T0134454

Studies in Computational Intelligence

Volume 799

The series "Studies in Computational Intelligence" (SCI) publishes new developments and advances in the various areas of computational intelligence—quickly and with a high quality. The intent is to cover the theory, applications, and design methods of computational intelligence, as embedded in the fields of engineering, computer science, physics and life sciences, as well as the methodologies behind them. The series contains monographs, lecture notes and edited volumes in computational intelligence spanning the areas of neural networks, connectionist systems, genetic algorithms, evolutionary computation, artificial intelligence, cellular automata, self-organizing systems, soft computing, fuzzy systems, and hybrid intelligent systems. Of particular value to both the contributors and the readership are the short publication timeframe and the world-wide distribution, which enable both wide and rapid dissemination of research output.

More information about this series at http://www.springer.com/series/7092

Boris Kryzhanovsky · Witali Dunin-Barkowski
Vladimir Redko · Yury Tiumentsev
Editors

Advances in Neural Computation, Machine Learning, and Cognitive Research II

Selected Papers from the XX International Conference on Neuroinformatics, October 8–12, 2018, Moscow, Russia

Editors
Boris Kryzhanovsky
Scientific Research Institute for System Analysis
Russian Academy of Sciences
Moscow, Russia

Vladimir Redko
Scientific Research Institute for System Analysis
Russian Academy of Sciences
Moscow, Russia

Witali Dunin-Barkowski
Scientific Research Institute for System Analysis
Russian Academy of Sciences
Moscow, Russia

Yury Tiumentsev
Moscow Aviation Institute (National Research University)
Moscow, Russia

ISSN 1860-949X ISSN 1860-9503 (electronic)
Studies in Computational Intelligence
ISBN 978-3-030-13170-8 ISBN 978-3-030-01328-8 (eBook)
https://doi.org/10.1007/978-3-030-01328-8

This Springer imprint is published by the registered company Springer Nature Switzerland AG
The registered company address is: Gewerbestrasse 11, 6330 Cham, Switzerland

Preface

The international conference "Neuroinformatics" is the annual multidisciplinary scientific forum dedicated to the theory and applications of artificial neural networks, the problems of neuroscience and biophysics systems, and adaptive behavior and cognitive studies.

The scope of the conference is wide, ranging from theory of artificial neural networks, machine learning algorithms, and evolutionary programming to neuroimaging and neurobiology.

Main topics of the conference cover theoretical and applied research from the following fields:

neurobiology and neurobionics: cognitive studies, neural excitability, cellular mechanisms, cognition and behavior, learning and memory, motivation and emotion, bioinformatics, computation, modeling, and simulation;
neural networks: neurocomputing and learning, architectures, biological foundations, computational neuroscience, neurodynamics, neuroinformatics, deep learning networks;
machine learning: pattern recognition, Bayesian networks, kernel methods, generative models, information theoretic learning, reinforcement learning, relational learning, dynamical models, classification and clustering algorithms, self-organizing systems;
applications: medicine, signal processing, control, simulation, robotics, hardware implementations, security, finance and business, data mining, natural language processing, image processing, and computer vision.

More than 100 reports were presented at the Neuroinformatics-2018 Conference. Of these, 42 papers were selected, including five invited papers, for which articles were prepared and published in this volume.

Boris Kryzhanovskiy
Witali Dunin-Barkowski
Vladimir Red'ko
Yury Tiumentsev

Editorial Board

Advisory Board

Program Committee of the XX International Conference "Neuroinformatics-2018"

General Chair

Kryzhanovskiy Boris	Scientific Research Institute for System Analysis, Moscow

Co-chairs

Dunin-Barkowski Witali	Scientific Research Institute for System Analysis, Moscow
Gorban Alexander Nikolaevich	University of Leicester, Great Britain
Red'ko Vladimir	Scientific Research Institute for System Analysis, Moscow

Program Committee

Abraham Ajith	Machine Intelligence Research Labs (MIR Labs), Scientific Network for Innovation and Research Excellence, Washington, USA
Anokhin Konstantin	National Research Centre "Kurchatov Institute," Moscow
Baidyk Tatiana	National Autonomous University of Mexico, Mexico
Balaban Pavel	Institute of Higher Nervous Activity and Neurophysiology of RAS, Moscow
Borisyuk Roman	Plymouth University, UK
Burtsev Mikhail	National Research Centre "Kurchatov Institute," Moscow

Cangelosi Angelo	Plymouth University, UK
Chizhov Anton	Ioffe Physical-Technical Institute, Russian Academy of Sciences, Saint Petersburg
Dolenko Sergey	Skobeltsyn Institute of Nuclear Physics Lomonosov Moscow State University
Dolev Shlomi	Ben-Gurion University of the Negev, Israel
Dosovitskiy Alexey	Albert-Ludwigs-Universität Freiburg, Germany
Dudkin Alexander	United Institute of Informatics Problems, Minsk, Belarus
Ezhov Alexander	State Research Center of Russian Federation "Troitsk institute for innovation & fusion research," Moscow
Frolov Alexander	Institute of Higher Nervous Activity and Neurophysiology of RAS, Moscow
Golovko Vladimir	Brest State Technical University, Belarus
Hayashi Yoichi	Meiji University, Kawasaki, Japan
Husek Dusan	Institute of Computer Science, Czech Republic
Ivanitsky Alexey	Institute of Higher Nervous Activity and Neurophysiology of RAS, Moscow
Izhikevich Eugene	Brain Corporation, San Diego, USA
Jankowski Stanislaw	Warsaw University of Technology, Poland
Kaganov Yuri	Bauman Moscow State Technical University
Kazanovich Yakov	Institute of Mathematical Problems of Biology of RAS, Pushchino, Moscow Region
Kecman Vojislav	Virginia Commonwealth University, USA
Kernbach Serge	Cybertronica Research, Research Center of Advanced Robotics and Environmental Science, Stuttgart, Germany
Koprinkova-Hristova Petia	Institute of Information and Communication Technologies, Bulgaria
Kussul Ernst	National Autonomous University of Mexico, Mexico
Litinsky Leonid	Scientific Research Institute for System Analysis, Moscow
Makarenko Nikolay	Central Astronomical Observatory of the Russian Academy of Sciences at Pulkovo, Saint Petersburg
Mishulina Olga	National Research Nuclear University (MEPhI), Moscow
Narynov Sergazy	Alem Research, Almaty, Kazakhstan
Nechaev Yuri	Institute of Experimental Medicine of the Russian Academy of Medical Sciences, Saint Petersburg
Pareja-Flores Cristobal	Complutense University of Madrid, Spain

Prokhorov Danil	Toyota Research Institute of North America, USA
Rudakov Konstantin	Dorodnicyn Computing Centre of RAS, Moscow
Rutkowski Leszek	Czestochowa University of Technology, Poland
Samarin Anatoly	A. B. Kogan Research Institute for Neurocybernetics Southern Federal University, Rostov-on-Don
Samsonovich Alexei Vladimirovich	George Mason University, USA
Sandamirskaya Yulia	Institute of Neuroinformatics, UZH/ETHZ, Switzerland
Shumskiy Sergey	P. N. Lebedev Physical Institute of the Russian Academy of Sciences, Moscow
Sirota Anton	Ludwig Maximilian University of Munich, Germany
Snasel Vaclav	Technical University Ostrava, Czech Republic
Terekhov Serge	JSC "Svyaznoy Logistics," Moscow
Tikidji-Hamburyan Ruben	Louisiana State University, USA
Tiumentsev Yury	Moscow Aviation Institute (National Research University)
Trofimov Alexander	National Research Nuclear University (MEPhI), Moscow
Tsodyks Misha	Weizmann Institute of Science, Rehovot, Israel
Tsoy Yury	Institut Pasteur Korea, Republic of Korea
Ushakov Vadim	National Research Centre "Kurchatov Institute," Moscow
Velichkovsky Boris	National Research Centre "Kurchatov Institute," Moscow
Vvedensky Viktor	National Research Centre "Kurchatov Institute," Moscow
Yakhno Vladimir	Institute of Applied Physics of the Russian Academy of Sciences, Nizhny Novgorod
Zhdanov Alexander	Lebedev Institute of Precision Mechanics and Computer Engineering, Russian Academy of Sciences, Moscow

Contents

Applications of Neural Networks

Cognitive Sciences and Adaptive Behavior

Neurobiology

Invited Papers

Quantumness and Irrationality

Alexandr A. Ezhov[1(✉)], Andrei G. Khromov[1], and Svetlana S. Terentyeva[2]

[1] SRC Troitsk Institute for Innovation and Fusion Research, Troitsk, Moscow, Russia
ezhov@triniti.ru
[2] Deutsche Bank Technological Center, Moscow, Russia

Abstract. Our interest to irrationality is mainly due to the agent modeling of unequal society and especially due to the study of critical phenomena in such models. Unfortunately, many similar models miss a point of agent irrational behavior. The last one dramatically change the model behavior. We argue that many current attempts to use quantum approach to describe irrational behavior deal in fact with the rational one. We also further elaborate the approach proposed by authors earlier and based on the definition of irrationality *per se*. It has been demonstrated that it is possible to define irrational action as the action which makes situation worser without any hope to improve it in future. It is shown that quantum approach is needed to describe this irrational behavior. Concretely, taking into account the analogy between decisions governed by classical implication function and simulated annealing it was shown that staying in classical domain it is impossible to receive decisions which are not rational. We also propose new different generalizations of the classical implication function to quantum domain which lead to the description of different kinds of irrationality including ones that suggest suicide-like behavior. These generalizations show analogy between modeling of irrational behavior and also of quantum simulated annealing.

Keywords: Quantumness · Irrationality · Annealing · Implication

1 Introduction

The key message of our report is as follows: *quantum approach is required for the description of irrational behavior* and this behavior can take different forms. Earlier we have demonstrated how quantum approach can be used for defining irrational strategies [1]. Here we repeat the essence of the approach based on the generalization of the implication function to the quantum domain and present some new types of such generalization. The goal of this paper may seem rather strange. It seems that now many examples of the use of the quantum approach to the description of non-rational human behavior are presented (see papers of Khrennikov [2], Sornette and Yukalov [3] and Bussemeyer and Wang [4] among others). Their approaches refer mainly to such phenomena as *quantum interference and entanglement*. But we will show that in fact these cases have relation to the *rational* behavior and suffer from the absence of the strong definition of irrationality. The last one connects irrationality with the quantum

B. Kryzhanovsky et al. (Eds.): NEUROINFORMATICS 2018, SCI 799, pp. 3–13, 2019.
https://doi.org/10.1007/978-3-030-01328-8_1

simulated annealing [5] and hence with some analogy of *quantum tunneling*. Our interest to irrationality is mainly due to the agent modeling of unequal society and especially due to the study of critical phenomena in such models [6, 7]. Unfortunately, many similar models miss a point of agent irrational behavior. Actually, agents considered by us in [6, 7] are completely rational and only in [1] we have introduced irrational agents. The last one dramatically change the model behavior. Let us now start from the definition of rationality and show that agents characterized by many authors as irrational are in fact rational ones.

2 Rational Agents

Neoclassical economics fails not only because it is based on physics of the 19th century with the idea of equilibrium and optimality but also because of the use of a wrong model of human behavior. It considers human as pure *rational* agent - *homo œconomicus* - which is a fully informed egoistic rational *utility maximizer* [8]. Now it is widely known that this is a wrong hypothesis because it contradicts to facts. For example let us consider a famous *ultimatum game* [9]. Alice (metavariable) will receive a rather serious amount of money if she proposes to Bob a part of this money and if he agrees to her proposal. Otherwise the both will receive nothing. According to a model of the rational agent Bob should accept any proposal because any amount of money is better than nothing. But in real experiments the probability of proposal acceptance drops dramatically if money sharing is unfair. In some sense Bob acts *non rationally* (if we accept the notion of rationality adopted in Neoclassical economics) when he rejects Alice's proposal. It is necessary to stress that this behavior is called by many researches as "irrational" or "non-rational" [10]. But this use of these terms is not justified properly. The Prisoner's Dilemma (PD) is another example of fiasco of rationality [11]. It was formulated by specialists of RAND to describe the nuclear confrontation between the Soviet Union and the USA. In our terms the dilemma is as follows. Alice and Bob are suspected of robbery. They are placed in different cameras and cannot communicate. If Alice confesses and if Bob also confesses then the both go to jail for five years. If Alice is silent while Bob confesses then Alice goes to jail for ten years and Bob goes free, and *vice versa*. However, if the both remain silent then they go to jail only for one year. Despite of Bob's decisions rational Alice has to betray Bob. Rational Bob will choose the same decision. In result the both agents will go to jail for 5 years. In case the both do not act rationally they will receive only one year of punishment. So, the rational decision (which corresponds to the Nash equilibrium) is worse than the non-rational (Pareto optimal one). One more example. It seems to be rational to risk a larger amount of money if the probability to win it is higher and *vice versa*. But surprisingly this behavior contradicts to the observed real wealth distribution [12]. It is interesting that human is not special. If you propose ape to choose between one high value banana and one high value banana *plus* one middle value tomato then the ape will choose only one banana [13]! This is again in contradiction with the utility maximization and rationality from the point of view of economics. Does it mean that these examples lead us to reject the hypothesis that in these cases agents act rationally? No, it doesn't. It is easy to recognize that these behaviors are really rational if we take into account that the human and the ape *can make prognosis* about the influence of their decisions on future possible proposals. For example, if Bob hopes to play the ultimatum game again he can suggest that after his rejection Alice's next proposal should be fair. This is obviously rational behavior. If people play Iterated Prisoner Dilemma without dramatic punishment they can think that next time their opponent will not confess. So, they decision to cooperate is obviously rational. The ape can also be afraid that if he agrees to take tomato along with a banana then next time only tomatoes will be proposed to him. Now let us consider the important case of the classical method of simulated annealing [14]. In searching for minimum at a finite temperature it is quite *rational* to make situation worse by performing hill climbing if in a future we hope to go downhill towards a deeper minimum. In general, we can recognize that there are a lot of such examples where implicit rationality is treated as a "*irrationality*" or "*non-rationality*" (for more examples see e.g. [15]). What is really irrational and how can we define it? To answer this question we must go deeper into logic and start from the consideration of *implication function*.

3 The Implication Function

The implication function is the key element of the Lefebvre algebra of conscience [16]. Implication is defined as "*b* implies *a*" and can be calculated as "*a* plus not *b*" $(a + \neg b)$. It can also be expressed in the exponential form a^b. There are a lot of applications of this function in logic, reflexive control, agent modeling and, what is important to us here, in simulated annealing. In logic *a* and *b* are the Boolean variables. Implication takes a false value for the wrong statement "*the truth implies false*" (Table 1). Otherwise it takes true value.

Table 1. The implication function $I(a, b) = a + \neg b = a^b$ $a, b = \{truth\ (1), false(0)\}$

a	b	a^b	
0	0	1	
0	1	0	The *truth* cannot imply *false*
1	0	1	
1	1	1	

In Lefebvre's algebra of conscience *a* and *b* take values *evil* and *good* and *a* is the proposal to person to make a good or bad thing and *b* is the awareness of this proposal by person. In this case the main axiom of algebra is as follows: an individual in a state of evil and aware of this, changes his state to good (Table 2).

In the classical simulated annealing when we search for minimum of the function, *a* and *b* take values which mean the move (1) uphill or (2) downhill (or stay) [1, 17]. In this case *a* means the proposal to go uphill or downhill (or stay) and *b* means a *prognosis* of the next proposal presented to the system by environment: $b = p$ (Table 3). The implication function here takes time reversed form: "*p* implies *a*" (prognosis implies current proposal). But exponential form a^p seems to be reasonable (it means that current

Table 2. The implication function I(a, b), a, b = {*good* (1), *evil* (0)}

a	b	a^b	
0	0	1	*evil* 'aware' of *evil* becomes good
0	1	0	*evil* 'aware' of *good* remains *evil*
1	0	1	*good* 'aware' of *evil* remains *good*
1	1	1	*good* 'aware' of *good* remains *good*

proposal a generated the guess of the future proposal, p). Taking into account this we can say, that the decisions defined by implication function correspond to the foundation of simulated annealing which is the suggestion that we go uphill if we hope that in future we shall go downhill. This is obviously rational behavior.

Table 3. Interpretation of I(a, p) in simulated annealing: a, p = {*stay or downhill* (1), *uphill* (0)}

a	p	a^p	
0	0	1	*stay* if *uphill* will be followed by *uphill*
0	1	0	go *uphill* if the following is *downhill*
1	0	1	*stay or go downhill* regardless the prognosis
1	1	1	*stay or go downhill* regardless the prognosis

In more details: consider 4 cases of proposal a and prognosis p *(see* Fig. 1)

If a = 1 (the so called "*mafia's proposal*" that cannot be rejected [1]) the proposal to go downhill (or stay) is accepted unconditionally regardless the prognosis. If it is proposed to go uphill (a = 0) then this proposal is accepted only if it is predicted that in future it will be proposed to go downhill. This obviously corresponds to thermal uphill movement. It is important to stress that all decisions are rational! It is useful to decompose the two-variable Implication function into two one-variable functions $I_{sym}(a)$ and $I_{asym}(a)$ taking into account that b can be equal to a or $\neg a$, correspondingly (Table 4). In the case of simulated annealing antisymmetric strategy defined by the function $I_{asym}(a)$ for which prediction differs from proposal corresponds to high temperature case when thermal uphill movement is permitted. The symmetrical strategy, defined by $I_{sym}(a)$, permits only to go downhill if the prediction corresponds to the proposal. This symmetrical strategy can be considered as low temperature one. Again, both strategies are obviously rational ones.

These two strategies also describe decision rules used by two types of agents in the model of an unequal society [6.7]. The world these agents occupy consists of discrete niches and agents have two lives: physical and mental. These agents have two rational strategies. The antisymmetric high temperature strategy describes the agent which accepts any proposal to survive physically by receiving food proposed in any niche (so it changes it freely). The symmetrical low temperature strategy describes the agent which prefers to hold niche (identity) or to survive mentally by neglecting the proposal to receive food in the other niche. The probability of a food proposal can be effectively

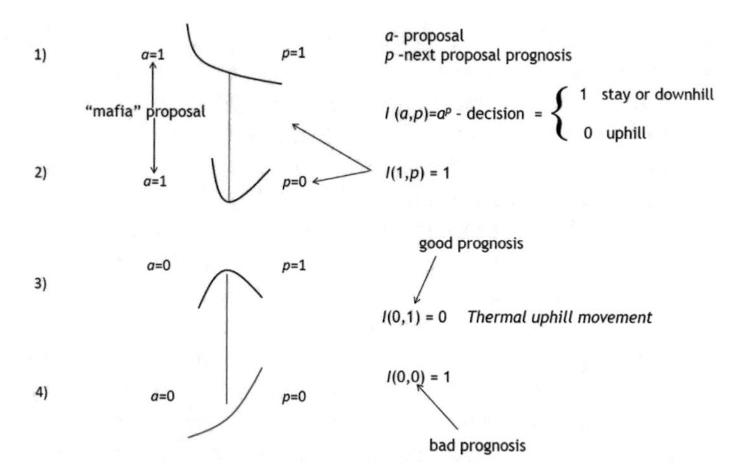

Fig. 1. Annealing interpretation of the the implication function. The first argument of the Implication function, a, denotes a proposal to *stay* or go *downhill* (a = 1) or to go *uphill* (a = 0), while the second argument, p, denotes a prognosis of the value of next proposal. The value of the Implication, $I(a, p)$, denotes agent's decision to *stay* or go *downhill* ($I(a, p)$ = 1) or to go *uphill*). The cases (1) and (2) correspond to "*mafia proposal*" (a = 1) the agent accepts despite on its prognosis p: $I(1, p)$ = 1. (3) An agent receives a proposal to go *uphill* (a = 0) but also has a good prognosis of the next proposal, p = 1. In this case it goes *uphill*. (4) An agent receives proposal to go *uphill* (a = 0) and also has bad prognosis of the next proposal, p = 0. In the classical annealing it rejects this proposal.

Table 4. One variable functions $I_{asym}(a)$ (left) and $I_{sym}(a)$ (right)

a	$\neg a$	$a^{\neg a}$
0	1	0
1	0	1

a	a	a^a
0	0	1
1	1	1

described by the efficient temperature. Higher temperatures correspond to fairer societies [6]. Two basic strategies of agents correspond to the two parts of implication function: $I_{asym}(a)$ and $I_{sym}(a)$. There are some reasons to call these strategies *left* and *right brain dominant*, respectively [6] (Table 5).

Table 5. The implication function $I(a, p)$ represents strategies of simulated annealing which become dominant at high and low temperature and correspond to strategies of left and brain dominant agents in the model of unequal society [6]

a	p	a^p	
0	0	1	Low temperature & right brain strategy/identity preference
0	1	0	High temperature left & brain strategy/physical life preference
1	0	1	High temperature left & brain strategy/physical life preference
1	1	1	Low temperature & right brain strategy/identity preference

It is important that the Lefebvre algebra permits to calculate rules of interaction of agents of these two types (this interaction permits, e.g., to right brain agents to change their niches). It turns, that left brain agents repulse each other while the right brain agents attract each other. It is not surprising that in equilibrium the distributions in homogeneous ensembles of such agents resemble the Bose and Fermi statistical distributions [6]. Using ultrametric data analysis [18] and by studying symmetry breaking phenomena it is possible to calculate critical levels of inequality in model agent societies [7]. Note that highly unequal societies are characterized by low effective temperatures so their study is closely related to low temperature physics. Note again that strategies of left and right brain dominant which corresponds to the agent decisions at high and low temperatures in classical simulated annealing and correspond to asymmetrical and symmetrical parts of the implication function are strictly rational. The only way to introduce irrational behavior is to generalize the implication function. Earlier [1] we have demonstrated that this can be done by it's generalization to quantum domain that is also equivalent to the transition from the classical simulated annealing to the quantum one [5].

4 Irrational Agents

The definition of irrational decision has been presented in [1]: *the irrational decision makes the situation worse without any hope to improve it in future*. This definition is out of the base of classical simulated annealing which implicitly suggests that any uphill movement is justified only if in future it will be followed by downhill one (see Fig. 2).

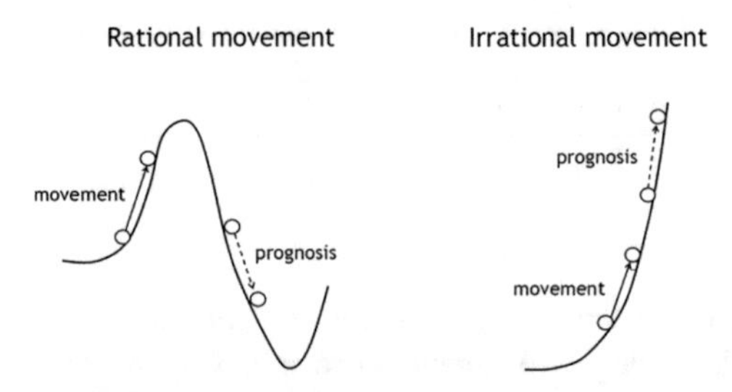

Fig. 2. In searching for the function minimum a rational movement (left) corresponds to uphill one connected with the prognosis to go downhill in future. Irrational decision (right) corresponds to move uphill with the bad prognosis to go uphill in future.

As we argued in previous section the irrational behavior can be obtained if we generalize implication function to quantum domain. One way to reach this goal is to start with the circuit implementation of classical implication function (see Fig. 3). The input to the first line, a, corresponds to the proposal in the agent model. The output value of the first line gives the decision of the right brain agent while the output value of the second line gives the decision of the left brain agent. Note, that the realization of the implication function by this circuit is easily verified.

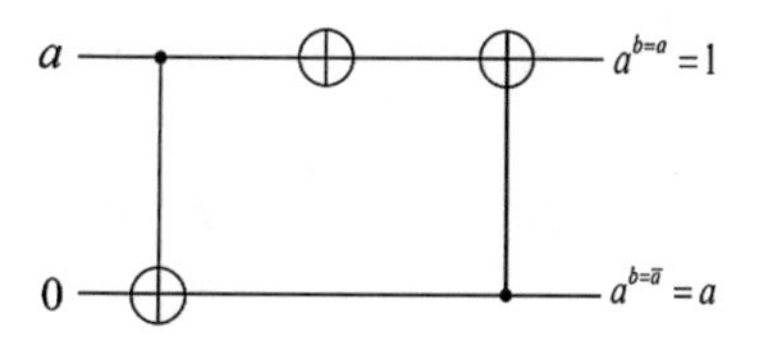

Fig. 3. The circuit implementation of the implication function

To change this scheme to explicitly quantum one we can change NOT gate by Hadamard gate and the second C-NOT gate by Control Hadamard gate [1]. We do not modify the first C-NOT because it simply transfers the value a to the second qubit (see Fig. 4). This generalization of implication function is special: it holds "mafia's proposal" as well as a logical behavior of the left brain agent [6]. In the agent model of unequal society mafia proposal is the proposal of food in the niche which is occupied by agents. Obviously, it should not be rejected because is profitable both from the point of view of mental and physical surviving.

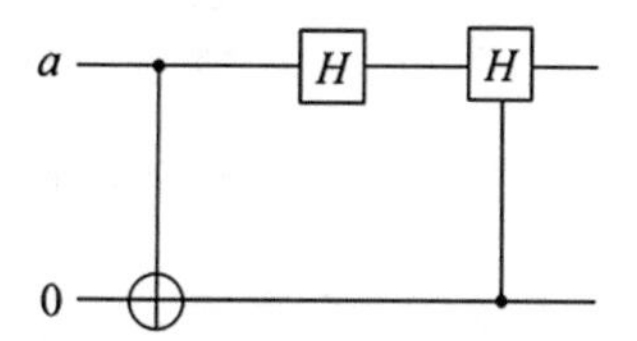

Fig. 4. Quantum generalization of the implication function

Now let's check how the new circuit works. It is easily to verify that only the decision of the right brain agent will be modified [1]. Corresponding state of the first qubit becomes the quantum superposition of 0 and 1 with equal amplitudes (Table 6). It means that the quantum right brain agent changes its decision when its classical analog rejects the proposal to change the identity and becomes slightly "left brain minded". Precisely, it will change it's niche (without any interaction with the other agent) with the probability of 50% if the food is proposed to it in another niche.

Table 6. Left and right brain agent strategies in the case of quantum generalization

a	$\neg a$	$a^{\neg a}$
0	1	0
1	0	1

a	a	a^a
0	0	$(\lvert 0\rangle + \lvert 1\rangle)/\sqrt{2}$
1	1	1

What does it mean in case of logic? Obviously, on contrast to the classical logic "the *false* can imply *false*" with the probability equal to 50% (Table 7). In the case of the Lefebvre algebra it will mean that "*evil* 'realizes' the *evil* becomes *good*" only in 50% of cases.

Table 7. Quantum generalization of implication function (see Fig. 4)

a	b	a^b	
0	0	$(\|0\rangle + \|1\rangle)/\sqrt{2}$	The *false* can imply *false* in 50%
0	1	0	
1	0	1	
1	1	1	

At last in the simulated annealing it will mean that *apart from thermal uphill movement* a *new uphill movement arises* in the case of the absence of the prediction that in future's downhill movement will happens (Fig. 5)! This can be considered as not an analog but anticipation of quantum tunneling in the nearest region surrounding deeper attractor! This is because environment also can react unexpectedly. It simply can be quantum. Also it will be an analog of the scheme of quantum annealing [5] and obviously will correspond to irrational behavior as it was defined in [1].

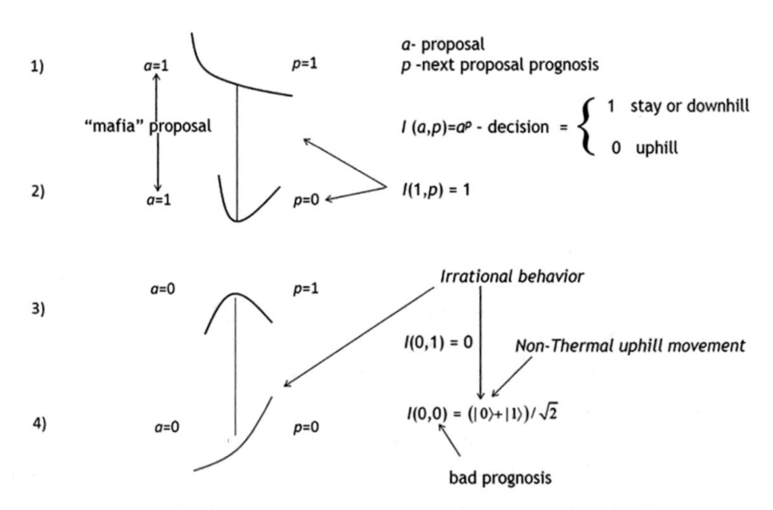

Fig. 5. Quantum annealing interpretation of the quantum implication. The cases (1), (2), (3) correspond to ones presented in Fig. 1 In the case (4) an agent receives proposal to go *uphill* ($a = 0$) and also has bad prognosis of the next proposal, $p = 0$. In the classical annealing an agent rejects this proposal (this corresponds to the low temperature decision of right brain agent). But in quantum case she has a chance to go *uphill* (with the probability 50%) even in this case (a tunneling possibility). This non-thermal *uphill* movement of an agent having bad prognosis can be interpreted as his *irrattional behavior*.

Note, that in Quantum Decision Theory [19] irrationality is attributed to the *intermediate stage* of decision making. In contrast, in our case the irrationality is the characteristics of the final decision.

Irrational behavior of agents corresponding to the generalization of implication function given above changes dramatically agent interactions and also equilibrium distributions by changing them from Bose and Fermi-like into intermediate statistics [1].

Now let us consider other generalizations of implication function which violate mafia's proposal.

5 Suicide-like Irrationality

Let us consider two cases when only one gate, NOT or second C-NOT, are changed by Hadamard and Controlled Hadamard gates, respectively. In the first case apart of the superposition of decisions of right brain agent to hold or to change its niche when the food is proposed in the other one (similar to its behavior in previous case of quantum generalization of implication function) an analogous superposition of decision arise for the case of mafia proposal (see Fig. 6).

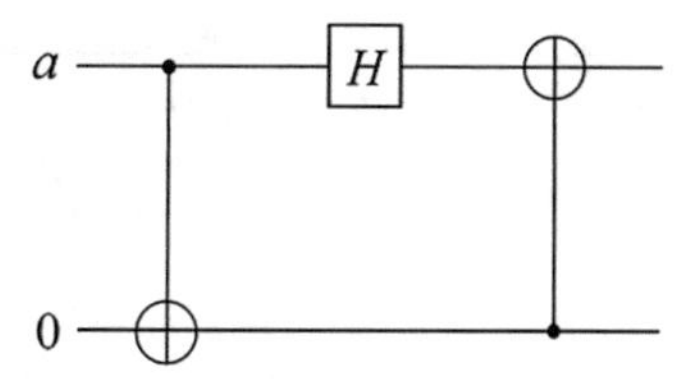

Fig. 6. Quantum generalization of the implication function (variant 2)

It means that the agent will reject the proposal to receive a food in a nice it occupies which is deadly for it from the point of view of physical and also of mental surviving (Table 8).

Table 8. Left and right brain agent strategies in the case of quantum generalization

a	$\neg a$	$a^{\neg a}$
0	1	0
1	0	1

a	a	a^a
0	0	$(\|0\rangle+\|1\rangle)/\sqrt{2}$
1	1	$(-\|0\rangle+\|1\rangle)/\sqrt{2}$

The same suicide-like decision is also valid for the second scheme for which right brain agent also loss the advantage to be slightly left-minded as in previous cases (see Fig. 7).

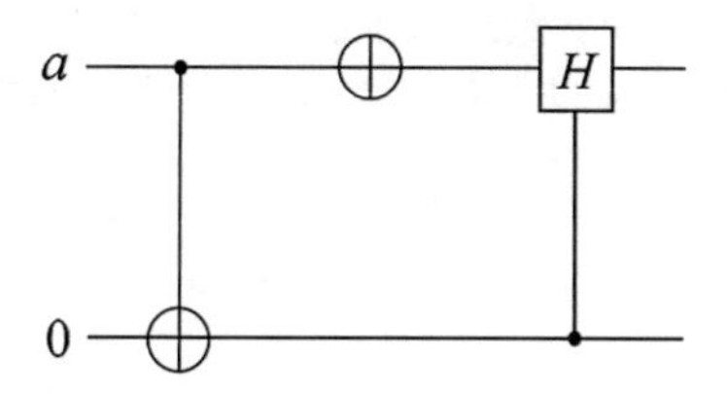

Fig. 7. Quantum generalization of the implication function (variant 3)

From the point of view of application of such agent model in sociophysics such suicide-like behavior is not completely unreasonable. People really make suicide decisions sometimes. According to Albert Camus "There is but one truly serious philosophical problem and that is suicide" [20]. But sometimes such decisions can be only illusory suicide-like. The point is that environment can be unpredictable. For example, irrational decision to go uphill without any prognosis to go downhill in future can be justified by the unrecognized quantum nature of the world which makes possible agent's quantum tunneling into the region with deeper minimum. This can happen also in the classical world (Table 9).

Table 9. Left and right brain agent strategies in the case of quantum generalization

a	$\neg a$	$a^{\neg a}$
0	1	0
1	0	1

a	a	a^a
0	0	1
1	1	$(-\vert 0\rangle + \vert 1\rangle)/\sqrt{2}$

For example, there is a version that if the helmsman kept the course and the Titanic would have collided with the iceberg forehead, then only two nasal compartments of the ship would be destroyed and the ship would not sink. So, this seemingly suicide-like decision could save thousands of lives. In some air accidents when plane falls down the suicide-like decision of pilot to release the steering wheel so that the plane itself comes out of the dive can be the only right decision. To avoid dangerous rip current the decision do not swim against the current to the shore also can be illusory seemed as suicide-like, but again this is the only decision which can save life of the swimmer.

6 Conclusions

We demonstrated here that it is possible to define irrationality as action which makes situation worser without a hope to improve it in future. Taking into account the analogy between decisions governed by implication function and simulated annealing it was shown that staying in classical domain it is impossible to receive decisions which are not rational. Using different quantum generalization of implication function it is possible to obtain irrational decisions which can be used to model different types of agent behaviors including suicide-like ones. These generalizations show analogy between modeling of irrational behavior and also of quantum simulated annealing.

References

1. Ezhov, A.A., Khromov, A.G., Terentyeva, S.S.: Artificial intelligence and machine learning for quantum computing: A Tutorial at the International Joint Conference on Neural Networks (IJCNN) 2015, 12 July 2015. http://webs.wichita.edu/depttools/depttoolsmemberfiles/quantumresearch/Australia%20conference/Ezhov.pdf
2. Asano, M., Ohya, M., Tanaka, Y., Khrennikov, I.A.: Quantum-like model of brain's functioning: Decision making from decoherence. J. of Theor. Biol. **281**(1), 56 (2011)
3. Yukalov, V.I., Sornette, D.: Quantitative Predictions in Quantum Decision Theory(2018). arXiv:1802.06348v1[physics.soc-ph]
4. Bussemeyer, J.R., Wang, Z.: What is quantum cognition, and how is it applied to psychology? Curr. Dir. Psychol. Sci. **24**(3), 163 (2015)
5. Das, A., Chakrabarti, B.K.: Colloquium: Quantum annealing and analog quantum computation. Rev. Mod. Phys. **80**, 1061 (2008)
6. Ezhov, A.A., Khrennikov, AYu.: Agents with left and right dominant hemispheres and quantum statistics. Phys. Rev. E **71**, 016138 (2005)
7. Ezhov, A.A., Khrennikov, A.Yu., Terentyeva, S.S.: Indications of a possible symmetry and its breaking in a many-agent model obeying quantum statistics. Phys. Rev. E **77**, 031126 (2008)
8. Ball, P.: News Feature. Nature **441**, 686 (2006)
9. Güth, W., Schmittberger, R., Schwarze, B.: J. of Economic Beh. Org. **3**, 367 (1982)
10. Fiala, J., Starý, O., Fialová, H., Holasová, A., Fialová, M.: Acta Oeconomica Pragensia **25** (1), 64 (2017)
11. Flood, M.M.: Some experimental games. Research memorandum RM-789. RAND Corporation, Santa Monica (1952)
12. Fuentesa, M.A., Kuperman, M., Iglesias, J.R.: Living in an irrational society: Wealth distribution with correlations between risk and expected profits. Physica A. **371**(1), 112–117 (2006)
13. Kralik, J.D., Xu, E.R., Knight, E.J., Khan, S.A., Levine, W.J.: When less is more: Evolutionary origins of the affect heuristic. PLoS ONE **7**(10), e46240 (2012)
14. Metropolis, N., Rosenbluth, A.W., Rosenbluth, M.N., Teller, A.H., Teller, E.: Equation of state calculations by fast computing machines. J. Chem. Phys. **21**(6), 1087 (1953)
15. Ariely, D.: Predictably Irrational, Harper Collins Publishers, New York (2008)
16. Lefebvre, V.: Algebra of Conscience. Kluwer Academic Publ., Dordrecht/Boston/London (2001)
17. Ezhov, A.A., Bur'yanitsa, A.V., Terentyeva, S.S., Khvalina, A.S.: In: Econophysics, MEPhI, Moscow, 529 (2007)
18. Murtagh, F.: Identifying the ultrametricity of time series. Eur. Phys. J. B **43**, 573 (2005)
19. Sornette, D., Yukalov, V.I.: Quantitative Predictions in Quantum Decision Theory, 18 Feb 2018. arXiv:1802.06348v1[physics.soc-ph]
20. Camus, A.: The Myth of Sisyphus. Hamish Hamilton, London (1955)

Planar Ising-Spin Models in Probabilistic Machine Learning

Iakov M. Karandashev(✉)

Center of Optical Neural Technologies, Scientific Research Institute for System Analysis RAS, Moscow, Russia
karandashev@niisi.ras.ru

Abstract. One of the approaches in machine learning is probabilistic models that operate with such concepts as binary neurons, the energy of a system, the normalization constant (partition function), the entropy, etc. Most of them are migrated to machine learning from physics. In solid-state physics, the most interesting are the results obtained using asymptotic approximations for crystal lattices of infinite dimensions. One of such most investigated objects is the infinite two-dimensional lattice of Ising spin model. In machine learning, on the contrary, lattices of finite sizes are of the greatest interest. In this paper we describe algorithms for the exact calculation of the partition function and other statistical quantities for the planar Ising model of finite dimensions (from $N = 5 \times 5$ to $N = 1000 \times 1000$). A polynomial algorithm for finding the partition function is described in detail, based on the calculation of the determinant of the dual graph of a planar lattice. Also, the results obtained by the classical Metropolis algorithm are discussed. The results obtained with the help of these algorithms allow us to draw some conclusions about the behavior of the model of the two-dimensional Ising model in the finite-size case and compare them with the asymptotic formulas.

Keywords: Planar graph · Ising model · Partition function · Binary model
Polynomial algorithm · Free energy · Critical point · Heat capacity
Normalizing constant · Two-dimensional grid · Nearest-neighbors interaction
Critical temperature · Noise of grid elements

1 Introduction

Today's tendencies in the field of neural networks mostly suggest the development of supervised learning methods. However, there are problems that involve few or no labeled data. In such cases the learning without a teacher is the only applicable approach. The unsupervised learning goes together with data dimensionality reduction and clustering. The demonstrative example of the approach is autoencoders [1–3]. Regrettably, the interest to them has dropped significantly with the advent of fast optimization methods for feed forward neural networks. Such networks generate a one-way function whose inverse is very difficult to build.

One of the first learning methods for deep autoencoders (e.g. the restricted Boltzmann machine) engages the maximization of the likelihood function of a

B. Kryzhanovsky et al. (Eds.): NEUROINFORMATICS 2018, SCI 799, pp. 14–38, 2019.
https://doi.org/10.1007/978-3-030-01328-8_2

Bayesian neural network. The computation of the likelihood function gradient comes down to the finding of the normalizing constant by approximate methods like contrastive divergence (CD) [1, 2].

The computation of the partition function (normalization constant) is an important problem in statistical physics as well in neural networks. For nearly a century, the Ising model and, particularly, the calculation of its partition function has been in the focused of active research. The solution of the problem for a finite system will allow a noticeable advance in the methods of deep learning and image processing [4, 5]. The use of graph models usually involves computation of two quantities: a posterior probability maximum estimate and marginal distributions. The calculation of the latter is closely related to the determination of the partition function [6]. The problem is to find some statistical properties (e.g. marginal probabilities), given a particular set of random variables in a certain graph model.

Unfortunately, it is known that there are very few problems for which the partition function can be calculated exactly [7, 8]. In particular, these are problems on trees, planar graphs, or general type graphs of small sizes. In most cases, when a problem has hundreds or thousands dimensions, the use of rough heuristic methods is the only possible approach. However, it is useful to have exact methods at hand (at least, for a limited range of problems) to develop heuristic methods for approximate calculations of partition functions. The approximate method based on the use of trees and known as the tree reweighting (TRW) method was offered in papers [9]. In [10–12] was proposed the n-vicinity method for approximate calculation of the partition function for an arbitrary graphs.

Our paper considers another class of solvable models – planar graphs. A graph is called the planar graph if it can be drawn on a plane without its branches intersecting. It was discovered that in a special type of planar Ising model without an external field the calculation of the partition function is polynomially reduced to the computation of the matrix determinant [13–18]. The planar Ising model presents a perfect object of investigation. First, a plane grid can be regarded as a collection of image pixels. Second, for this model the statistical physics offers the exact analytical solution known as Onsager's solution [19] for constant couplings.

A significant progress has been made in the development of numerical algorithms allowing successful investigation of critical characteristics [20–23] and energy spectra of spin systems [24]. The Monte Carlo approach [25, 26], which permits rough estimations, is mostly used in this kind of algorithms. However, algorithms [17, 18] that make it possible to exactly calculate the free energy of a finite planar spin lattice have been developed.

In the paper we show the relationship between system parameters and dimensionality N of the problem and find analytical expressions suitable for finite N. The results given below make it possible to understand how large the dimensionality of the problem should be for the simulation results to give a satisfactory description of properties of real models. Besides, the analytical expressions suggest their use in further development of image processing algorithms and another machine learning applications.

Also, in this paper, we have calculated the free energy for planar graphs, when the interaction between neighbors has a constant average value with a small noise.

We found that with the increasing noise of interactions the peak of the second derivative of the free energy goes farther and farther from Onsager's prediction until it disappears at all. The analysis of the heat capacity showed that when noise variations are small, the critical temperature shifts linearly with the growing noise dispersion. We used the n-vicinities method [11, 12] to compute the free energy, which allowed us to find an approximate relationship between the shift of critical temperature and the growth of noise dispersion, which agrees well with the experimental data.

This lecture is an overview of materials on the papers [18, 33–36]. Section 2 of the paper offers the problem of partition function calculation statement. Section 3 based on the paper [18] in detail describes the Kasteleyn-Fisher algorithm. Section 4 based on the paper [33–36] gives the experimental results and analytic approximations and compare them with Onsager's analytic solution for the large dimensionality limit. In Sect. 5 some conclusions are reported.

2 Problem Statement and Onsager's Solution

It is best to use the terms of physics for our purposes. Let there be a set of N binary spins described by the Hamiltonian

$$E = -\frac{1}{2N}\sum_{i,j=1}^{N} J_{ij}s_i s_j, \quad s_i = \pm 1. \tag{2.1}$$

This functional is often used in problems of machine learning and image processing. Quantities $s_i = \pm 1$ mean the belongingness to one of two pixel classes (background or object) or the neuron activity in a Bayesian neural network. Given a particular temperature, the possibility of the system staying in state s is defined by the formula

$$P(s) = \frac{1}{Z}e^{-\beta E(s)} \tag{2.2}$$

where β is the inverse temperature, and $Z = \sum_s e^{-N\beta E(s)}$ is the normalizing constant (partition function), the sum over all possible configurations s. As seen from (2), to evaluate the probability of a particular configuration, we need to know the normalizing constant, which is not so simple. It is the usual practice to find the specific free energy of the system rather than the constant Z itself:

$$f = -\ln Z/N, \tag{2.3}$$

The knowledge of the free energy allows us to compute the major parameters of the system such as internal energy $U = \bar{E}$, energy variance $\sigma^2 = \overline{E^2} - \bar{E}^2$ and heat capacity $C = \beta^2\sigma^2$:

$$U = \frac{\partial f}{\partial \beta}, \sigma^2 = -\frac{\partial^2 f}{\partial \beta^2}, C = -\beta^2\frac{\partial^2 f}{\partial \beta^2} \tag{2.4}$$

In the paper we study a planar model where spins are positioned in a square lattice and only four nearest neighbors interact $J_{ij} = J$. Onsager's solution [19] found for $N \to \infty$ for this sort of system with periodical boundary conditions has the form:

$$
\begin{aligned}
f(\beta) &= -\frac{\ln 2}{2} - \ln(\cosh 2\beta J) - \frac{1}{2\pi}\int_0^{\pi} \ln\left(1+\sqrt{1-k^2\cos^2\theta}\right) d\theta \\
U &= -\coth 2\beta \cdot \left[1+\frac{2}{\pi}E_1(k)\left(2\tanh^2 2\beta - 1\right)\right] \\
C &= \frac{4\beta^2}{\pi\,\tanh^2 2\beta}\cdot\left\{E_1(k) - E_2(k) - \left(1-\tanh^2 2\beta\right)\left[\frac{\pi}{2} + \left(2\tanh^2 2\beta - 1\right)E_1(k)\right]\right\}
\end{aligned}
\tag{2.5}
$$

where

$$
k = \frac{2\sinh 2\beta}{\cosh^2 2\beta} \tag{2.6}
$$

and $E_1 = E_1(k)$ and $E_2 = E_2(k)$ are the complete elliptic integrals of the first and the second kinds, defined as:

$$
E_1(k) = \int_0^{\pi/2} (1-k^2\sin^2\phi)^{-1/2}\, d\phi, \quad E_2(k) = \int_0^{\pi/2} (1-k^2\sin^2\phi)^{1/2}\, d\phi. \tag{2.7}
$$

It is known, that in the framework of the Onsager solution the heat capacity diverges logarithmically ($C \to \infty$) when $\beta \to \beta_{ONS}$, where the critical value of inverse temperature and critical values of free energy and internal energy are determined from the condition $k = 1$ in the form:

$$
\begin{aligned}
\beta_{ONS} &= \frac{1}{2}\ln\left(1+\sqrt{2}\right) \\
f_{ONS} &= -\frac{\ln 2}{2} - \frac{2G}{\pi} \approx -0.9297 \\
U_{ONS} &= -\sqrt{2}
\end{aligned}
\tag{2.8}
$$

where $G \approx 0.915966$ is Catalan's constant.

The solution (2.5) describes the logarithmic divergence of heat capacity when $\beta \to \beta_{ONS}$, where the critical temperature is determined from condition $k = 1$ as:

$$
\beta_{ONS} = \frac{1}{2}\ln\left(1+\sqrt{2}\right). \tag{2.9}
$$

3 Kasteleyn-Fisher Algorithm

In this Section we thoroughly explain an algorithm[1] for exact calculation of partition function for planar graph models with binary variables. The complexity of the algorithm is $O(N^2)$. This Section repeats materials of [18].

The calculation of the partition function on a planar graph amounts to finding the number of perfect matchings adjusted for their weights by using linear algebra methods as was suggested by P. Kasteleyn and M. Fisher in 1961. Some portions of the algorithm were borrowed from book [27] and papers [28, 29].

Let there be a planar graph G for which the partition function should be found. In short, the algorithm includes the following steps:

1. Dual graph D is built for initial planar graph G.
2. The nodes of graph D of degree greater than 2 are unfolded in a planarity-retaining manner to produce extended graph R with coupling matrix $W = \{w_{ij}\}$.
3. Skew-symmetric matrix $B = \{b_{ij}\}$ corresponding to the Pfaff orientation of graph R is constructed.
4. The sought-for partition function is equal to the Pfaffian of matrix $A = \{a_{ij} = b_{ij}w_{ij}\}$, which, in turn, is the square root of the determinant, i.e. $Z = \mathrm{Pf}(A) = \sqrt{\det A}$.

Steps 3–4 are known as the FKT (Fisher-Kasteleyn-Temperley) algorithm. Though each of the four steps of the algorithm is known, for first thing there haven't been so far any consistent description of all four steps solving the partition function problem, for another there haven't been their realization except for [15].

Below thorough consideration of each of the four steps is given.

3.1 Dual Graph Construction

Let there be original graph G with coupling matrix J_{ij}. Let each node hold a particle whose spin can take either of two values $s_i = \pm 1$. We recognize these values as the node states. It is necessary to compute the partition function over all possible node states:

$$Z = \sum_{s} e^{-\beta E(s)}, \tag{3.1}$$

where β is the inverse temperature, and the energy is determined by pair interaction between graph nodes (please note we do not use factor ½ in energy definition as in (2.1) for simplicity):

$$E(s) = -\sum_{i,j}^{N} J_{ij} s_i s_j. \tag{3.2}$$

[1] The code is publicly available at https://github.com/Thrawn1985/2D-Partition-Function.

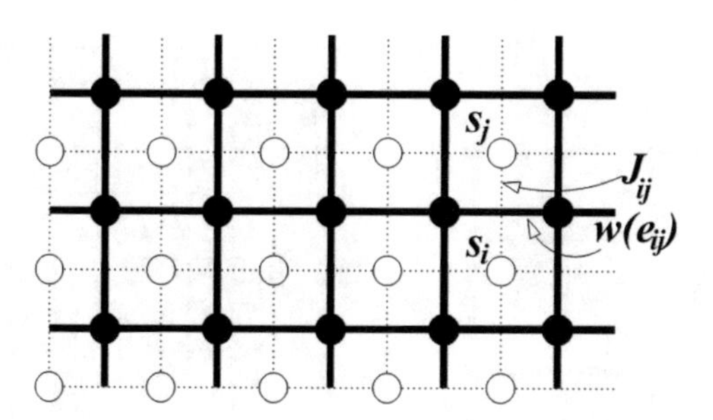

Fig. 1. The bold lines and black circles designate the dual graph, and the thin lines and white circles stand for the initial graph.

Let $V = \{i: \; s_i = +1\}$, i.e. V is a collection of nodes whose states are +1. Then the expression for the energy can be rewritten in the following manner:

$$E(s) = -\sum_{i,j}^{N} J_{ij}s_is_j = 2\sum_{i\in V}\sum_{j\notin V} J_{ij} - \sum_{i,j\in V}^{N} J_{ij} - \sum_{i,j\notin V}^{N} J_{ij} = 4\sum_{i\in V}\sum_{j\notin V} J_{ij} - \sum_{i,j}^{N} J_{ij}. \quad (3.3)$$

It is seen that the second term $\sum_{i,j}^{N} J_{ij}$ is a constant, and the first term $\sum_{i\in V}\sum_{j\notin V} J_{ij}$ is defined only by node pairs $<i,j>$ with opposite states. It is particularly evident with a planar graph where such node pairs are situated on the boundary of set V.

When we deal with planar graph G, it is possible to build dual graph D whose nodes are the faces of the original graph and edge weights are determined as follows (see Fig. 1):

$$w_{ij} = e^{-4\beta J_{ij}}. \quad (3.4)$$

For a dual graph it is possible to say that (see Fig. 2) any configuration s of node states in original graph G (Fig. 2a) corresponds to a set of Eulerian cycles in dual graph D (Fig. 2b). Eulerian cycles are closed curves going along edges w_{ij} and constraining nodes s_i in state +1. It follows that finding partition function amounts to summation over all possible Eulerian subgraphs in dual graph D:

$$Z = C\sum_{V} e^{-4\beta\sum_{i\in V}\sum_{j\notin V} J_{ij}} = C \sum_{\substack{\emptyset - Eulerian \\ subgraphs\ of\ D}} \prod_{e\in\emptyset} w_e, \quad (3.5)$$

where

$$C = \exp\left(\beta\sum_{i,j}^{N} J_{ij}\right). \quad (3.6)$$

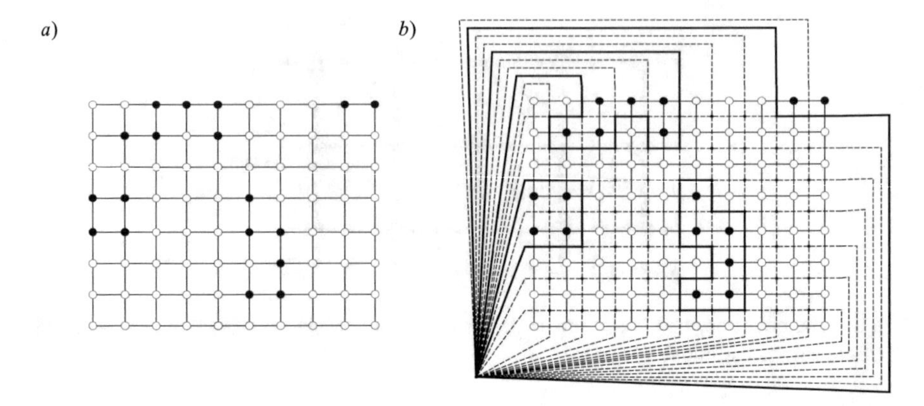

Fig. 2. (a) The original configuration in which filled circles corresponds to state +1, and blank circles stand for state −1; (b) The dual graph (dashed lines and small circles). The Eulerian cycles corresponding to the original configuration are drawn by bold lines.

3.2 Unfolding the Graph Nodes

Let dual graph D be built. We are going to do the two operations:

1. Let us first make all nodes of graph D have degree three by unfolding the nodes of degree greater than three (as shown in Fig. 3a).
2. Then let us replace each node (which is of degree three now) as shown in Fig. 3b. This node pattern is chosen because it always allows perfect matching.

The result of the transformation is a new graph, which we call graph R. The distinction of the graph is that it remains planar for one thing. For another, it has the even number of nodes and always permits perfect matching. Moreover, it is possible to show that a subgraph of Eulerian cycles in graph D will correspond to each perfect matching in graph R after the reverse operation of node contraction.

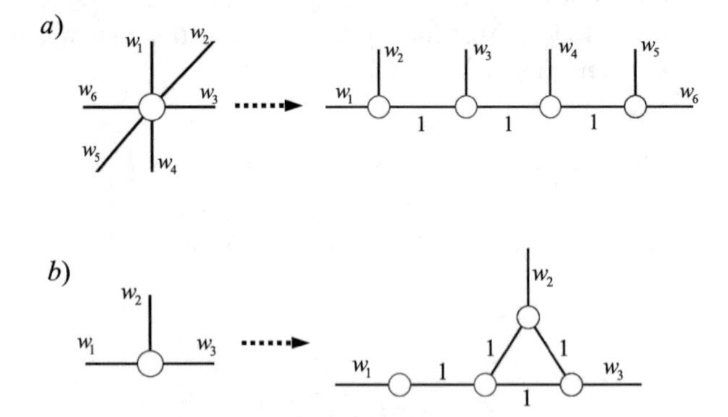

Fig. 3. Unfolding graph nodes. (a) Stage one: A node of degree greater than three is replaced by the chain of nodes. (b) Stage two: A node of degree three is replaced by the four nodes.

3.3 Perfect Matching and Search for the Pfaffian Orientation in the Graph

It is known that for any skew-symmetric matrix $A = (a_{ij} : a_{ij} = -a_{ji})$ of size $2n \times 2n$ the Pfaffian can be determined by the formula:

$$\mathrm{pf}(A) = \sum_{\pi} sign\begin{pmatrix} 1 & 2 & \dots & 2n \\ i_1 & j_1 & \dots & j_n \end{pmatrix} a_{i_1 j_1} a_{i_2 j_2} \dots a_{i_n j_n}, \tag{3.7}$$

where the summation is made over all pair combinations $\pi = \{\{i_1, j_1\}, \dots, \{i_n, j_n\}\}$ of set $\{1, \dots, 2n\}$, and *sign* denotes the sign of the substitution $\begin{pmatrix} 1 & 2 & \dots & 2n \\ i_1 & j_1 & \dots & j_n \end{pmatrix}$.

Graph R has been undirected so far. Let us assume that in graph R we have chosen a particular edge orientation and designated it as $\overrightarrow{R}$. We define matrix A as

$$a_{ij} = b_{ij} w_{ij}, \quad \text{where } b_{ij} = \begin{cases} 1, & \text{if } (i,j) \in e(\overrightarrow{R}) \\ -1, & \text{if } (j,i) \in e(\overrightarrow{R}) \\ 0, & \text{else} \end{cases} \tag{3.8}$$

Note that for graph $\overrightarrow{R}$ each term of sum (3.7) belonging to pairing $\pi = \{\{i_1, j_1\}, \dots, \{i_n, j_n\}\}$ corresponds to a certain matching. If a particular edge is missing in the matching (i.e. its weight is zero $a_{i_k j_k} = 0$), then the whole product $a_{i_1 j_1} a_{i_2 j_2} \dots a_{i_n j_n}$ is zero. It means that the summation (3.7) is equivalent to the summation only over available perfect matchings of graph $\overrightarrow{R}$. Since each perfect matching in graph R corresponds to a collection of Eulerian cycles in graph D (see previous paragraph), the partition function can be calculated by trying all perfect matchings:

$$Z = C \sum_{\substack{\emptyset - Eulerian \\ subgraphs\ in\ D}} \prod_{e \in \emptyset} w_e = C \sum_{\substack{PM - perfect \\ matchings\ in\ R}} \prod_{e \in PM} w_e \geq C \cdot |\mathrm{Pf}(A)|. \tag{3.9}$$

The inequality in the right side is caused by the fact that the summation in Pfaffian (3.7) holds terms of opposite signs. On the other hand, if the edge orientation in graph $\overrightarrow{R}$ was chosen so that matrix $B = \{b_{ij}\}$ would compensate the signs of permutations π, then all non-zero addends in sum (3.7) would appear with the plus sign and we would get

$$Z = C \cdot \mathrm{Pf}(A). \tag{3.10}$$

The corresponding orientation $B = \{b_{ij}\}$ is called Pfaffian orientation and, what is most important, according to Kasteleyn's theorem it really exists in planar graphs and can be found in polynomial time.

The criterion that can be used to check the Pfaffian orientation is the following theorem:

If $\overrightarrow{R}$ is a connected directed planar graph whose faces (perhaps, except for the infinite face) have an odd number of clockwise directed edges, this directed graph is a Pfaffian graph $\overrightarrow{R}$.

The algorithm of finding the Pfaffian orientation is as follows (see Fig. 4). We apply induction on the number of edges in graph R. If graph R is a tree, any orientation suits. Let us now assume that it is not a tree and select an edge belonging to the cycle and lying on the boundary of the infinite face. Let F_0 be a finite face that holds this edge e.

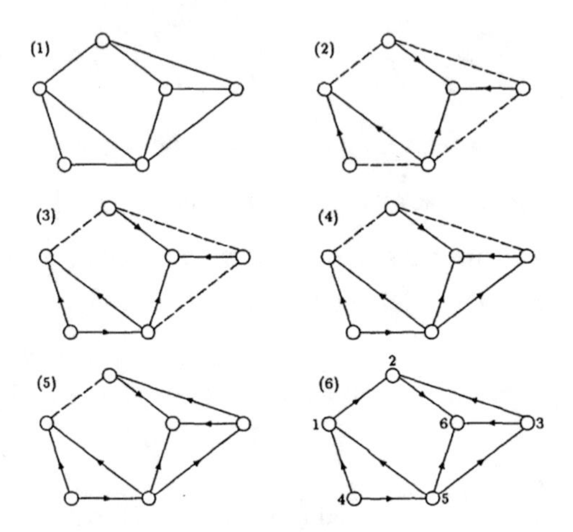

Fig. 4. Giving the Pfaffian orientation to a planar graph. (1) The original undirected graph. (2)–(6) The Pfaffian orientation constructing stages.

Following the induction, graph $(R - e)$ has the direction where the boundary of each finite face holds an odd number of clockwise directed edges. Let us return edge e to the graph and direct it so that the boundary of face F_0 has an odd number of clockwise directed edges. Since all boundaries of finite faces different from F_0 have not changed, the resulting orientation of graph R will have necessary properties.

3.4 Calculating the Pfaffian and Partition Function

After the Pfaffian orientation of graph $\overrightarrow{R}$ is found, formula (8) is used to determine matrix A. Then it is no problem to compute the partition function:

$$Z = C \cdot \text{Pf}(A) = C \cdot \sqrt{\det(A)}, \tag{3.11}$$

because the square of the Pfaffian of a skew-symmetric matrix is equal to the determinant of this matrix.

3.5 Discussion of Kasteleyn-Fisher Algorithm

We have described the algorithm [18] of exact computation of the partition function for two-dimensional binary-variable Ising models. The C/C++-based realization of the algorithm uses the Boost and Csparse libraries [30]. The algorithm complexity is $O(N^2)$ and is limited mostly by the time needed for computing the determinant of the sparse matrix. Due to optimization of procedures for working with matrices, the computational complexity can be reduced to $O(N^{3/2})$ [17].

To analyze the behavior of more complex than planar systems, the standard is the Monte Carlo method, which allows the most complete study of the measured quantities and establishes the critical parameters of the system [26]. Unfortunately, this method requires a large amount of calculations and does not allow direct calculation of free energy, but with the help of this method we obtained magnetization distributions and correlation lengths, which cannot be obtained with the help of the Kasteleyn-Fisher algorithm.

4 Results

In this Section, we show the results obtained with algorithms described above. The thermodynamic characteristics of the two-dimensional Ising model as a function of the number of spins N are examined. We show how to generalize Onsager's solution to a finite-size lattice. Experimentally validated analytical expressions for the free energy and its derivatives are computed. The heat capacity at the critical point is shown to grow logarithmically with N. Due to the finite extent of the system the critical temperature can only be determined to some accuracy. This Section combines the materials of the papers [33–35]. Some new results on Ising model with noisy couplings can be found here [36].

4.1 Experimental Results

We make an intensive use of the Kasteleyn-Fisher algorithm here to compute the free energy of the 2D square spin system. The algorithm gives exact results because the finding of the partition function is reduced to computation of the determinant of a matrix generated in accordance with the model under consideration. The algorithm permits us to exactly calculate the free energy of a spin system for an arbitrary planar graph with arbitrary links in a polynomial time. In the paper we use the realization [17] of the algorithm that can give the same results in a shorter time. Using this algorithm, we were able to examine the behavior of free energy and its derivatives (internal energy U and heat capacity C) for a few lattices of different dimensions $N = L \times L$. The length of the lattice varied from $L = 25$ to $L = 10^3$. Let us point out that the algorithm we use is only applicable to planar lattices. It means that we considered only lattices with free boundary conditions because lattices with periodic boundary conditions do not belong to a planar graph. Correspondingly, the basic state energy is

$$E_0 = -2\left(1 - \frac{1}{\sqrt{N}}\right). \tag{4.1}$$

In the experiment we computed free energy $f = f(\beta)$ and its derivatives. As expected, the peak of curve $C = C(\beta)$ is shifted to the right from the peak of curve $\sigma = \sigma(\beta)$. The position of the heat capacity peak is used determine critical temperature β_c and critical values $f_c = f(\beta_c)$, $U_c = U(\beta_c)$, $\sigma_c = \sigma(\beta_c)$ and $C_c = C(\beta_c)$. The position of the energy variance peak is used to find the second critical point β_c^* and corresponding critical values $f_c^* = f(\beta_c^*)$, $U_c^* = U(\beta_c^*)$, $\sigma_c^* = \sigma(\beta_c^*)$ and $C_c^* = C(\beta_c^*)$. All of these values are given in Table 1.

The results of the experiment and data analysis are presented graphically in Figs. 5, 6 and 7. As seen from Fig. 5, the experimentally found values of free energy and internal energy approach Onsager's solution with growing dimensionality. The figure gives the curves only for small lengths $L = 25,\ 50,\ 100$. When $L > 100$, the curves practically repeat Onsager's solution and are not shown in the Figure for this reason. According to (4.1), the asymptotical behavior of free energy for large β is described as

$$f \approx -2\beta\left(1 - \frac{1}{\sqrt{N}}\right). \tag{4.2}$$

It is the presence of term $\sim 1/\sqrt{N}$ that causes the curves representing small linear dimensions not to follow Onsager's solution.

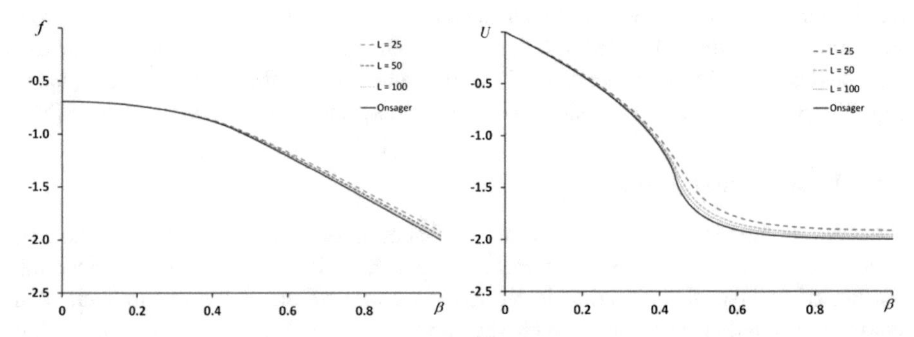

Fig. 5. Free energy $f = f(\beta)$ (the left plot) and internal energy $U = U(\beta)$ (the right plot) for small-dimension lattices and asymptotic Onsager's solution ($L \to \infty$)

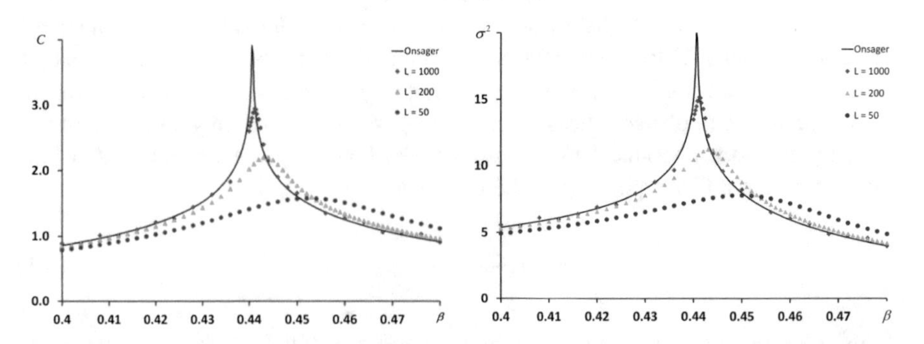

Fig. 6. Dependences of heat capacity C (the left plot) and variance σ^2 (the right plot) from β for some lattice dimensions from $N = 50 \times 50$ to $N = 10^3 \times 10^3$ (dots). Osanger's solution is drawn in solid.

For the finite-extent lattice the difference between Osanger's solution and $f = f(\beta)$ is best seen in the plot of the second derivative (heat capacity C). According to Osanger's solution the heat capacity exhibits logarithmic divergence when $\beta \to \beta_{ONS}$. In the case of finite lattice it does not happen; however, we can observe the peak in the heat capacity curve $C = C(\beta)$, the peak becoming sharper with the growing lattice dimension (see Fig. 6). A closer examination shows that the peak height increases logarithmically with the lattice dimension and the peak itself is slightly shifted to the right from β_{ONS}, the distance between the two points shortening as the lattice dimension grows.

The examination of the data of Table 1 shows that the position of the peak (critical value β_c) and the dimension dependencies of the critical values of free energy and heat capacity can be approximated well by the following expressions:

$$\begin{aligned} \beta_c &= \beta_{ONS}\left(1 + \frac{5}{4\sqrt{N}}\right) \\ U_c &= -\sqrt{2}\cdot\left(1 - \frac{1}{2\sqrt{N}}\right) \\ C_c &= \frac{4\beta_c^2}{\pi}(\ln N - 1.7808) \end{aligned} \tag{4.3}$$

The approximation (4.3) of dependency $\beta_c = \beta_c(N)$ gives a rather small relative error: the largest error of $\sim 0.3\%$ is at $L = 25$, the error decreases rapidly with growing L (to 0.01% at $L = 10^3$). The relative error of the approximation of U_c is less than 0.4% and C_c less than 0.8%. Figure 7 shows how close expressions (4.3) follow experimental data.

The position of the energy variance peak (critical value β_c^*) and the corresponding values of free energy and heat capacity are well approximated (see Table 1) by the expressions:

$$\begin{aligned} \beta_c^* &= \beta_{ONS}\left(1 + \frac{1}{\sqrt{N}}\right) \\ U_c^* &= -\sqrt{2}\cdot\left(1 - \frac{1}{\sqrt{N}}\right) \\ C_c^* &= 1.197\,\beta_{\mathrm{ONS}}^2\cdot(\ln N - 1) \end{aligned} \tag{4.4}$$

These formulae give good agreement with experimental data: β_c^*, U_c^* and C_c^* have the greatest relative error 0.6%, 2.1% and 1.2% correspondingly at $L = 25$. The relative errors fall rapidly with L and at $L = 10^3$ become 0.02%, 0.03% and 0.08%.

Table 1. Critical values at the peaks of heat capacity and energy variance.

L	β_c/β_c^*	f_c/f_c^*	U_c/U_c^*	σ_c/σ_c^*	C_c/C_c^*
25	0.4642/0.4556	0.9467/0.9351	1.3808/1.3288	2.4444/2.4678	1.2875/1.2642
50	0.4522/0.4494	0.9382/0.9344	1.3985/1.3768	2.7762/2.7849	1.5760/1.5664
100	0.4462/0.4454	0.9337/0.9326	1.4054/1.3978	3.0767/3.0782	1.8846/1.8797
200	0.4436/0.4432	0.9320/0.9314	1.4120/1.4075	3.3491/3.3502	2.2072/2.2046
300	0.4428/0.4422	0.9315/0.9306	1.4152/1.4078	3.4990/3.5001	2.4005/2.3955
400	0.4422/0.4418	0.9309/0.9304	1.4143/1.4091	3.6050/3.6052	2.5413/2.5369
500	0.4418/0.4418	0.9305/0.9305	1.4131/1.4131	3.6832/3.6832	2.6479/2.6479
600	0.4414/0.4414	0.9301/0.9301	1.4104/1.4104	3.7525/3.7525	2.7435/2.7435
700	0.4414/0.4414	0.9302/0.9302	1.4124/1.4124	3.8141/3.8141	2.8344/2.8344
800	0.4414/0.4414	0.9302/0.9303	1.4139/1.4139	3.8544/3.8544	2.8945/2.8945
900	0.4414/0.4414	0.9303/0.9303	1.4152/1.4152	3.8702/3.8702	2.9184/2.9184
1000	0.4412/0.4412	0.9301/0.9301	1.4132/1.4132	3.8914/3.8914	2.9477/2.9477

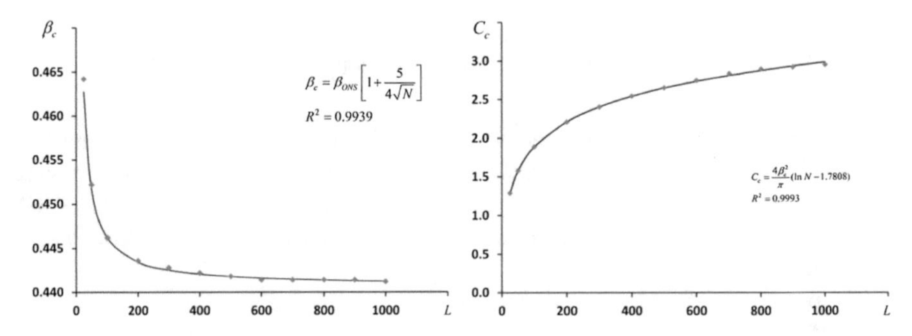

Fig. 7. Critical temperature β_c (the left plot) and heat capacity C_c (the right plot) as functions of dimension L: the dots represent experimental data, the solid lines approximation formulae (7).

4.2 Generalization of Onsanger's Solution

The analysis shows that it is possible to get the analytical expressions that can describe experimental data and the above approximation formulae quite well. It is sufficient to make substitutions $2\beta J \rightarrow z$ and $k \rightarrow \kappa$ in (2.5), where

$$\kappa = \frac{2\sinh z}{(1+\delta)\cosh^2 z}. \tag{4.5}$$

Then for free energy, internal energy and heat capacity we get

$$f(\beta) = -\frac{\ln 2}{2} - \ln(\cosh z) - \frac{1}{2\pi}\int_0^{\pi} \ln\left(1+\sqrt{1-\kappa^2\cos^2\theta}\right)\, d\theta z = \frac{2\beta J}{1+\Delta}$$

$$U = -\frac{1}{1+\Delta}\left\{2\tanh z + \frac{\sinh^2 z - 1}{\sinh z \cdot \cosh z}\left[\frac{2}{\pi}K_1 - 1\right]\right\}$$

$$C = \frac{z^2}{\pi \tanh^2 z} \cdot \left\{ a_1(K_1 - K_2) - \left(1 - \tanh^2 z\right) \left[\frac{\pi}{2} + \left(2a_2 \tanh^2 z - 1\right) K_1 \right] \right\} \quad (4.6)$$

where $K_1 = K_1(\kappa)$ and $K_2 = K_2(\kappa)$ are full elliptical integrals of first and second type correspondingly and

$$a_1 = p\,(1+\delta)^2,\; a_2 = 2p - 1,\; p = \frac{\left(1 - \sinh^2 z\right)^2}{(1+\delta)^2 \cosh^4 z - 4 \sinh^2 z} \quad (4.7)$$

As could be expected, when $N \to \infty$, formulae (4.7) give $p \to 1$, $a_{1,2} \to 1$ and expressions (4.6) turn into well-known ones (2.5). If we compare resulting expressions (4.6) with experimental data, we can see that the best agreement occurs when the adjustment parameters take the form:

$$\Delta = \frac{5}{4\sqrt{N}},\; \delta = \frac{\pi^2}{N} \quad (4.8)$$

Expressions (4.6) give good approximation of experimental results even if the lattice dimension is small. By way of illustration Fig. 8 gives the curves of energy variance and heat capacity for a $N = 25 \times 25$ lattice. It is seen that there is good agreement between the theory and experiment. This agreement becomes better with the growing lattice dimension.

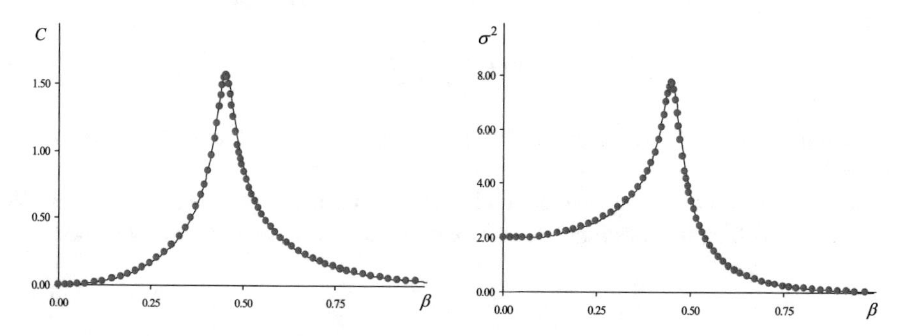

Fig. 8. Dependencies of heat capacity (the left plot) and variance (the right plot) on β for a $N = 25 \times 25$ lattice: solid-line curves are produced by formulae (4.6), dots are experimental data.

The examination of expressions (4.6) shows that the introduction of the correction for a finite lattice dimension does not change behavior of free and internal energy much. On the other hand, in the heat capacity formula the logarithmic divergence at the critical point disappears. Indeed, the examination shows that the maximum of heat capacity occurs at $\sinh z = 1$, which corresponds to the critical temperature from $\beta_c = \beta_{ONS}\,(1+\Delta)$. Borrowing Δ from (4.8), we find that the expression agrees with empirically defined expression (4.4) fully.

Expanding function $C(\beta)$ in a series about critical point β_c and omitting the terms that are polynomial in $(\beta - \beta_c)$, we

$$C(\beta) \approx \frac{4\beta_c^2 J^2}{\pi}\left\{3\ln 2 - \frac{\pi}{2} - \ln\left[4J^2(\beta-\beta_c)^2 + \frac{\pi^2}{N}\right]\right\}. \tag{4.9}$$

This expression yields the following expression for the critical heat capacity:

$$C_c = \frac{4\beta_c^2 J^2}{\pi}\left(\ln N + 3\ln 2 - 2\ln\pi - \frac{\pi}{2}\right), \tag{4.10}$$

which corresponds to (4.4) because $2\ln\pi + \pi/2 - 3\ln 2 \approx 1.7808$.

4.3 Metropolis Algorithm Simulations

Our Eq. (4.5) allow us to examine the dependence of the correlation length ξ on N. In the limit $N \to \infty$, in accordance with the well-known expression from [7] this dependence is $\xi = -1/2\ln\eta$, where $\eta = k/(1+\sqrt{1-k^2}\,)$. To use this expression for a lattice of a finite size we make the substitution (4.5) $k \to \kappa$. Then we obtain the correlation length in the form

$$\xi = -\frac{1}{2\ln\bar{\eta}},\ \bar{\eta} = \frac{\kappa}{1+\sqrt{1-\kappa^2}} \tag{4.11}$$

At the critical point $\beta = \beta_c$, the value $\kappa = \kappa(z)$ reaches its maximum equal to $\kappa_{\max} = 1/(1+\delta)$ and the correlation length is minimal and equal to

$$\xi_{mzx} = \frac{L}{2\pi\sqrt{2}}. \tag{4.12}$$

Approximately this value one order of magnitude less than the linear size of the lattice L.

The correlation length $r_d = \exp(-d/\xi)$ characterize the correlation between spins at a distance d. Taking into account Eq. (4.11), we obtain the expression for r_d in the form:

$$r_d = \bar{\eta}^{2d} \tag{4.13}$$

Figures 9 and 10 shows the experimental values of the correlation between four nearest spins ($d = 1, 2, 3, 4$), as well as the theoretical curves constructed from this formula. The results are obtained by the Metropolis Monte Carlo method.

In the same way, we substitute Eq. (4.5) in the expression for the spontaneous magnetization obtained by Yang [31]. Then in our case the spontaneous magnetization is

$$M_0 = \left[1 - \left(\frac{\bar{\eta}}{\bar{\eta}_0}\right)^2\right]^{1/8}, \tag{4.14}$$

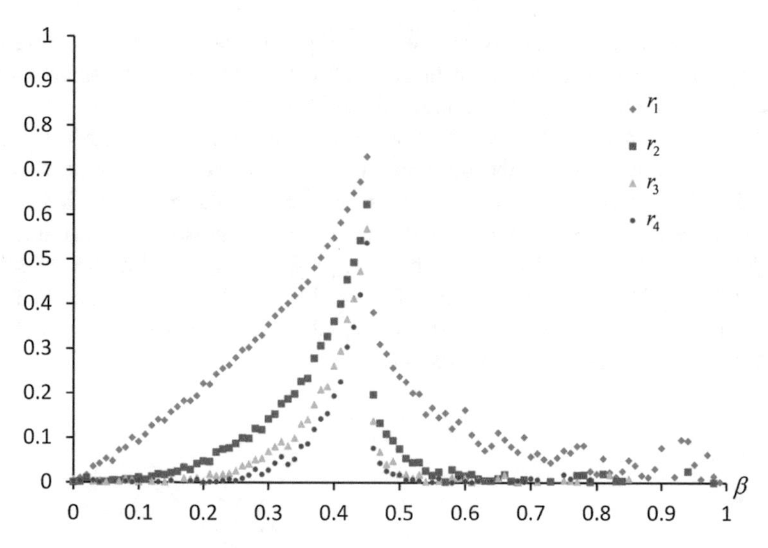

Fig. 9. Correlation r_d as a function of temperature: lattice size $L = 50$ and $d = 1, 2, 3, 4$, counting from the center of the lattice.

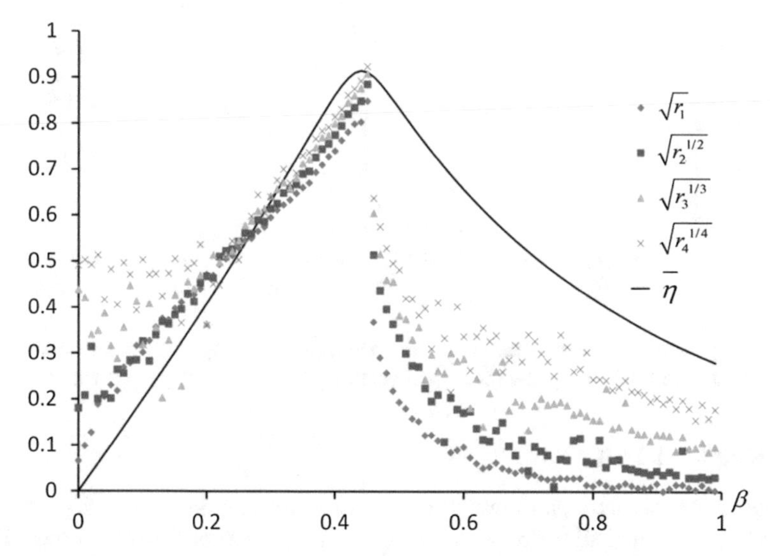

Fig. 10. Graphs $\sqrt[2d]{r_d}$ vs β for $d = 1, 2, 3, 4$ for square lattice of size $L = 50$. The solid line shows the theoretical curve $\bar{\eta}(\beta)$.

where $\bar{\eta}_0 \approx 1/(1+\sqrt{2\delta})$ is the value of the function $\bar{\eta} = \bar{\eta}(\kappa)$ for $\beta = \beta_c$ ($M_0 = 0$ when $\beta < \beta_c$). It must be emphasized, that when the size of the system is finite the interpretation of the expression (4.14) is completely different than in the case $N \to \infty$. As it was pointed out in [7], in the absence of magnetic field mean values of magnetizations of finite systems are equal to zero. The reason is that for each configuration with $s_i = +1$ there is the equally probable configuration with $s_i = -1$. Consequently Eq. (4.14) states that at different instants of time the magnetization obtained in the framework of simulations takes on any value between M_0 and $-M_0$.

To verify the obtained theoretical values of the magnetization for each temperature we examined experimentally the distribution of the magnetization by means of the Metropolis Monte Carlo method. Since the mean value of the magnetization is always equal to zero we checked the value of magnetization at the peak of distribution only. When β is not so large the distribution has one peak at $M = 0$. When β is close to the critical value or larger the peak splits. In Fig. 11, we show the coordinates of the magnetization peak as function of the temperature. As we see, they are in good agreement with the obtained theoretical curve.

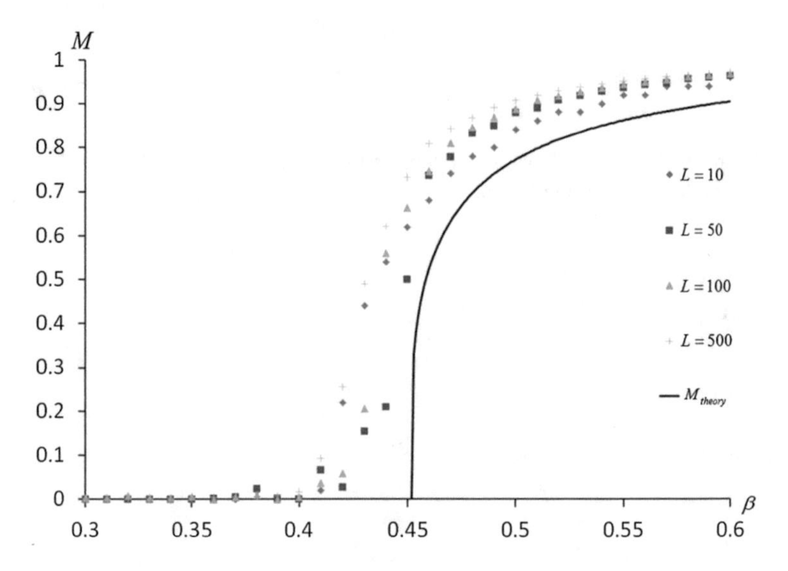

Fig. 11. Dependence of the magnetization M on the reciprocal temperature β for different values $l = 10...500$. The solid line is the theoretical curve (4.14) for $l = 50$ ($M = 0$ when $\beta < \beta_c$).

4.4 Ising Model with Noise

It can be assumed that logarithmic growth with increasing dimensionality is a consequence of the fact that all matrix elements J_{ij} in the grid are equal to each other. Thus, a kind of coherence arises, leading to an unlimited increase in the specific heat with increasing grid size. To avoid this kind of coherence, we investigate the case in which the value of the interaction coefficients varies about unit just slightly:

$$J_{ij} = 1 + \varepsilon,\ \varepsilon \in (-\eta, \eta) \tag{4.15}$$

ε stands for the additive noise, which is uniform over the interval and whose maximum amplitude is equal to parameter η.

In computational experiments we used the Kasteleyn-Fisher algorithm. Enabling the calculation of free energy for any planar models, the algorithm allowed us to analyze the behavior of free energy and its derivatives (internal energy U and heat capacity C) for different grids of dimension $N = L \times L$. It is to be noted that the algorithm is applicable to planar grids only. It means that we consider only grids with free boundary conditions because a grid with periodic boundary conditions, as well as a two-dimensional model subjected to an external field, is not a planar graph.

We carried out a few experiments with grids whose dimension ranged from $L = 100$ to $L = 400$, noise amplitude η varying from 0 to 1.5.

Since our major interest was the behavior of the critical temperature, its second derivative (energy dispersion) σ^2 rather than the free energy itself provided us most information. Figure 12 shows the results of our experiments.

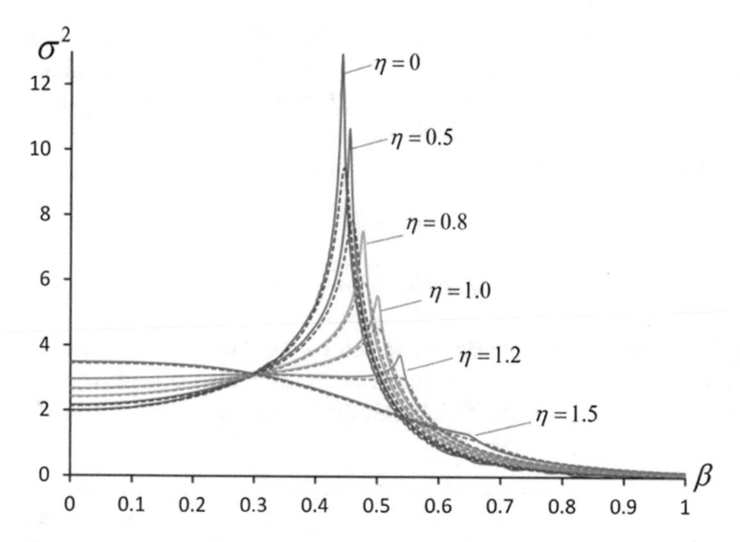

Fig. 12. Energy dispersion σ^2 as a function of inverse temperature β for different noise amplitudes η. The dashed and solid lines represent data for $L = 100$ and $L = 400$, correspondingly.

It is seen from the Fig. 12 that the dispersion peak moves smoothly to the right, i.e. towards greater values of β, with the noise amplitude η. The height of the peak decreases until it becomes almost invisible at $\eta = 1.5$ ($L = 400$) and disappears at all with small grid dimensions ($L < 100$).

It is worth noticing that there is a point near $\beta \approx 0.3$ in Fig. 12 where all the curves intersect. We don't know if this observation bears any physical significance, but the intersection looks like a sharp point only at small noise amplitudes ($\eta < 1.5$), it spreads out with higher noise amplitudes.

Figure 12 permits the conclusion that the peak disappears at all starting from a particular noise amplitude. It is probable that the phase transition, which is typical of the two-dimensional Ising model, disappears in the same manner as the dispersion peak does.

Besides the temperature dependence of thermodynamic properties, we investigated the effect of added noise on the spectral characteristics of the system. Let's define the probability of the system being in a particular energy state as:

$$P(E) = \frac{e^{N\Psi(E)}}{2^N}, \tag{4.16}$$

where $\Psi(E)$ is the spectral density of the energy state distribution. It is simple to show [10] that spectral density $\Psi(E)$ is related to the free energy and its derivatives by the following implicit formulae:

$$E = \frac{df(x)}{dx}, \ \Psi(E) = xE - f(x) \tag{4.17}$$

where $f(x)$ results from $f(\beta)$ by substituting β by x. Here we introduce a variable x to point out that the spectral density is not related to the temperature variable. Varying x from 0 to ∞, we go over the whole energy spectrum: expressions (13) allow energy E and density $\Psi(E)$ to be found for each value of x. The first and second derivatives of the spectral density are also determined by implicit formulae:

$$\partial\Psi/\partial E = -x, \tag{4.18}$$

$$\partial^2\Psi/\partial E^2 = -\sigma^{-2}(x). \tag{4.19}$$

The plot of the spectral density itself holds little information because it is almost independent of noise of matrix elements. More information can be obtained from the plot of the second derivative of $\Psi(E)$ with respect to energy (see Fig. 13).

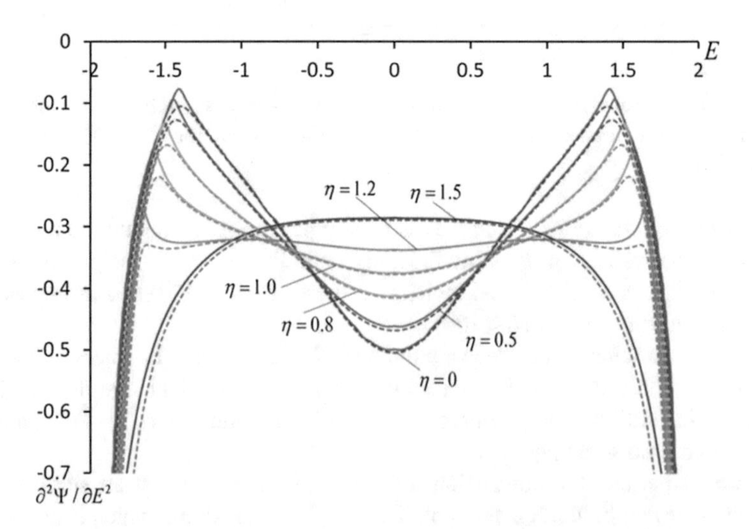

Fig. 13. The second derivative of the distribution density as a function of energy E for different noise amplitudes η. The dashed and solid lines represent data for $L = 100$ and $L = 400$, correspondingly.

It is clear that the higher the dispersion peak σ^2, the nearer $\partial^2\Psi/\partial E^2$ approaches zero (see Eq. (4.19)). When $N \to \infty$, dispersion $\sigma_c^2 \to \infty$ at the critical point, and the curve $\partial^2\Psi/\partial E^2$ comes close to zero. With finite dimensions, the dispersion is always finite. For this reason, $\partial^2\Psi/\partial E^2$ also stays a finite distance off zero.

When the noise is added, the structure of the spectrum changes gradually, which is clearly seen in Fig. 13. When $\eta < 1.2$, we can see a noticeable sag in the middle of the plot. When $\eta > 1.5$, there is no sag, which may mean a qualitative change of the system.

4.5 n-Vicinities Method Estimates

Let us consider the relationship between the critical temperature and scatter of matrix elements. Since Onsager's solution doesn't hold such information, let us use the approximation based on so-called n-vicinities method [11, 12]. After some simplifications the equation of state from [11, 12] takes the form:

$$\ln\frac{1+m}{1-m} = 8mb\left[\frac{2-b}{2+\sigma_\eta^2} - b\left(\frac{1}{2} - m^2\right)\right] \tag{4.20}$$

where variable m is the magnetization, σ_η^2 is the noise dispersion of matrix elements, b is the dimensionless parameter related to β as:

$$\beta = \frac{\pi b}{\sqrt{2}\,(2+\sigma_\eta^2)} \tag{4.21}$$

Finding the critical point is reduced to the numerical solution of (4.20) for given parameters b and σ_η^2. When b is small, Eq. (4.20) has a unique solution $m = 0$. When b is greater than critical value b_c, there is another solution $m \neq 0$. Substituting b_c in (4.21) gives the critical value of inverse temperature $\beta = \beta_c$, at which the phase transition occurs.

Table 2 gives β_c for different σ_η^2: the first row holds experimental values of β_c, and the second row carries the values obtained by solving Eq. (4.20) numerically.

Table 2. The relationship between β_c and σ_η^2: experiment and theory.

η	0.0	0.2	0.5	0.8	1.0	1.2
$\sigma_\eta^2 = \eta^2/3$	0.000	0.013	0.083	0.213	0.333	0.480
β_c experiment	0.442	0.444	0.454	0.476	0.500	0.536
β_c theory	0.440	0.442	0.452	0.477	0.501	0.534

It is seen that theoretical and experimental values differ by less than 0.5%, which corresponds to the experimental accuracy (see also Fig. 14). It can be noticed that

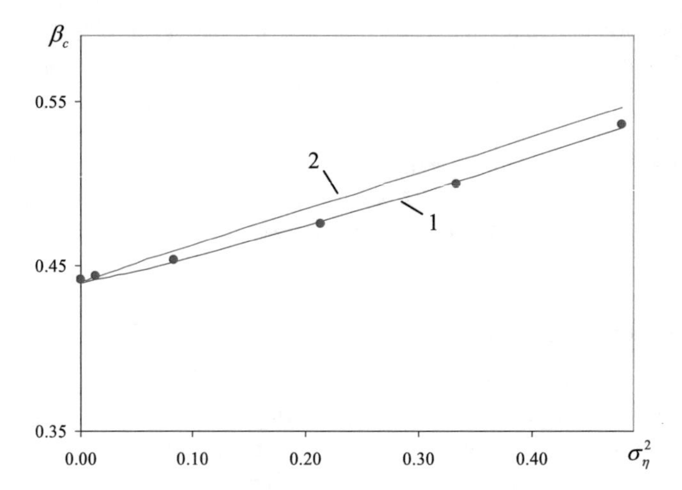

Fig. 14. Dependence $\beta_c = \beta_c(\sigma_\eta)$: the rhombi stand for experimental points; 1 is the theoretical curve; 2 is the approximation curve (4.22).

within the given range of dispersion σ_η^2 dependence $\beta_c = \beta_c(\sigma)$ can be approximated accurately enough by the expression:

$$\beta_c = \beta_{\text{ONS}}\left(1 + \frac{1}{2}\sigma_\eta^2\right), \tag{4.22}$$

where β_{ONS} is Onsager's critical value (2.9). The curves $\beta_c = \beta_c(\sigma_\eta)$ are presented in Fig. 14. It is seen that critical value β_c grows with σ_η^2 almost linearly.

5 Conclusion

We have described the algorithm of exact computation of the partition function for two-dimensional binary-variable Ising models. The algorithm has polynomial complexity and is limited mostly by the time needed for computing the determinant of the sparse matrix.

Basing on our experimental data, we obtained simple expressions (4.4) allowing us to estimate the critical values that are in good agreement with experiments. If we choose the fitting parameters in the form (4.8) the error of approximation in the limit $\beta \to \infty$ is rather small and we can describe the coordinates and heights of peaks almost precisely. Of course, one may try to define the fitting parameters more accurately and then write down the estimates for critical values to higher accuracy. However, it was not our goal since we seek for the simplest expressions providing the high accuracy when $l \geq 40$. The objective of our work was to find how the critical values behave themselves as functions of N.

We introduced two fitting parameters (4.8) and this allows us to derive analytic expressions (4.6) generalizing the Onsager solution to the case of lattices of finite sizes.

With great accuracy, these expressions describe the behavior of spin systems even when sizes of their lattices are small ($N \sim 25 \times 25$). When $N \geq 50 \times 50$ the differences between our analytical solution and the experimental data become less than experimental errors. Moreover, introduction of these fitting parameters allowed us to derive approximate expressions for the magnetization and the correlation length as functions of the temperature that provide good qualitative agreement with the dynamic Metropolis Monte Carlo simulations.

Basing on our analysis, we can conclude that:

First, computer simulations allows one to determine correctly distinctive features of behavior of spin systems even when the values of N are comparatively small. Increase of N only allows one to determine more accurately the values of the critical parameters. However, this refinement is not very important: from Eq. (4.4) we see that the accuracy of the critical values β_c and U_c is determined to within $\sim 1/\sqrt{N}$.

Second, a system of a finite size has no logarithmic divergence of the heat capacity in the critical point predicted by Onsager. The same is also true for the energy dispersion. It was to be expected from the most general considerations. Instead we see that the heat capacity in the critical point increases logarithmically, that is $C_c \sim \ln N$. One would think that when $N \to \infty$ we return to the Onsager solution ($C \to \infty$ for $\beta \to \beta_c$). However, it is difficult to realize this transition since if we increase the size of the lattice even up to the Avogadro number $N \sim 10^{23}$, the heat capacity C_c increases only 4 times comparing with the case $N = 10^6$. Moreover, the dependence of the heat capacity on N means violation of the additivity concept for a classical system that is clearly seen even when $N \sim 10^{23}$: if we double the size of the system the value of C_c is up by 1.3%. Of course, we examined the system with the free boundary conditions and such systems are non-additive by definition. In the same time, non-additivity is present also in the models with the periodic boundary conditions since according Eq. (4.12) in the critical point correlations increase proportional to linear sizes of the lattices.

The logarithmic dependence of the critical heat capacity on the dimension was predicted as far in the Onsager paper [19]. He examined a rectangular lattice of the size $l \times l'$ and when $l' \to \infty$ obtained the dependence:

$$C_c \approx 0.4945 \cdot \ln l + 0.1879 \tag{5.1}$$

We examined a somewhat different system (a square lattice) and obtained

$$C_c \approx 0.4945 \cdot \ln l - 0.4403 \tag{5.2}$$

Comparing Eqs. (5.1) and (5.2) we see that only the constant but not the dependence of C_c on l changed. We suppose that this is a result of different boundary conditions: Onsager used the periodic boundary conditions, while we explore the free boundary conditions. The experiment performed for a square lattice with the periodic boundary conditions [32] confirms this conclusion.

Finally, our simulations showed that the peaks of the curves $\sigma_E^2 = \sigma_E^2(\beta)$ and $C = C(\beta)$ do not coincide. At that the heat capacity reaches its maximum at larger values of β. This is an expected result since the relation between the heat capacity and the energy dispersion is $C(\beta) = \beta^2 \sigma_E^2(\beta)$. However, there is a question: which of the peaks we have to use to define the critical temperature? Indeed, the first peak corresponds to the maximum of the energy dispersion and the second to the maximum of the correlation length. They both are characteristics of a phase transitions. We (mostly through a habit) defined the critical temperature as the coordinate of the peak of the heat capacity. When the size of the lattice was large, this approach was justified: when $N \geq 400 \times 400$ the distance between these peaks was so small that we did not see it. However, at smaller sizes the distance between the peaks was noticeable. Most likely when interpreting the results of computer simulations we have to state that the phase transition is smeared over the interval from β_c^* to β_c. Consequently, numerical experiments allow one to define the critical temperature within the accuracy of the length of this interval. This means that the absolute error has to be of the order of $\pm\beta_{ONS}/4\sqrt{N}$.

We analyzed the dependences of the critical parameters on the size of the lattice by example of the 2D Ising model. However, we hope that our basic conclusions are also true for other models.

It was interesting for us to know how the adding of noise to interaction elements changes the system and its thermodynamic properties. The conventional Onsager model, in which all elements are strictly units, seems to be rather far from practice. We have showed that a small addition of noise does not bring about sharp changes. Rather it leads to smooth changes of thermodynamic properties such as heat capacity and spectral density. At the same time, starting with a certain value of the noise amplitude, a qualitative rearrangement of the system is likely, perhaps even a disappearance of the phase transition. However, judging about the presence of phase transition in finite-dimension systems is rather a problem.

The primary goal of further research is the addition of an external field to the model. We believe that the described methods for estimating and calculating the partition function for the case of random connections will be useful in training Bayesian neural networks. We compared experimental results with ones obtained with the earlier developed n-vicinities method. This approximation method showed excellent agreement with the results of the numerical experiment, which gives us reasons to look forward to its further development and application in machine learning and image processing.

Acknowledgements. The research was done as a part of program “35.14. Investigation of the multiple-extremum quadratic functional energy surface topology in the q-ary variable state space (0065-2018-0001)”.

References

1. Hinton, G.E., Salakhutdinov, R.R.: Reducing the dimensionality of data with neural networks. Science **313**(5786), 504–507 (2006)
2. Krizhevsky, A., Hinton, G.E.: Using very deep autoencoders for content-based image retrieval. In: European Symposium on Artificial Neural Networks, ESANN 2011, Bruges, Belgium (2011)

3. Zeiler, M.D., Taylor, G.W., Fergus, R.: Adaptive deconvolutional networks for mid and high level feature learning. In: Proceedings of the 2011 International Conference on Computer Vision (ICCV 2011), Washington, DC, USA, pp. 2018–2025. IEEE Computer Society (2011) https://doi.org/10.1109/iccv.2011.6126474
4. Wainwright, M.J., Jaakkola, T., Willsky, A.S.: A new class of upper bounds on the log partition function. IEEE Trans. Inf. Theory **51**, 2313–2335 (2005)
5. Yedidia, J.S., Freeman, W.T., Weiss, Y.: Constructing free-energy approximations and generalized belief propagation approximations. IEEE Trans. Inf. Theory **51**, 2282–2312 (2005)
6. Wainwright, M.J., Jordan, M.I.: Graphical models, exponential families, and variational inference. Technical report, UC Berkeley, Department of Statistics (2003)
7. Baxter, R.J.: Exactly Solved Models in Statistical Mechanics. Academic Press, London (1982)
8. Stanley, H.: Introduction to Phase Transitions and Critical Phenomena. Clarendon Press, Oxford (1971)
9. Wainwright, M.J., Jaakkola, T., Willsky, A.S.: Tree-based reparameterization framework for analysis of sum-product and related algorithms. IEEE Trans. Inf. Theory **49**(5), 1120–1146 (2003)
10. Kryzhanovsky, B., Litinskii, L.: Approximate method of free energy calculation for spin system with arbitrary connection matrix. J. Phys. Conf. Ser. **574**, 012017 (2015). http://arxiv.org/abs/1410.6696
11. Kryzhanovsky, B., Litinskii, L.: Generalized approach to energy distribution of spin system. Opt. Mem. Neural Netw. **24**, 165 (2015). http://arxiv.org/abs/1505.03393
12. Kryzhanovsky, B., Litinskii, L.: Applicability of n-vicinity method for calculation of free energy of Ising model. Phys. A Stat. Mech. Appl. http://dx.doi.org/10.1016/j.physa.2016.10.074. Accessed 3 Nov 2016. ISSN 0378-4371
13. Kac, M., Ward, J.: A combinatorial solution of the two-dimensional Ising model. Phys. Rev. **88**(6), 1332 (1952)
14. Sherman, S.: Combinatorial aspects of the Ising model for ferromagnetism. I. A conjecture of Feynman on paths and graphs. J. Math. Phys. **1**(3), 202–217 (1960)
15. Kasteleyn, P.: Dimer statistics and phase transitions. J. Math. Phys. **4**(2), 287–293 (1963)
16. Fisher, M.: On the dimer solution of planar Ising models. J. Math. Phys. **7**(10), 1776–1781 (1966)
17. Schraudolph, N., Kamenetsky, D.: Efficient exact inference in planar Ising models. In: NIPS (2008). https://arxiv.org/abs/0810.4401
18. Karandashev, Ya.M., Malsagov, M.Yu.: Polynomial algorithm for exact calculation of partition function for binary spin model on planar graphs. Opt. Mem. Neural Netw. (Inf. Opt.) **26**(2) (2017). https://arxiv.org/abs/1611.00922
19. Onsager, L.: Crystal statistics. I. A two-dimensional model with an order–disorder transition. Phys. Rev. **65**(3–4), 117–149 (1944)
20. Blote, H.W.J., Shchur, L.N., Talapov, A.L.: The cluster processor: new results. Int. J. Mod. Phys. C **10**(6), 1137–1148 (1999)
21. Häggkvist, R., Rosengren, A., Lundow, P.H., Markström, K., Andren, D., Kundrotas, P.: On the Ising model for the simple cubic lattice. Adv. Phys. **56**(5), 653–755 (2007)
22. Lundow, P.H., Markstrom, K.: The critical behavior of the Ising model on the 4-dimensional lattice. Phys. Rev. E. **80**, 031104 (2009). Preprint: arXiv:1202.3031v1
23. Lundow, P.H., Markstrom, K.: The discontinuity of the specific heat for the 5D Ising model. Nucl. Phys. B **895**, 305–318 (2015)

24. Dixon, J.M., Tuszynski, J.A., Carpenter, E.J.: Analytical expressions for energies, degeneracies and critical temperatures of the 2D square and 3D cubic Ising models. Phys. A **349**, 487–510 (2005)
25. Lyklema, J.W.: Monte Carlo study of the one-dimensional quantum Heisenberg ferromagnet near T ¼ 0. Phys. Rev. B. **27**(5), 3108–3110 (1983)
26. Binder, K., Luijten, E.: Monte Carlo tests of renormalization-group predictions for critical phenomena in Ising models. Phys. Rep. **344**, 179–253 (2001)
27. Lovas, L., Plammer, M.: Applied Problems of the Graph Theory. The Pair Matching Theory in Mathematics, Physics, Chemistry. Mir (1998)
28. Middleton, A., Thomas, C.K.: Matching Kasteleyn cities for spin glass ground states. Physics. Paper 180 (2007) http://surface.syr.edu/phy/180
29. Liers, F., Pardella, G.: A simple MAX-CUT algorithm for planar graphs. Technical report, 16 p. (2008)
30. Davis, T.A.: Direct Methods for Sparse Linear Systems. SIAM, Philadelphia (2006)
31. Yang, C.N.: The spontaneous magnetization of a two-dimensional Ising model. Phys. Rev. **65**, 808 (1952)
32. Häggkvist, R., Rosengren, A., Andrén, D., Kundrotas, P., Lundow, P.H., Markström, K.: Computation of the Ising partition function for two-dimensional square grids. Phys. Rev. E **69**, 046104 (2004)
33. Karandashev, I.M., Kryzhanovsky, B.V., Malsagov, M.Yu.: The analytical expressions for a finite-size 2D Ising model. Opt. Mem. Neural Netw. (Inf. Opt.) **26**(3), 165–171 (2017)
34. Kryzhanovsky, B.V., Malsagov, M.Yu., Karandashev, I.M.: Dependence of critical parameters of 2D Ising model on lattice size. Opt. Mem. Neural Netw. **27**(1), 10–22 (2018)
35. Kryzhanovsky, B.V., Karandashev, I.M., Malsagov, M.Y.: Dependence of critical temperature on dispersion of connections in 2D Grid. In: Huang, T., Lv, J., Sun, C., Tuzikov, A. (eds.) Advances in Neural Networks, ISNN 2018. Lecture Notes in Computer Science, vol. 10878, Minsk, Belarus, 25–28 June 2018, Proceedings, pp. 695–702. Springer, Cham (2018). https://doi.org/10.1007/978-3-319-92537-0_79
36. Kryzhanovsky, B., Malsagov, M., Karandashev, I.: Investigation of finite-size 2D Ising model with a noisy matrix of spin-spin interactions. Entropy, Special issue: Entropy and Complexity of Data (2018, in press)

Intellectual Agents Based on a Cognitive Architecture Supporting Humanlike Social Emotionality and Creativity

Alexei V. Samsonovich(✉)

National Research Nuclear University "MEPhI" (Moscow Engineering Physics Institute), Kashirskoe Shosse 31, Moscow 115409, Russian Federation
asamsono@gmu.edu

Abstract. Human-friendly virtual and physical collaborative robots, or cobots, will work side-by-side with users as helping minds and hands in a variety of creative cognitive tasks, including design, invention, creation of art, or goal setting in unexpected situations in unpredictable environments. These tasks require autonomous reasoning and engage social-emotional attitudes, because the cobot needs to maintain mutual trust with the team or the user. In addition, the cobot needs to understand the global context to be able to determine its role and specific task in a joint mission. All this can be achieved based on a cognitive architecture, supporting social-emotional and narrative reasoning. A general concept of such architecture is presented here, together with an overview of evaluations of prototypes and potential practical applications.

Keywords: Affective computing · Cognitive architectures · Creative assistants Human-friendly AI · Semantic mapping

1 Introduction

Today, in many areas of human activity, there is a growing need for intelligent systems that are accepted by humans at the social level, as partners and assistants. As a consequence, the number of software tools on the market is growing rapidly. And this happens at all levels: from large systems such as IBM Watson to various kinds of assistants and bots: including chatbots, virtual characters, search engines, navigators, etc.

At the same time, a plateau is identified in the development of existing platforms, which should provide such opportunities. For example, virtual agents such as Siri or Cortana are not perceived by users as "reasonable" entities, and are not used as intended by their creators. The most popular operating systems of computers and smartphones, creating new functionality, do not evolve in the direction of their "animation". The situation resembles a "conceptual ceiling" in neural network technologies based on reinforcement learning: despite their impressive progress, the highest human cognitive capacities remain beyond their limits.

In this situation, there is a need to create a new generation of platforms, built on different principles, different from the means and traditions that dominate today in the development of software. This requires a revolutionary rather than evolutionary

B. Kryzhanovsky et al. (Eds.): NEUROINFORMATICS 2018, SCI 799, pp. 39–50, 2019.
https://doi.org/10.1007/978-3-030-01328-8_3

approach. Today, many people pin their hopes on the future of AI with cognitive architectures built on the principles of psychology and physiology of the human brain. The concept of a platform of this sort is a technological line on the basis of the cognitive model of the human mind, which is outlined in this work.

A strength of modern AI, as compared to natural intelligence, is in its ability to quickly find rational solutions to a broad variety of specific cognitive tasks. However, this strength could be reduced to nothing, if the very selection of the task was wrong in a given situation. Until recently, it was assumed that goals for AI should be set by a human, but things are changing rapidly [25–28, 38, 52]. The challenge for creators of AI today is to make a machine that not only can reason logically, but also can think like a human. In particular, this machine should be able to:

- Conceive and set its own, unforeseen goals.
- Think outside the logical or statistical inference box.
- Understand, evaluate and create art, music, dance, design, etc.
- Develop and maintain personal relationships, mutual understanding and trust.
- Understand the meaning of what is happening in the global context and its role in it.
- Learn and grow up like a human, developing dreams, values, motives, and higher ideals.

With all or some of these capabilities, new-generation intelligent machines will become partners and assistants to individuals and teams in various kinds of scenarios, while interacting with humans on equal. Examples range from military missions to creative works, from virtual managers and assistants to personal companions. This is how AI will become integrated into the human society. This meta-goal can be achieved with an approach based on cognitive architectures supporting the above capabilities. The present work brings together several recently developed interrelated concepts, combining them into one emergent picture of a unifying human-like cognitive architecture of this sort. The framework can serve as a theoretical infrastructure for the development of intelligent agents: collaborative robots (cobots) and autonomous actors, both physical and virtual.

2 The Common Model of Cognition

The study of cognitive architectures has a history almost as long as the history of AI [31–36], yet this field is still considered young and emergent. Recently it was noted that the variety of models, systems and frameworks, known as cognitive architectures under different names, converge to one and the same computational Common Model of Cognition (CMC) [5] (see also http://sm.ict.usc.edu). CMC is the common part of the majority of cognitive architecture frameworks, from the classical ones, such as Soar [39] and Act-R [41–44], to their newest descendants, such as Sigma [5]. Essentially, CMC can be characterized by a set of building blocks, primarily including (1) cognitive cycles, underlying system dynamics at all levels – from individual operations to the whole system action, (2) the standard set of memory systems (Fig. 1), each known under different names in different sources, and (3) standard elements of representations of knowledge. According to the terminology accepted in cognitive sciences, the standard set of memory systems includes working memory, where the active processing of currently actual information occurs, semantic memory, where all general knowledge is stored in the form of schemas, chunks, frames, operators, rules, etc. (here the term schema will be used); procedural memory, that stores automated skills, reflexes and biases, implemented as specialized rules and automata, and finally, the input-output buffer, where symbolic processing of information begins and ends. These four memory systems, marked by blue color in Fig. 1, belong to CMC, according to [5].

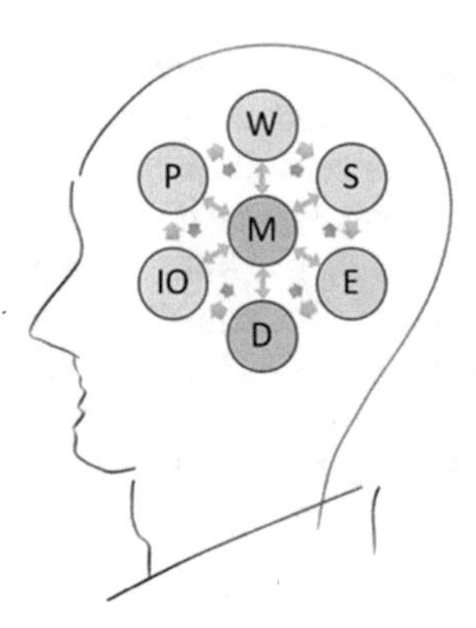

Fig. 1. A bird's-eye view of the unifying cognitive architecture. Blue: CMC, red: value system, green: narrative system. W: working memory, P: procedural memory, S: semantic memory, IO: input-output buffer, M: semantic map, D: drives and motives, E: episodic memory.

Representations used in memory systems (Fig. 1) are endowed with semantics related to the environment where the system operates, and in this sense they are cognitive. Elementary building blocks of these representations have different names in different models. Here they are called schemas, understood in a generalized sense: the term "schema" here unifies the notions borrowed from popular cognitive architectures – a chunk, a rule, an operator, a frame, etc. Dynamics of these representations is governed by a hierarchy of cognitive cycles: from the cognitive cycle of a schema - to the cognitive cycle of a voluntary act of the system as a whole. The latter is sometimes called the OODA Loop (Observe, Orient, Decide, Act) [37]. This cognitive cycle can be expressed as follows:

$$\begin{aligned}&\text{Perceive} \rightarrow \text{Understand} \rightarrow \text{List sensible actions} \rightarrow \text{Predict outcomes} \rightarrow \\&\text{Intend} \rightarrow \text{Commit} \rightarrow \text{Perform}\end{aligned} \quad (1)$$

The implementation of this cycle relies on a variety of standard computer science techniques of search, pattern matching, binding, instantiating, rule application, logical inference, constraint satisfaction, planning, optimization, and so on, with possible usage of specialized systems and tools, such as neural networks and genetic algorithms.

Without going into further details, one can note that the scheme outlined so far is missing essential human functionalities: for example, emotional stimuli and episodic memory. They are added below step by step. The first step is to add social-emotional motivations and biases to cognitive processes.

3 A Cognitive-Architecture Model of Affective Social Cognition

To be favorable for human users, cognitive cobots will need to possess social-emotional intelligence at a human level, at least in some limited sense [4, 10]. This implies not only the ability to recognize and express emotions during social interactions and collaboration with users. An emotionally-intelligent cobot needs to understand emotional attitudes of others in the context of social interaction, and to know how to react to them. This ability is necessary in order to deserve trust, to establish and maintain an emotional contact leading to lasting personal relationships, built based on the believability of cobot behavior [40]. Achieving this capability in a virtual actor or cobot is feasible today and may lead to a major technological breakthrough. A common approach to this challenge involves machine learning (ML) techniques, e.g., training neural network classifiers using deep learning [11]. Indeed, a deep neural network can be trained to recognize emotions from behavior and to select an appropriate reaction to them in a given paradigm. General problems with this approach are that

(i) the amount of available annotated data for a selected domain may not be sufficient;
(ii) reinforcement learning does not necessarily yield an understanding of the entrained mechanisms, that would be necessary for a generalization to other domains, and
(iii) a particular system can be trained to do only the tasks for which it was designed, and remains clueless outside of their scope.

Therefore, an alternative to the traditional ML approach is desirable. An alternative approach is to build a computer model of emotional cognition based on a scientific understanding of the principles that support emotional cognition in vivo. This approach should be based on a cognitive architecture that supports socially-emotional cognition and decision making, as explained below.

Recent progress in understanding of cognitive aspects of affects [3, 7, 8, 12, 13] suggests that fundamental cognitive mechanisms underlying emotional biases in reasoning, decision making and behavior generation must be common across domains and modalities. In other words, a general model capturing the 'magic' of human emotional intelligence is possible. This model can be formulated as a standard-model-based cognitive architecture [5, 39], a part of which supports higher-level emotional cognition, while being abstracted from details of domain-specific intelligence (Fig. 1). In principle, this cognitive architecture can be used as a template to expand virtually any intelligent agent, giving it the ability to 'understand' emotional values of its actions and exhibit 'feelings' naturally. One can expect that the addition of this affective module to a cobot will increase its believability, human-compatibility and social acceptability, resulting in a higher productivity and favorability in interactions with humans (individuals or teams). Furthermore, certain tasks (e.g., creation of art) may only be solvable collaboratively, if the virtual partner possesses the aforementioned qualities.

Conceptual model of this approach illustrated in Fig. 2 can be explained in simple terms. The cognitive architecture, in addition to all its basic components (working, procedural, semantic, episodic memory systems), has a semantic map [47, 48] representing affective appraisals of all elements of cognition, including representations of events, actors and schemas.

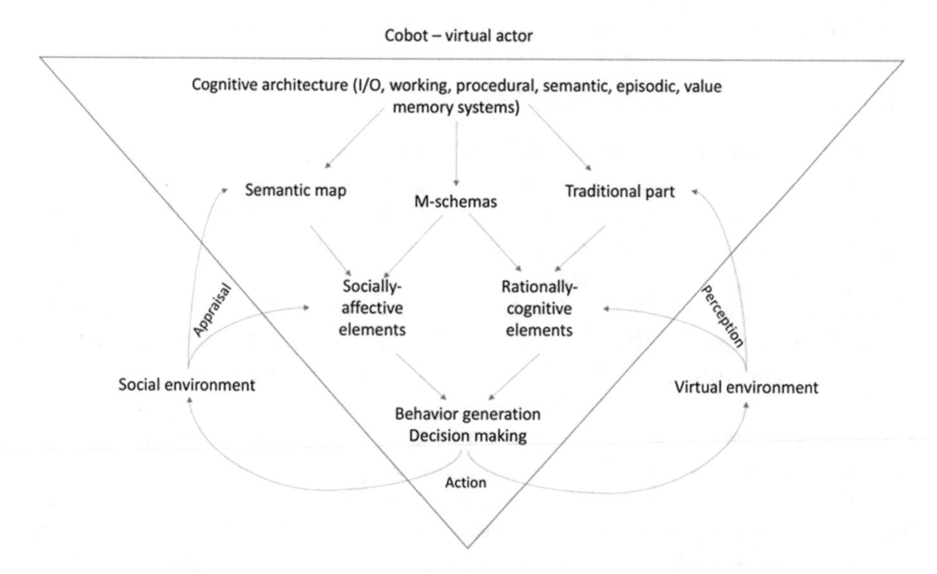

Fig. 2. Conceptual model of a cobot operating as a virtual actor.

3.1 Semantic Mapping

Semantic mapping becomes a new universal paradigm at the intersection of the fundamental and applied sciences from linguistics to biology. Here the main role is played by an abstract metric space, the purpose of which is to present the semantic or functional characteristics and relations of concepts geometrically. Such a representation allows us to classify knowledge, find solutions to abstract problems using methods of geometry and topology. It is no coincidence that cognitive maps are ubiquitous in the brain, from sensory to higher associative areas (a popular example is the hippocampus, embodying a cognitive map of physical space and time, and indexing memory distributed in the associative cortex). Apparently, semantic maps should play a central role in artificial cognitive architectures as well.

Understood in the broad sense, semantic maps can be divided into discrete and continuous, as well as "strong" and "weak" [47]. Discrete methods of representing knowledge based on graphs, such as the semantic network, ontology, or conceptual lattice, are not sufficient to isolate semantic features. The same shortcoming is characteristic of semantic maps created within the framework of a "strong" approach, i.e. based on the metrics of dissimilarity: these include latent semantic analysis and related methods: hyperspace analogy for language, (HAL), Eigenwords, multidimensional scaling (MDS) and other methods. Previous attempts to construct feature spaces of "weak" type, with coordinates having a certain semantics, were mainly limited to estimates made by test subjects or experts manually. This includes the ANEW databases [1], WordSim-353, Simlex-999, and others. The method of automated construction of "weak" semantic maps [55] overcomes the above-mentioned drawback, and has been successfully applied to the dictionary of English. The quality of the results exceeds the known analogues, such as SenticNet [2]. The same approach can be extended to annotated databases of images and elements of behavior.

3.2 Operation Principles of eBICA

Here eBICA stands for emotional BICA (biologically inspired cognitive architecture). In addition to semantic maps, the general eBICA framework [45, 46] implies a set of so-called M-schemas (originally called "moral schemas" [46]), that determine how affects should modulate behavior and decision making. Then, it has a system of drives and motives, that create motivation for selection of M-schemas, goals and actions. These affective cognitive factors work together with traditional, 'rational' mechanisms, within the same system. The principle of their combination is that whenever a choice should be made probabilistically, affective factors bias the probabilities in favor of the likelihoods given by the currently active M-schema. The general principle of calculation of these likelihoods can be explained intuitively as follows. An M-schema plays the role of an agency on its own, for which a certain state of affairs is taken as the normal condition, regardless of whether it is achievable or not. Higher likelihoods are assigned to those behaviors that help approaching the normal condition. This principle leads to equations for the likelihoods L of action selection, as functions of the current affective appraisals A of the author and the target (or recipient) of the action, and the action itself. In general,

$$L(\text{action}) = F(A(\text{action}), A(\text{author}), A(\text{target}), \qquad (2)$$

where affective appraisals A are variables that take values on the semantic map. The latter is implemented as a two-dimensional vector space, with semantics of the main coordinates (determined by principal components) interpreted as Valence and Dominance. This is a simplified version of traditional semantic space models [1, 2, 6, 9]. Dynamics of the appraisal variables depend on the history of interactions among actors and are determined by the currently active M-schema M:

$$A(\text{actor}) = f(M, A(\text{actor}), \{\text{action1}, .., \text{ action}\}). \qquad (3)$$

When no M-schema has been bound to actors present in the environment, then the default M-schema is used.

These general principles lead to two algorithms (Tables 1 and 2), defining interactions between a virtual cobot and a user. Which one of them should be used depends on whether the cobot is an autonomous virtual actor (i.e., a complete virtually embodied agent on its own) – or merely an extension of the human mind-and-embodiment in the virtual environment (i.e., a "virtual assistant")

Table 1. *Algorithm 1.* Affective part of the cobot interaction cycle for a virtual actor.

Input: active M-schemas, states of actors, history of events, a priori probabilities of cobot actions. Output: biased probabilities of cobot actions.
For each active M-schema in working memory,
1. Update affective appraisals of all actors bound to it, using the history of interactions (Eq. 3);
2. Evaluate affective appraisals of all sensible in the current context actions of the cobot;
3. Given Eq. (2) and the a priori computed probabilities of action selection, compute biased probabilities of cobot actions.

After executing the affective part (Algorithm 1 or Algorithm 2), the cobot will proceed with the traditional part, selecting and performing actions, and then will return to the beginning of the cycle.

4 Believable Character Reasoning

The next step is to add to the framework narratives [18, 29, 30], including reasoning in terms of characters and their motives [14, 15, 17, 19, 20]. This approach is closely related to goal-driven autonomy [21, 25–28, 38] and relies on believable narrative reasoning [23, 24]. The concept of a character and its motive [19, 20] extends ideas of an agent goal, and agent role or behavior pattern [16]. One of the central notions in believable character reasoning (BCR: [53, 54]) is that of a hierarchical narrative network [22, 53], which includes sets of possible worlds, of connecting them events, of characters and character arcs involved in those events. The resultant general architecture is represented in Fig. 3.

Table 2. *Algorithm 2.* Affective part of the cobot interaction cycle for a virtual assistant.

Input: active M-schema, state of the user, history of events, a priori probabilities of cobot actions. Output: biased probabilities of cobot actions.
For each active M-schema in working memory,
1. Update affective appraisals of current states of the user and the cobot (Eq. 3);
2. Evaluate affective appraisals of all sensible in the current context actions of the cobot;
3. Given Eq. (2) and the a priori computed probabilities of action selection, compute biased probabilities of cobot actions.

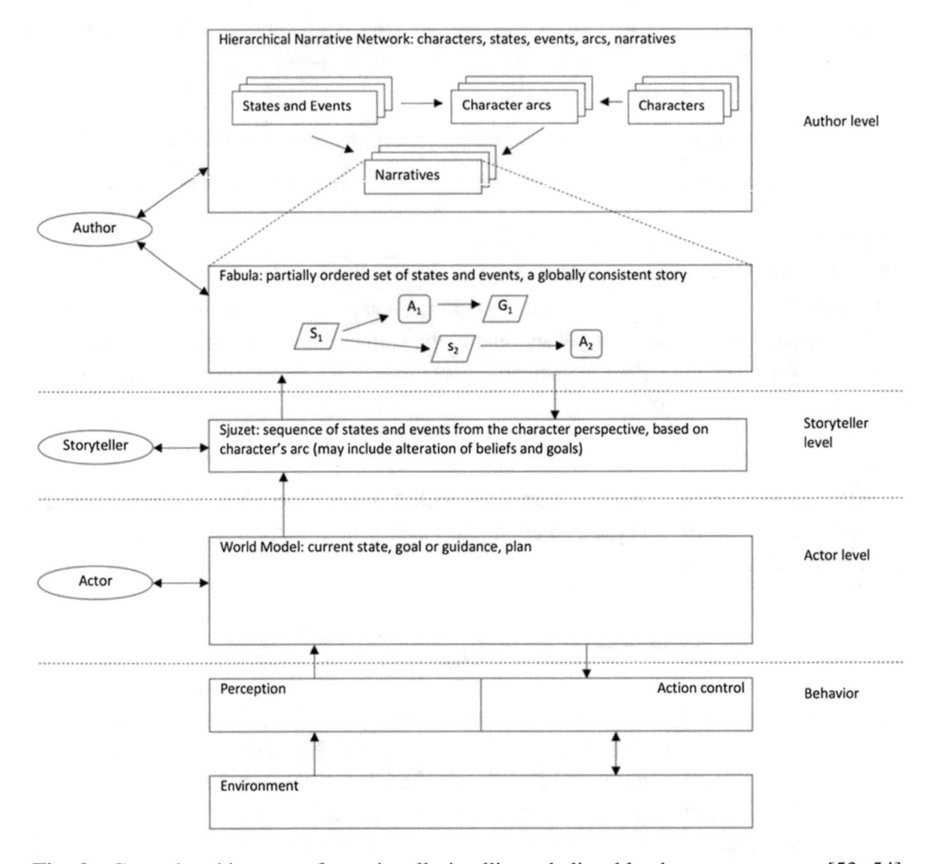

Fig. 3. General architecture of emotionally-intelligent believable character reasoner [53, 54].

5 Concluding Remarks

It is shown that human-friendly virtual cobots can be achieved on the basis of a cognitive architecture, in which the current emotional state of the user is represented on a two-dimensional semantic map. Cobot action selection is determined by this state, the action appraisal, and the currently active M-schema. This approach contrasts popular machine learning techniques [11], and at the same time can be combined with them.

Main results of evaluation of rapid prototypes [45] include a significantly higher quality of the outcome, compared to the control condition, without a semantic map. In selected paradigms, the cobot demonstrates human-level believability and social acceptability. These preliminary results suggest many potential practical applications [49–51]. Examples include creative assistants of a choreographer, a composer, a designer, an insight problem solver.

Speaking generally, on the one hand, today there is a growing need for intelligent agents and devices that would be accepted by man as partners and assistants, while traditional means reach the plateau without reaching a social level (this is indicated, for example, by the data on the limited use of agents Siri and Cortana). In this situation, there is a need to create platforms for production of new generation software that will supplant traditional software. The key qualities of the new software should be a social-emotional AI of the human level and the ability to learn like a human being.

On the other hand, the latest developments of AI trace the tendency of convergence and integration of different approaches based on cognitive architectures similar to the structure of the human mind (that is, BICA). The previous work [45] resulted in a general approach, which should form the basis for the development of creative assistants in a wide range of practical fields.

The concept of an emotional biologically inspired cognitive architecture (eBICA) covers approaches to the creation of intelligent agents based on ideas, concepts and theories of neuroscience and cognitive psychology. Biological intellectual systems, and especially humans, have many qualities that are lacking in artificial intelligence agents. The main of these qualities are reliability, flexibility and adaptability to the environment. Today, when advances in neuroscience and information technology put forward the task of reproducing in the computer all the significant aspects of the human mind, BICA are at the center of attention. Initially, the interdisciplinary nature of the direction of BICA makes the results obtained in it relevant to both biological (neurobiology, psychology) and computer science, primarily AI.

Because emotions are intimately involved in creativity, this approach could be the key to creation of human-level intelligent creative assistants. The presented new approach may also be crucial for the development of the ability of artificial agents to grow cognitively by developing new systems of values and learning goals.

Acknowledgments. This work was supported by the Russian Science Foundation Grant # 18-11-00336.

References

1. Bradley, M.M., Lang, P.J.: Affective Norms for English Words (ANEW): Stimuli, Instruction Manual and Affective Ratings. Technical report C-1. University of Florida, Gainesville, FL (1999)
2. Cambria, E., Speer, R., Havasi, C., Hussain, A.: SenticNet: a publicly available semantic resource for opinion mining. In: Commonsense Knowledge: Papers from the AAAI Fall Symposium FS-10-02, pp. 14–18, AAAI Press, Menlo Park (2010)
3. Gratch, J., Marsella, S.: A domain-independent framework for modeling emotion. Cogn. Syst. Res. **5**, 269–306 (2004)
4. Hudlicka, E.: Affective game engines: motivation and requirements. In: Proceedings of the 4th International Conference on Foundations of Digital Games, pp. 299–306. ACM (2009)
5. Laird, J.E., Lebiere, C., Rosenbloom, P.S.: A Standard model of the mind: toward a common computational framework across artificial intelligence, cognitive science, neuroscience, and robotics. AI Mag. **38**(4), 13–26 (2017)
6. Lövheim, H.: A new three-dimensional model for emotions and monoamine neurotransmitters. Med. Hypotheses **78**(2), 341–348 (2012)
7. Marsella, S.C., Gratch, J.: EMA: a process model of appraisal dynamics. Cogn. Syst. Res. **10** (1), 70–90 (2009)

8. Ortony, A., Clore, G.L., Collins, A.: The Cognitive Structure of Emotions. Cambridge University Press, New York (1988)
9. Osgood, C.E., Suci, G., Tannenbaum, P.: The Measurement of Meaning. University of Illinois Press, Urbana (1957)
10. Picard, R.: Affective Computing. The MIT Press, Cambridge (1997)
11. Shum, H.-Y., He, X.-D., Li, D.: From Eliza to XiaoIce: challenges and opportunities with social chatbots. Front. Inf. Technol. Electron. Eng. **19**(1), 10–26 (2018)
12. Sloman, A.: How many separately evolved emotional beasties live within us? In: Trappl, R., Petta, P., Payr, S. (eds.) Emotions in Humans and Artifacts, pp. 35–114. Vienna (2002)
13. Sun, R., Wilson, N., Lynch, M.: Emotion: a unified mechanistic interpretation from a cognitive architecture. Cogn. Comput. **8**(1), 1–14 (2016)
14. Abell, P.: A case for cases: comparative narratives in sociological explanation. Sociol. Methods Res. **38**(1), 38–70 (2009)
15. Abell, P.: Singular mechanisms and Bayesian narratives. In: Demeulenaere, P. (ed.) Analytical Sociology and Social Mechanisms, pp. 121–135. Cambridge University Press, Cambridge (2011). ISBN 9780521154352
16. Campbell, A., Wu, A.S.: Multi-agent role allocation: issues, approaches, and multiple perspectives. Auton. Agent. Multi-Agent Syst. **22**, 317–355 (2010). https://doi.org/10.1007/s10458-010-9127-4
17. Finlayson, M.A., Corman, S.R.: The military interest in narrative. Sprache und Datenverarbeitung **37**(1–2), 173–191 (2013)
18. Freytag, G.: Die Technik des Dramas (1863). Published by S. Hirzel. https://archive.org/details/dietechnikdesdr01freygoog
19. Haven, K.: Story Smart: Using the Science of Story to Persuade, Influence, Inspire, and Teach. ABC-CLIO, LLC, Santa Barbara, CA (2014). ISBN 9781610698115
20. Haven, K.: Story Proof: The Science Behind the Startling Power of Story. Libraries Unlimited, Westport (2007). ISBN 978-1-59158-546-6
21. Klenk, M., Molineaux, M., Aha, D.W.: Goal-driven autonomy for responding to unexpected events in strategy simulations. Comput. Intell. **29**(2), 187–206 (2013). https://doi.org/10.1111/j.1467-8640.2012.00445.x
22. Pentland, B.T., Feldman, M.S.: Narrative networks: patterns of technology and organization. Organ. Sci. **18**(5), 781–795 (2007). https://doi.org/10.1287/orsc.1070.0283
23. Riedl, M.O., Stern, A., Dini, D., Alderman, J.: Dynamic experience management in virtual worlds for entertainment, education, and training. Int. Trans. Syst. Sci. Appl. (Special Issue on Agent Based Systems for Human Learning) **3**(1), 23–42 (2008)
24. Riedl, M.O., Young, R.M.: Narrative planning: balancing plot and character. J. Artif. Intell. Res. **39**, 217–268 (2010)
25. Roberts, M., Vattam, S., Aha, D.W., Wilson, M., Apker, T., Auslander, B.: Iterative goal refinement for robotics. In: Finzi, A., Orlandini, A. (eds.) Planning and Robotics: Papers from the ICAPS Workshop. AAAI Press, Portsmouth (2014)
26. Roberts, M., Vattam, S., Alford, R., Auslander, Apker, T., Johnson, B., Aha, D.W.: Goal reasoning to coordinate robotic teams for disaster relief. In: Finzi, A., Ingrand, F., Orlandini, A. (eds.) Planning and Robotics: Papers from the ICAPS Workshop. AAAI Press, Jerusalem (2015)
27. Samsonovich, A.V.: Goal reasoning as a general form of metacognition in BICA. Biol. Inspired Cogn. Arch. **9**, 105–122 (2014). https://doi.org/10.1016/j.bica.2014.07.003

28. Samsonovich, A.V., Aha, D.W.: Character-oriented narrative goal reasoning in autonomous actors. In: Aha, D.W. (ed.) Goal Reasoning: Papers from the ACS Workshop. Technical Report GT-IRIM-CR-2015-001, pp. 166–181. Georgia Institute of Technology, Institute for Robotics and Intelligent Machines, Atlanta, GA (2015). https://smartech.gatech.edu/bitstream/handle/1853/53646/Technical%20Report%20GT-IRIM-CR-2015-001.pdf#page=169
29. Schmid, W.: Narratology: an introduction. Walter de Gruyter GmbH & Co. KG, Berlin and New York (2010). ISBN 978-3-11-022631-7
30. Ware, S.G., Young, R.M.: Glaive: a state-space narrative planner supporting intentionality and conflict. In: Proceedings of the 10th Conference on Artificial Intelligence and Interactive Digital Entertainment (AIIDE 2014), Raleigh, NC, pp. 80–86. (2014)
31. Newell, A.: Unified Theories of Cognition. Harward University Press, Cambridge, MA (1990)
32. SIGArt: Special section on integrated cognitive architectures. Sigart Bulletin, vol. 2(4) (1991)
33. Pew, R.W., Mavor, A.S. (eds.): Modeling Human and Organizational Behavior: Application to Military Simulations. National Academy Press, Washington, DC (1998). books.nap.edu/catalog/6173.html
34. Ritter, F.E., Shadbolt, N.R., Elliman, D., Young, R.M., Gobet, F., Baxter, G.D.: Techniques for Modeling Human Performance in Synthetic Environments: a Supplementary Review. Human Systems Information Analysis Center (HSIAC), Wright-Patterson Air Force Base, OH (2003)
35. Gluck, K.A., Pew, R.W. (eds.): Modeling Human Behavior with Integrated Cognitive Architectures: Comparison, Evaluation, and Validation. Erlbaum, Mahwah (2005)
36. Gray, W.D. (ed.): Integrated Models of Cognitive Systems. Series on Cognitive Models and Architectures. Oxford University Press, Oxford (2007)
37. Osinga, F.: Science, Strategy and War: The Strategic Theory of John Boyd. Routledge, Abingdon (2007). ISBN 0-415-37103-1
38. Aha, D.W.: Goal reasoning: foundations, emerging applications, and prospects. AI Mag. **39** (2), 3–24 (2018)
39. Laird, J.E.: The Soar Cognitive Architecture. MIT Press, Cambridge (2012)
40. Johnson, W.L., Lester, J.C.: Pedagogical agents: back to the future. AI Mag. **39**(2), 33–44 (2018)
41. Anderson, J.: Language, Memory and Thought. Erlbaum Associates, Hillsdale (1976)
42. Anderson, J.R., Lebiere, C.: The Atomic Components of Thought. Lawrence Erlbaum Associates, Mahwah (1998)
43. Anderson, J.R., Bothell, D., Byrne, M.D., Douglass, S., Lebiere, C., Qin, Y.L.: An integrated theory of the mind. Psychol. Rev. **111**(4), 1036–1060 (2004)
44. Anderson, J.R.: How Can the Human Mind Occur in the Physical Universe? Oxford University Press, New York (2007)
45. Samsonovich, A.V.: On semantic map as a key component in socially-emotional BICA. Biol. Inspired Cogn. Arch. **23**, 1–6 (2018). https://doi.org/10.1016/j.bica.2017.12.002
46. Samsonovich, A.V.: Emotional biologically inspired cognitive architecture. Biol. Inspired Cogn. Arch. **6**, 109–125 (2013). https://doi.org/10.1016/j.bica.2013.07.009
47. Samsonovich, A.V., Goldin, R.F., Ascoli, G.A.: Toward a semantic general theory of everything. Complexity **15**(4), 12–18 (2010). https://doi.org/10.1002/cplx.20293
48. Gärdenfors, P.: Conceptual Spaces: The Geometry of Thought. MIT Press, Cambridge (2004)
49. Augello, A., Infantino, I., Manfrè, A., Pilato, G., Vella, F., Chella, A.: Creation and cognition for humanoid live dancing. Robot. Auton. Syst. **88**, 107–114 (2016)

50. Augello, A., Infantino, I., Pilato, G., Rizzo, R., Vel-la, F.: Creativity evaluation in a cognitive architecture. Biol. Inspired Cogn. Arch. **11**, 29–37 (2015)
51. Mohan, V., Morasso, P., Zenzeri, J., et al.: Teaching a humanoid robot to draw 'shapes'. Auton. Robot. **31**(1), 21–53 (2011)
52. Parker, L.: Creation of the national artificial intelligence research and development strategic plan. AI Mag. **39**(2), 25–32 (2018)
53. Samsonovich, A.V.: Believable character reasoning and a measure of self-confidence for autonomous team actors. In: Nisar, A., Cummings, M., Miller, C. (eds.) Self-Confidence in Autonomous Systems: Papers from the AAAI Fall Symposium. AAAI Technical Report FS-15-05. AAAI Press, Palo Alto, CA (2015)
54. Bortnikov, P.A., Samsonovich, A.V.: A simple virtual actor model supporting believable character reasoning in virtual environments. In: Advances in Intelligent Systems and Computing, vol. 636, pp. 17–26. Springer Nature, Cham (2017)
55. Ascoli, G.A., Samsonovich, A.V.: Semantic Cognitive Map. US Patent No. 8,190,422 B2, issued on May 29, 2012. U.S. Patent and Trademark Office, Washington, DC (2012)

A Novel Methodology for Simulation of EEG Traveling Waves on the Folding Surface of the Human Cerebral Cortex

Vitaly M. Verkhlyutov[1](✉), Vladislav V. Balaev[1], Vadim L. Ushakov[2,3], and Boris M. Velichkovsky[2,4]

[1] Institute of Higher Nervous Activity and Neurophysiology of RAS, Moscow, Russia
verkhliutov@ihna.ru
[2] National Research Center "Kurchatov Institute", Moscow, Russia
[3] National Research Nuclear University "MEPhI", Moscow, Russia
[4] Russian State University for the Humanities, Moscow, Russia

Abstract. There is an ample evidence on the existence of traveling waves in the cortex of subhuman animals such as rats, ferrets, monkey, and even birds. These waves have been registered invasively by electrical and optical imaging techniques. Such methodology is not possible in healthy humans. Non-invasive EEG recordings show scalp waves propagation at rates two orders greater than the data obtained invasively in animal experiments. At the same time, it has recently been argued that the traveling waves of both local and global nature do exist in the human cortex. In this article, we report a novel methodology for simulation of EEG spatial dynamics as produced by depolarization waves with parameters taken from animal models. Our simulation of radially propagating waves takes into account the geometry of the surface of the gyri and sulci in the areas of the visual, motor, somatosensory and auditory cortex. The dynamics of the electrical field distribution on the scalp in our simulations is fully consistent with the experimental EEG data recorded in humans.

Keywords: Traveling waves · Propagating waves · Spontaneous alpha-rhythm Human cerebral cortex · EEG

1 Introduction

Traveling waves (TW) on the cerebral cortex surface were discovered in the 30s of the 20th century [1], and were then detected in human via the EEG mapping [2, 3] and then were investigated minutely in 50–60 s [4–11] and less intensively in subsequent years [12]. At present, interest to this phenomenon returned for reasons such as the registration techniques improvement [13] and owing to a number of physiological discoveries [14–16], which may explain the TW.

Electronic supplementary material The online version of this chapter (https://doi.org/10.1007/978-3-030-01328-8_4) contains supplementary material, which is available to authorized users.

B. Kryzhanovsky et al. (Eds.): NEUROINFORMATICS 2018, SCI 799, pp. 51–63, 2019.
https://doi.org/10.1007/978-3-030-01328-8_4

TW on the surface of the scalp have a speed of about 5–15 m/s [17], while lower speed was recorded in electrocorticography (ECoG) experiments [18]. Moreover, when recorded using microelectrode matrices placed on the cortical surface, the velocity of the TW appeared to be lower than 0.5 m/s [19] which is similar to the speed detected in animals by means of electrical [20–23] or optical recordings [24]. Importantly, when registered within a small area about 16 mm^2, simpler TW configurations appear, which in many cases might be interpreted as radial waves propagating from one epicenter [19].

Previously, we proposed an elementary model for the TW spread within the human visual cortex, which explained all known to us at that time phenomena of EEG TW in alpha-range on the head [25]. However, the primitiveness of this model made it necessary to search ways of calculating TW in a more realistic model. This became possible with the appearance of the Boundary Element Method (BEM) based on the high quality MRI images [26] and algorithms for geodetic distances calculation on complex surfaces [27].

In this report we propose the EEG data generation method consistent with the intracortical hypothesis [15], which suggests that the main activity recorded in the EEG is related to the TWs of electrical potentials propagated by intra-cortical short fibers [28–30]. We limit our attention to only radially propagating omni-directional waves in the 10 Hz range.

We also assume that spontaneous TWs are not a constant process on the cortex surface and we believe that TW are more related to rest, sleep, pathological and epileptic activity [31]. The spontaneous TWs are suppressed during sensory stimulation [24]. At the same time the evoked and motor potentials exhibit TW-like nature [20].

2 Material and Methods

TW were modeled in the form of concentric waves propagating over a complex folded surface according to the algorithm developed by authors. The algorithm is implemented as a package of procedures in the Matlab environment which aim to model the electrical potential on a complex folded surface mimicking the human cerebral cortex as a result of the wave excitation process of electrical activity propagating from a single epicenter.

The procedure package consists of the following functions: meshm_dist, meshm_wave, meshm_dipl, meshm_pot. Two matrices define triangulated surfaces: Vertices of size 3xN with coordinates of surface nodes and Faces of size 3xL with numbers of three nodes forming L triangles of the surface. All metric values for calculations are specified in mm. The functions work in the following order: (1) the localization of the epicenter as a node of a triangulated surface, (2) the calculation of the geodetic distances to all other nodes of the surface, (3) the calculation of the amplitudes A at each node for the propagating wave, according to the function f(xj, t) basing on the distance from the epicenter xj at each moment time t, (4) calculation of dipole parameters, (5) calculation of potentials.

The epicenter of TW was specified as the target node of the triangulated surface. The user can specify the node directly or the nearest node is determined to a point with user-defined coordinates. The operation is part of the meshm_dist function, which takes a vector with three Cartesian coordinates of the starting point and an array of Vertices, and returns the index of the starting node.

To calculate the geodetic distance from the selected node to all other nodes of the triangulated surface, the latter is represented as a graph, and the graph adjacency matrix is calculated using the Brainstorm toolbox tess_vertconn function [32]. This function takes two arguments: the vertices and faces of the triangulated surface. As a result, tess_vertconn returns the adjacency matrix of the nodes of the triangulated surface. The adjacency matrix is used for breadthisfirst search across the graph. In the first iteration, the matrix is multiplied by the vector x = [0 0 0 … 1 … 0 0], where 1 corresponds to the index of the initial node, resulting in the vector y = [0 0 0 … 1 … 0 … 1 … 0], where 1 corresponds to the indexes of nodes adjacent to the first, i.e. its first level of adjacency. The distance between the initial node and its first level of adjacency is calculated. At the next iteration, the adjacency matrix is multiplied by the vector y and the second adjacency level for the starting node is calculated.

The distance to the nodes of the second level of adjacency is calculated by adding the distances to the connected nodes of the first level of adjacency and the distances between the nodes of the first level of adjacency and the initial node. As a result of each iteration, the vector of distances from the initial node to the already passed nodes and a vector with the indexes of these nodes are stored. The procedure is performed by the meshm_dist function, which takes the following arguments: a structure type variable with Faces and Vertices fields specifying the triangulated surface, and a second variable specifying the index of the starting node, or a vector 1x3 specifying the Cartesian coordinates of the starting point. As a result, the function returns a 1xN vector, where each node number corresponds to the geodetic distance to it from the node specified by the user.

Calculation of the amplitudes of the electrical activity wave that propagates along the surface of the brain from the initial node is performed by the function meshm_-wave. The 1xN vector with distances from the initial node to all nodes of the surface, the variable specifying the maximum propagation distance of the wave, the wavelength λ, the number of time readings M, the frequency of the wave oscillations ω in Hz, and the sampling rate (SR) in Hz are taken as arguments. The amplitude was calculated by the formula 1,

$$A(j,t) = \sin\left(2\pi\left(\lambda x_j - \omega t/\mathrm{SR}\right)\right) \tag{1}$$

where A (j, t) is the amplitude of the wave at the j-th node point, n is the node index, t is the time moment, and xj is the distance to the j-th node. As a result, the function returns the amplitude matrix NxM for N nodes and M time samples.

The parameters of the elementary dipoles vectors located at the nodes of the surface are calculated by the function meshm_dipl. The function takes a structure type variable with Vertices and Faces fields of the triangulated surface and an NxM amplitude matrix for N nodes and M time samples as arguments. Using the functions of the Brainstrorm software package (http://neuroimage.usc.edu/brainstorm/) tess_normals, the values of unit vectors normal to the surface in each node are calculated. This function takes the Vertices and Faces matrices and returns an array of 3 × N with three Cartesian coordinates of the normal vectors for each of the N nodes. Then, in mesm_dipl, the lengths of the surface normal vectors along which the propagation takes place are multiplied with the amplitudes of the propagating wave.

In addition, the coordinates of the equivalent dipole are calculated as the vector sum of elementary dipoles. Next, the coordinates of the origin of the equivalent dipole are calculated. To do this, cosines are first calculated between each elementary dipole and the equivalent dipole, and then we obtain the radius vectors drawn to the beginnings of elementary dipoles, projected using these cosines. The values of the projections are summed, returning the coordinates of the origin of the dipole. The meshm_dipl function returns a structure type variable with Loc fields of 3xM coordinates of the location of the equivalent dipole, Amp with the size of 3xM amplitudes of the equivalent dipole, and elem with similar Loc and Amp fields of 3xNxM coordinates specifying the position of the elementary dipoles.

Calculation of the electric field on the surface of the head or at selected points simulating the electrodes of the EEG is performed by the function meshm_pot. The function takes the surface of the cerebral cortex, the inner and outer surface of the skull, and the surface of the head, the structure-type variable that specifies the location of the sensors, the vector that specifies the indexing of the sensors, the structure-type variable that specifies the coordinates of the elementary dipoles or the total equivalent dipole as the arguments. The variable specifying the location of the sensors is a structure variable of size 1xK with Name and Loc fields. Name is the name of the electrode, for example Cz. Loc - location coordinates of electrodes. The method of calculation is given by a string and can be 'elem' or 'equiv'. The first method 'elem' calculates the electric field basing on the elementary dipoles, and the second 'equiv' basing on the equivalent dipole. Using the bst_openmeeg function of the Brainstorm software package, gain matrix Kx3N (K is the number of channels) is calculated for the first time point. Next, for each time point, the value of the electric or magnetic field potential on the electrodes is calculated by multiplication of elementary dipoles vectors with the gain matrix and written into a matrix of the size KxM, which is returned by a function meshm_pot. In the case of the 'equiv' method, the gain matrix is calculated for each time point and multiplied by the coordinates of the equivalent dipole corresponding to this time point.

An equivalent dipole was calculated from a set of dipoles perpendicular to the model surface. The algorithm allowed to calculate the electrical potentials on the model surface of the scalp at the locations of the 129 electrodes of the Geodesics system (Electrical Geodesics Inc., USA), including the 129th reference electrode Cz. In order to do this, the boundary element method was implemented in the OpenMEEG software [33, 34]. As a result a 100-msec model EEG was generated, reproduced by the Brainstorm program [32], and the interpolated dynamical distribution of the electric field on the scalp was mapped. At the same time, EEG sources were localized using the standard wMNE method implemented in the Brainstorm toolbox [32].

In the computer model of the wave propagating along the complex folded structure of the human cerebral cortex, the current density was set by a unit vector V which was collinear to the normals of the triangulated surface and varied sinusoidally from -1 to $+1$ depending on the geodetic distance from the epicenter. To define the current density in nA/mm^2, a special procedure was used to calculate the total area S occupied by the propagating wave process. Then the number of normals Nv located on this area and the area per one normal $Sv = S/Nv$ were calculated, and then we estimated the current at the normal CDv = V * CD/Sv, where CD is the current density in A/m^2 (nA/mm^2). The current density in different models was set to vary from 50 to 250 nA/mm^2, according to measurements from the visual evoked response registration from the monkey cortex [35]. Basing on the model of the cortical surface ICBM152 from the Brainstorm toolbox (http://neuroimage.usc.edu/brainstorm/) the distribution of vectors and current density values estimated on them was computed in a geodesic radius of 4 cm from the epicenter for 50 time points, i.e. every 2 ms, which at a sampling rate of 500 Hz was 100 ms. Thus a complete unit wave with a frequency of 10 Hz was formed, propagating at a velocity of 0.2 m/s [25].

We modeled 28 wave patterns per 100 ms of the model EEG and the dipole distributions in the cortex were calculated. According to the described procedure we calculated an equivalent dipole and simulated EEGs.

The estimated EEG temporal patterns in the form of three-dimensional matrices were compared with each other using the corr2 procedure from the Matlab standard library. We calculated mean, minimal, maximal values (except a value of 1) and first and third quartiles of mutual correlation of the simulated EEG from radial waves propagating from 28 epicenters on the ICBN152 cortical surface model (Fig. 1).

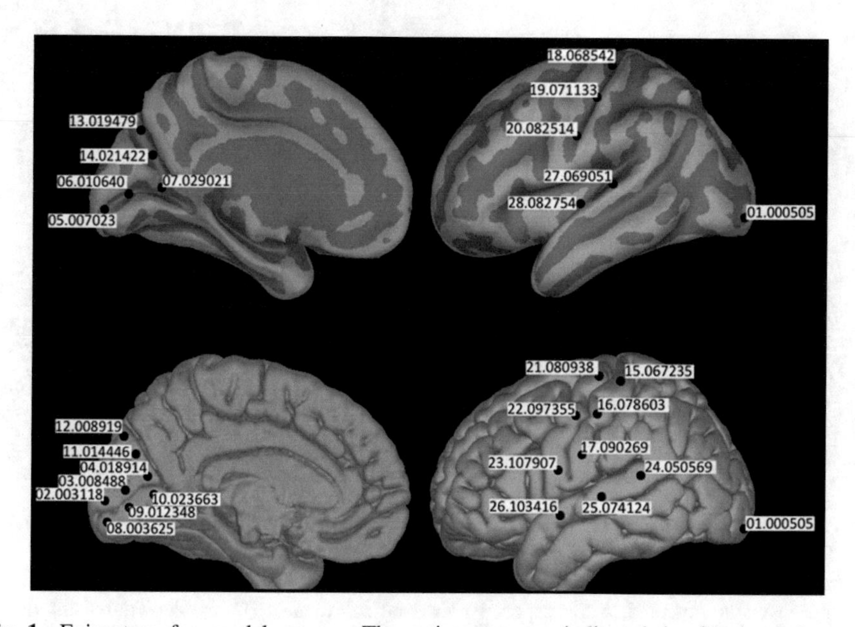

Fig. 1. Epicenters for model waves. The epicenters are indicated by black circles and correspond to the order numbers and to the indices of the vertices of the left hemisphere of the cortical surface ICBM152 from Brainstorm data base [32]. Sagittal and lateral inflated surfaces are at the top, sagittal and lateral no inflated surfaces are at the bottom.

3 Results

Dynamic distributions of dipoles (Fig. 2C) from concentrically propagating potential waves (Figs 2D, 3C) from 28 epicenters located at various cortical points (Fig. 1) were calculated. We simulated 28 128-channel EEG tracks (Fig. 2A) and calculated the distribution of EEG potentials on the scalp (Fig. 2B). In all of the cases the standard localization methods did not manage to reproduce the wave-like potential distribution on the cortical surface (Fig. 2D). We also reproduced traveling waves on the scalp.

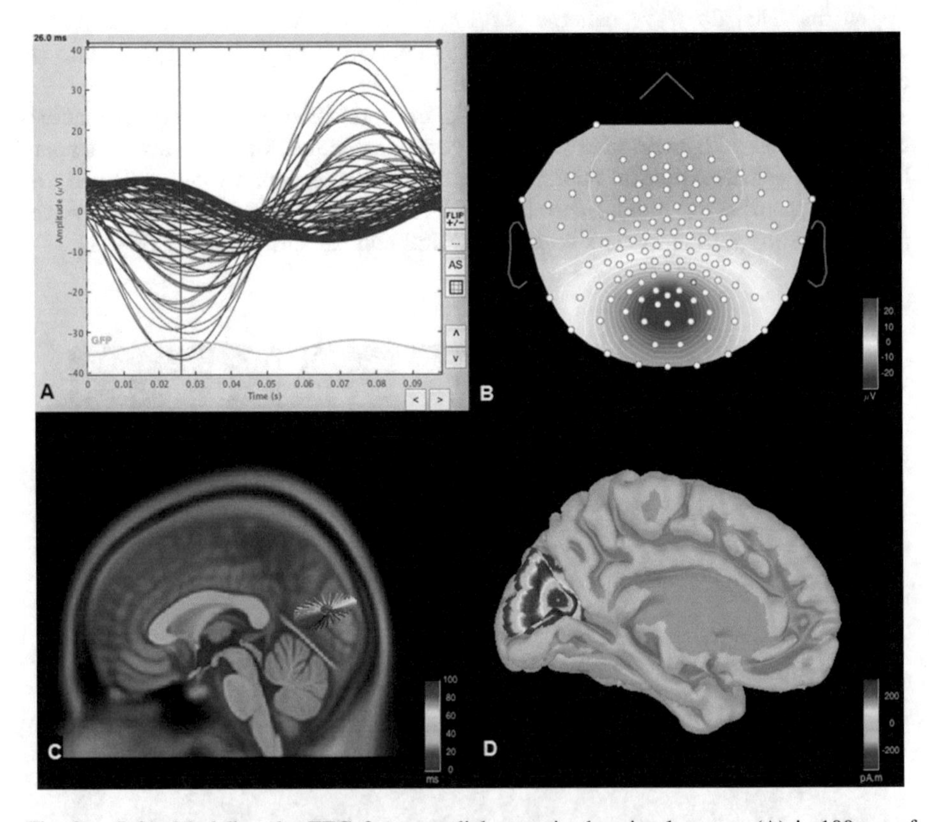

Fig. 2. (left). Modeling the EEG from a radial wave in the visual cortex. (A) is 100 ms of simulated EEG, (B) is electric field distribution on the "scalp" from two symmetric waves in both hemispheres with an epicenter at the vertex 04.18914 of the cortical surface model ICBN152 (the second wave is mirrored with respect to the sagittal plane), (C) is the equivalent dipole corresponding to the time moment marked by the cursor in (A), (D) is the current density distribution from model radial wave on the cortical surface.

Table 1 summarizes the data: the epicenters of modeled waves, the anatomical structures their pertain to, the functional cortical networks in which the wave process is assumed to propagate, the EEG rhythms, the maximum given current density in the

Table 1. The maximum amplitude of the modeled EEG and the type of moving waves on the scalp from radial waves propagating from the epicenters located in the visual, sensorimotor and auditory cortex of the ICBN152 model of human brain.

№	Vertex	MNI	Structure	RSN	Rhythm	CD	SPL	STWL	SPB	STWB
1	01.000505	−18.8 −102.7 −3.2	Pol.occipital.	VIS	alpha	±50	±40	RL	±50	2OF
2	04.018914	−0.0 − 73.3 15.3	Cuneus	VIS	alpha	±50	±20	RR	±40	FO
3	03.008488	0.5− 84.4 7.3	Cuneus	VIS	alpha	±50	±25	L	±45	OF
4	02.003118	0.6 − 93.8 2.6	Cuneus	VIS	alpha	±50	±40	RR	±80	FO
5	11.014446	0.4 − 76.9 24.4	Cuneus	VIS	alpha	±50	±45	RR	±80	FO
6	12.008919	−0.0 −83.2 34.1	Cuneus	VIS	alpha	±50	±15	RR	±30	FO
7	10.023663	−0.9 −66.8 3.0	G.lingualis	VIS	alpha	±50	±20	RA	±30	OF
8	09.012348	−0.0 −80.2 −2.1	G.occ.temp.	VIS	alpha	±50	±20	R	±35	OF
9	08.003625	−0.6 −93.0 −10.7	G.occ.temp.	VIS	alpha	±50	±20	OF	±35	OF
10	07.029021	−28.1 −62.8 5.7	S.calcarinus	VIS	alpha	±50	±7	RP	±9	2FO
11	06.010640	−12.3 −80.5 4.2	S.calcarinus	VIS	alpha	±50	±20	L	±7	FO
12	05.007023	−4.5 − 87.7 −0.8	S.calcarinus	VIS	alpha	±50	±15	RR	±30	OF
13	14.021422	−15.4 − 68.5 26.5	S.p.occip.	VIS	alpha	±50	±20	RP	±40	FO
14	13.019479	−14.6 −71.3 34.7	S.p.occip.	VIS	alpha	±50	±20	RP	±35	FO
15	15.067235	−47.6 −28.9 64.0	G.postcentr.	SSM	mu	±50	±15	RL	±15	2FO
16	16.078603	−59.1 −18.8 49.3	G.postcentr.	SSM	mu	±50	±15	LP	±25	2FO
17	17.090269	−65.6 −9.9 25.3	G.postcentr.	SSM	mu	±50	±25	L	±30	2OF
18	21.080938	−37.8 −17.3 69.7	G.precentr.	SSM	mu	±50	±15	RP	±25	FO
19	22.097355	−56.2 −4.6 47.6	G.precentr.	SSM	mu	±50	±25	RR	±25	2OF
20	23.107907	−62.4 4.3 20.3	G.precentr.	SSM	mu	±50	±20	RL	±20	2OF
21	18.068542	−22.0 −27.4 55.1	S.Rolandic	SSM	mu	±50	±30	RP	±55	2FO
22	19.071133	−34.4 −25.7 49.3	S.Rolandic	SSM	mu	±50	±30	FO	±50	FO
23	20.082514	−41.8 −15.7 33.7	S.Rolandic	SSM	mu	±50	±40	RA	±30	2OF
24	24.050569	−67.1 −41.3 14.8	G.temp.sup.	AUD	tau	±50	±15	R	±20	2OF
25	25.074124	−67.9 −23.1 4.9	G.temp.sup.	AUD	tau	±50	±15	RR	±25	2FO
26	26.103416	−61.0 0.6 −7.1	G.temp.sup.	AUD	tau	±50	±25	RR	±30	2FO
27	27.069051	−61.0 0.6 − 7.1	F.Sylvian	AUD	tau	±50	±25	L	±40	2OF
28	28.082754	−51.3 −17.0 9.9	F.Sylvian	AUD	tau	±50	±40	RL	±45	2OF

Vertex is index of the vertex in the ICBN152 model of the cerebral cortex coinciding with the epicenter of the propagating cortical wave, **MNI** is epicenter coordinates in mm, **Structure** is anatomical structures of the epicenter location, **RSN** is functional (resting) brain network, where the epicenter is placed (VIS is visual, SSM is sensorimotor, VIS is auditory), **Rhythm** is assumed rhythm **10 Hz**, CD is predetermined maximal current density in the cortex in nA/mm^2, **SPL** is estimated maximum amplitude of potentials on the scalp in μV of the model EEG in the case of one wave in the left hemisphere, **STWL** are dynamics of the wave on the scalp in the case of one wave in the left hemisphere (OF is occipito-frontal, RL is left rotation, RR is right rotation, R is right, L is left, RA is right forward (anterior), RP is right backward (posterior), LP is left backward (posterior), CO is from center to occipital lobe), **SPB** is maximum amplitude of potentials on the scalp in μV is scale modeled EEG for symmetric propagating waves in both hemispheres, **STWB** is wave dynamics on the scalp in the case of symmetric propagating waves in both hemispheres (2 is bifurkated extrema, FO is fronto-occipital, OF is occipito-frontal, CO is from center to occipital).

cortex, the maximum potential amplitude on the scalp surface, dynamics of the distribution of potentials on the surface of the scalp.

The models of alpha rhythm and alpha-like rhythms differed in localization only. The epicenters for the alpha rhythm were placed in the regions of the calcarinus and parieto-occipital sulcuses, the ones for the mu-rhythm placed in the area of the central sulcus, and for the tau rhythm, the epicenters were set in the vicinity of the Silvian fissure.

With an equal wave source current density of ± 50 nA/mm^2 in the cortex, the potential amplitudes on the scalp were higher for the alpha rhythm reaching 160 μV from peak to peak. The same value for the mu-rhythm was 110 μV, and 90 μV for the tau rhythm. Such values agree with those observed experimentally for these rhythms [36].

The most frequent dynamics is TW from frontal to occipital lobe - 10 epicenters out of 14 are observed for alpha sources. The total amplitude in this case was 802 μV. For TW from the occipital to the frontal regions was 290 μV.

TWs from the frontal to the occipital lobe and in the opposite direction for mu-and tau-rhythms are approximately equal for different epicenters. In most cases, for Rolandic and Temporal rhythms, a bifurcation of the field extrema on the scalp surface is observed for symmetrically placed sources in both hemispheres.

The rotation of the patterns in the visual cortex is predominantly clockwise for the source in the left hemisphere. Left rotation is observed for the epicenter at the pole of the occipital lobe. In the other cases, transverse and diagonal TWs are detected. Rotational, diagonal and transverse wave movements are observed in case of an asymmetric source location for mu and tau rhythms without noticeable predominance for any type.

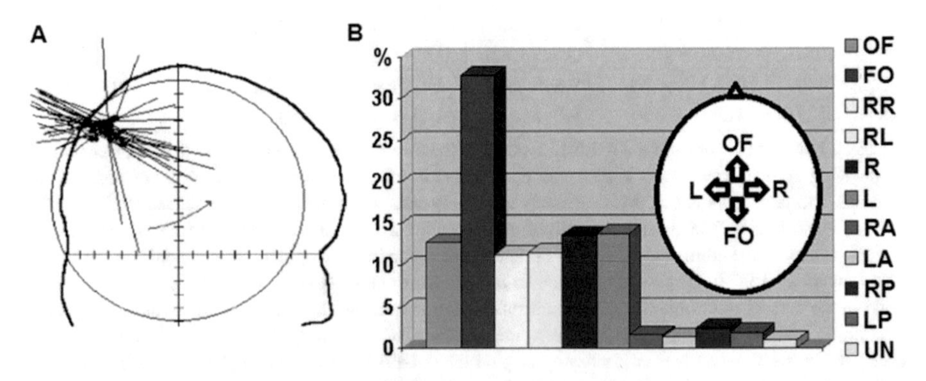

Fig. 3. EEG traveling waves. (A) are dynamics of an equivalent dipole in the occipital lobe of the human brain. (B) is a diagram illustrating the predominance of frontal-occipital EEG TW on a scalp in an experiment. Data were obtained from 12 healthy subjects at rest with closed eyes (Verkhlyutov [37]). Types of TW on the scalp: FO is fronto-occipital, OF is occipito-frontal, RR is right rotation, RL is left rotation, R is right movement, L is left movement, RA is right-anterior, LA is left-arterior, RP is right-posterior, LP is left-posterior, UN is unknow.

The predominance of the TW movements from the frontal to the occipital regions in the simulation coincides with the experimental occurrence frequency of such patterns for EEG (Fig. 3B). Since in our model the form of the EEG distribution dynamics on the scalp is related to the location of the epicenters and their symmetry in the hemispheres, the important objective is to recognize such patterns and relate them to the localization of the wave process.

To demonstrate the degree of similarity/difference within the dynamics of 128-channel EEG patterns, we used the corr2 procedure from the MATLAB standard library, which is intended for raster images analysis. The dynamics of the simulated EEG patterns were compared from 28 points in asymmetric and symmetric cases. For both asymmetric and symmetric cases, the average correlation values were below 0.08 and did not differ statistically except for epicenter No. 6. The maximum correlation values for most epicenters were close to 1 (Fig. 4).

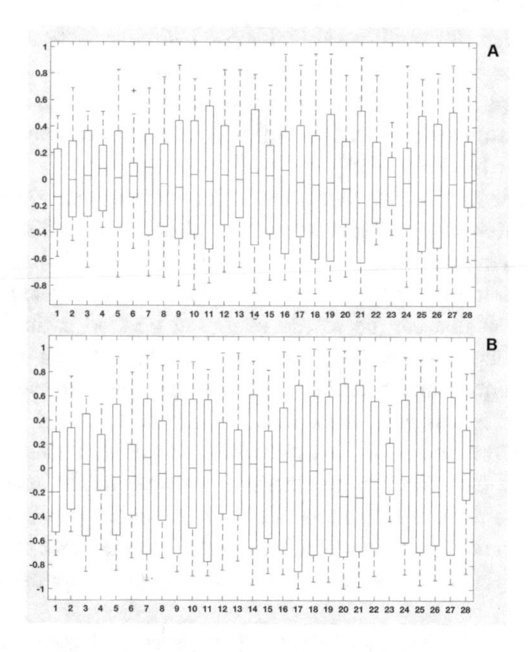

Fig. 4. Comparison of mean, minimal and maximal values, first and third quartiles of the correlation of dynamic patterns of simulated EEGs from radial waves with epicenters 1.-28. (Fig. 1) when compared to each other for the left hemisphere (A) and for both hemispheres (B).

4 Discussion

According to our model, the dynamics of EEG potentials can point out the epicenter of wave propagation. While modeling, we found that the types of the wave movement occur most often in the direction from the frontal to the occipital regions (FO) and this predominance coincides with the experimental data [37, 38] (Fig. 3). It has also been shown experimentally that the TW from the frontal to the occipital lobe is accompanied by the rotation of an equivalent dipole in the region of the occipital lobe [37], which we reproduced in simulation (Fig. 2C).

Thus we can conclude that the propagation of radial waves from certain epicenters can describe the experimental data. In our case, as it can be seen from the Table 1, 8 epicenters for the visual cortex and 1 for the precentral gyrus fulfill this task. We are also able to eliminate all asymmetric cases that give rotational, transverse and diagonal dynamic patterns on the surface of the scalp. In our case, we can conclude that the most likely symmetrical epicenters for fronto-occipital TW on the scalp are located in the visual cortex of both hemispheres in the region of the calcarine and parieto-occipital sulcus. The same dynamics is possible for symmetric epicenters located in the precentral gyrus near the interhemispheric sulcus.

Occipito-frontal TW also may have epicenters in the visual cortex. For other locations of the epicenter, the dynamics of TW has bifurcated extrema of the equal sign (see Table & Supplementary materials).

All the TW trajectories on the scalp are closed, similar to those that were obtained in a work of Hindriks et al. confirming our assumptions about the intracortical origin of electroencephalographic TWs [15]. Similar results were also obtained in a work on 27 subjects, where the authors however did not make assumptions about the origin of the TW observed on the EEG data [39].

According to our model, the dynamics of the electric field on the surface of the scalp (EEG) for different epicenters should be clearly distinguishable. However, in some cases problems may arise in their identification due to the similarity of the dynamic patterns, as indicated by the high correlation level for some pairs (Fig. 4).

The very first task that can be solved using our method is the localization of the epicenter of the epileptic focus basing on ECoG data [40]. Two types of electrode matrices - macroelectrode and microelectrode are considered for solving this problem. The results obtained in this case show that TW can be recorded throughout the convectional surface of the brain [18], but not within the sulci. Therefore, it is difficult to say what amount of data from macroelectrode matrices might arise from volumetric currents, and what amount is related to local field potential (LFP), because these data do not take into account the activity inside the sulci. On the other hand, microelectrode matrices register fragments of radial waves in the parietal, temporal and occipital areas [19].

Our modeling can fill the gap between the macro and micro scales of leads by recording the potential distribution fragments in ECoG and LFP for microelectrode matrices and comparing them with simulation data.

The outlined considerations allow us to a new approach to the solution of the inverse problem in EEG and MEG. The new solution can be based on the propagating wave prior. To this aim, one would need to formulate a specific function which maximization will be equivalent to solving this problem. This will also require the use of individual head geometry and exploit of the advanced techniques for solving the forward problem. The developed methodology is of importance in studying the wave bases of brain neural network integration in cognitive processes [41].

Acknowledgements. The work was supported by the NRC "Kurchatov Institute" grant to the last author (№ 1378 from 23.08.2017) on studying the multilevel cognitive organization of the human brain for brain-computer interfaces, in part by RFBR Grants 17-04-02211 to the first author (traveling waves in the human brain) and by ofi-m grant 17-29-02518 (the cognitive-effective structures of the human brain). Raw data for this paper are available at https://github.com/BrainTravelingWaves and http://braintw.org/. The authors have no competing financial interests.

References

1. Adrian, E.D., Matthews, B.H.C.: The interpretation of potential waves in the cortex. J. Physiol. **81**, 440–471 (1934)
2. Adrian, E.D., Yamagiwa, K.: The origin of the Berger rhythm. Brain **58**, 323–351 (1935)
3. Lindsley, D.B.: Foci of activity of the alpha rhythm in the human electroencephalogram. J. Exp. Psychol. **23**, 159–171 (1938)
4. Walter, W.G.: Toposcopy. In: Third International EEG Congress 1953, Symposium I on Recent Developments in Electroencephalographic Techniques, pp. 7–16. Elsevier, Amsterdam (1953)
5. Petsche, H., Marko, A.: Toposkopische Untersuchungen zur Ausbreitung des Alpharhythmus. Wien. Z. Nervenheilk. **12**, 87–100 (1955)
6. Shipton, H.W.: An improved electrotoposcope. Electroencephalogr. Clin. Neurophysiol. **9**, 182 (1957)
7. Anan'ev, V.M., Livanov, M.N., Bekhtereva, N.P.: Electroencephaloscopic studies on bioelectric map of the cerebral cortex in cerebral tumors and injuries. Zh. Nevropatol. Psikhiatr. Im. SS Korsakova **56**, 778–790 (1956)
8. Livanov, M.N., Anan'ev, V.M.: Electroencephaloscopy. Medgiz, Moscow (1960)
9. Monakhov, K.K.: "Overflows" as a special form of the spatial distribution of electrical activity of the brain. Proc. IHNA **6**, 279–291 (1961)
10. Dubikaytis, Yu.V, Dubakitis, V.V.: On the potential field and alpha rhythm on the surface of the human head. Biophysics **7**, 345–350 (1962)
11. Remond, A.: Integrated and topographical analysis of the EEG. Electroencephalogr. Clin. Neurophysiol. **20**, 64–67 (1961)
12. Hughes, J.R.: The phenomenon of travelling waves: a review. Clin. Electroencephalogr. **26**, 1–6 (1995)
13. Ferrea, E., et al.: Large-scale, high-resolution electrophysiological imaging of field potentials in brain slices with microelectronic multielectrode arrays. Front. Neural. Circuits. **6**, 80 (2012)
14. Biswal, B., Yetkin, F.Z., Haughton, V.M., Hyde, J.S.: Functional connectivity in the motor cortex of resting human brain using EP MRI. Magn. Reson. Med. **34**, 537–541 (1995)
15. Hindriks, R., van Putten, M.J., Deco, G.: Intra-cortical propagation of EEG alpha oscillations. Neuroimage **103**, 444–453 (2014)
16. Matsui, T., Murakami, T., Ohki, K.: Transient neuronal coactivations embedded in globally propagating waves underlie resting-state functional connectivity. Proc. Natl. Acad. Sci. U.S. A. **113**, 6556–6561 (2016)

17. Alexander, D.M., et al.: Traveling waves and trial averaging: the nature of single-trial and averaged brain responses in large-scale cortical signals. Neuroimage **73**, 95–112 (2013)
18. Zhang, H., Andrew, J., Watrous, A.J., Patel, A., Jacobs, J.: Theta and alpha oscillations are traveling waves in the human neocortex. BioRxiv (2017)
19. Martinet, L.E., et al.: Human seizures couple across spatial scales through travelling wave dynamics. Nat. Commun. **8**, 4896 (2017)
20. Rubino, D., Robbins, K.A., Hatsopoulos, N.G.: Propagating waves mediate information transfer in the motor cortex. Nat. Neurosci. **9**, 1549–1557 (2006)
21. Ferezou, I., Bolea, S., Petersen, C.C.: Visualizing the cortical representation of whisker touch: Voltage-sensitive dye imaging in freely moving mice. Neuron **50**, 617–629 (2006)
22. Reimer, A., Hubka, P., Engel, A.K., Kral, A.: Fast propagating waves within the rodent auditory cortex. Cereb. Cortex **21**, 166–177 (2011)
23. Takahashi, K., Saleh, M., Richard, D., Penn, R.D., Hatsopoulos, N.G.: Propagating waves in human motor cortex. Front. Hum. Neurosci. **5**, 40 (2011)
24. Muller, L., Reynaud, A., Chavane, F., Destexhe, A.: The stimulus-evoked population response in visual cortex of awake monkey is a propagating wave. Nat. Commun. **5**, 3675 (2014)
25. Verkhliutov, V.M.: A model of the structure of the dipole source of the alpha rhythm in the human visual cortex. Zh. Vyssh. Nerv. Deiat. Im. I P Pavlova **46**, 496–503 (1996)
26. Fuchs, M., Drenckhahn, R., Wischmann, H.A., Wagner, M.: An improved boundary element method for realistic volume-conductor modeling. IEEE Trans. Biomed. Eng. **45**, 980–997 (1998)
27. Siek, J.G., Lee, L.-Q., Lumsdaine, A.: The Boost Graph Library User Guide and Reference Manual. Pearson Education, Upper Saddle River (2002)
28. Ermentrout, G.B., Kleinfeld, D.: Traveling electrical waves in cortex: insights from phase dynamics and speculation on a computational role. Neuron **29**, 33–44 (2001)
29. Han, F., Caporale, N., Dan, Y.: Reverberation of recent visual experience in spontaneous cortical waves. Neuron **60**, 321–327 (2008)
30. Zheng, L., Yao, H.: Stimulus-entrained oscillatory activity propagates as waves from area 18 to 17 in cat visual cortex. PLoS ONE **7**, 41960 (2012)
31. Wu, J.-Y., Xiaoying, H., Chuan, Z.: Propagating waves of activity in the neocortex: what they are, what they do. Neuroscientist **14**, 487–502 (2008)
32. Tadel, F., Baillet, S., Mosher, J.C., Pantazis, D., Leahy, R.M.: Brainstorm: a user friendly application for MEG/EEG analysis. Comput. Intell. Neurosci. **2011**, 879716 (2011)
33. Kybic, J., Clerc, M., Abboud, T., Faugeras, O., Keriven, R., Papadopoulo, T.: A common formalism for the integral formulations of the forward EEG problem. IEEE Trans. Med. Imaging **24**, 12–28 (2005)
34. Gramfort, A., Papadopoulo, T., Olivi, E., Clerc, M.: OpenMEEG: opensource software for quasistatic bioelectromagnetics. BioMed. Eng. OnLine **45**, 9 (2010)
35. Hämäläinen, M., Hari, R., Ilmoniemi, R., Knuutila, J., Lounasmaa, O.V.: Magnetoencephalography - theory, instrumentation, and applications to noninvasive studies of the working human brain. Rev. Mod. Phys. **65**, 413–497 (1993)
36. Markand, O.N.: Alpha rhythms. J. Clin. Neurophysiol. **2**, 163–189 (1990)
37. Verkhlyutov, V.M.: "Overflows" and traveling waves of the human cerebral cortex. Dissertation for the degree of a Candidate of Medical Sciences, Institute of Higher Nervous Activity and Neurophysiology of RAS, Moscow, (1999)
38. Patten, T.M., Rennie, C.J., Robinson, P.A., Gong, P.: Human cortical traveling waves: dynamical properties and correlations with responses. PLoS ONE **7**, 38392 (2012)

39. Manjarrez, E., Vázquez, M., Flores, A.: Computing the center of mass for traveling alpha waves in the human brain. Brain Res. **1145**, 239–247 (2007)
40. Hindriks, R., et al.: LFP and CSD phase-patterns: a forward modeling study. Front. Neural. Circuits **10**, 51 (2016)
41. Velichkovskiy, B.M., Kovalchuk, M.V., Ushakov, V.L., Sharaev, M.G.: The study of consciousness by natural science methods: on a possible role of wave-like integration processes. RFBR J. N **3**(91), 61–69 (2016)

Neural Network Theory

Comparative Test of Evolutionary Algorithms to Build an Approximate Neural Network Solution of the Model Boundary Value Problem

A. R. Galyautdinova, J. S. Sedova, D. A. Tarkhov, E. A. Varshavchik, and A. N. Vasilyev(✉)

Peter the Great St. Petersburg Polytechnic University, Saint Petersburg, Russia
dtarkhov@gmail.com, a.n.vasilyev@gmail.com

Abstract. In this paper, we continue to study our neural network (NN) approach to the construction of approximate solutions of boundary value problems for partial differential equations. We test the NN learning algorithms on the example of the Dirichlet problem for the Laplace equation in the unit square. We do not use linearity and other features of the Laplace equation; therefore we expect the similar behavior of the algorithms under study for other problems. In the first algorithm, we use a constant number of neurons and train the entire network at once. The second and third algorithms are evolutionary ones, i.e., the structure of the NN (the number of neurons) changes in the learning process. In the second algorithm, we add neurons one by one through a certain number of training steps and conduct additional training of the entire resulting NN. In the third algorithm, we add neurons in the same way, but the learning process is only for the last added neuron. In all series, we use the maximum number of neurons equal to 10, which was enough to achieve an acceptable quality of training. We estimate the quality of NN training by the error functional consisting of two terms which are responsible for satisfying the equation and boundary conditions. These summands are calculated on the set of trial points (TP). We compared the quality of NN training in the case of constant TP and the application of the proposed procedure of re-generation of TP. Computational experiments have shown that the procedure of TP re-generation with a small number of them allows reducing the error by order of magnitude.

Keywords: Artificial neural network · Error functional · Test point set Laplace's equation · Boundary value problem · Neural network model

1 Introduction

Currently, artificial neural networks are one of the standard methods for solving differential equations, successfully applied to a wide range of problems [1–11]. However, some important moments remain still outside the scope of research. This article is devoted to two issues of them.

First, the structure of the neural network (the number of neurons and how they communicate with each other) is usually given by the researcher, and the weights are

B. Kryzhanovsky et al. (Eds.): NEUROINFORMATICS 2018, SCI 799, pp. 67–76, 2019.
https://doi.org/10.1007/978-3-030-01328-8_5

selected in the learning process. In this paper, two ways to “grow” a neural network in the learning process are compared with this algorithm. Both methods are associated with the addition of neurons one by one, only in the first method after adding the entire network is trained (the process of optimizing the error functional continues for all weights), and in the second method, only the last added neuron is trained (its weights are selected).

Secondly, the most significant difference between our approach (see works [6–11]) and the approach of other researchers (see works [1–5]) is the regeneration of test points, which makes the learning process more stable (it does not get stuck in local extrema) and reduces the number of test points without loss of accuracy.

2 Problem Statement and Methods of Solution

In this article, we present the results of computational experiments for a model problem – the Dirichlet problem for the Laplace equation $\frac{\partial^2 u}{\partial x^2} + \frac{\partial^2 u}{\partial y^2} = 0$ in the unit square [0; 1] × [0; 1]. As boundary conditions, we have chosen conditions on the sides of square $u = 0$ at $x = 0$ and $y = 0$ and $u = 1$ at $x = 1$ and $y = 1$.

We are looking for a solution in the form of RBF-network $u = \sum_{j=1}^{m} w_j \varphi_j(\| \mathbf{x} - \mathbf{c}_j \|)$.

We used standard Gaussians as basis functions $\varphi_i(x, y, a_i, x_i, y_i) = exp\{-a_i[(x-x_i)^2 + (y - y_i)^2]\}$.

We select an approximate solution by adjusting neural network parameters (weights) with the help of global optimization of an error functional $J = J_1 + \delta J_2$.

As the first summand J_1, which is responsible for the equation fulfillment we used $\sum_{j=1}^{M} \left(\Delta u(x'_j, y'_j)\right)^2$, where (x'_j, y'_j) are trial (test) points, randomly evenly distributed inside the square [0; 1] × [0; 1]. We re-generate the trial points after every 5 steps of nonlinear optimization of the error functional J. At the same time, a certain proportion d_t of these points, in which $\left(\Delta u(x'_j, y'_j)\right)^2$ is maximal, are not re-generated. In the works of other researchers [1–5] trial points are not re-generated, i.e. $d_t = 1$. In our works [6–11] all points were re-generated.

As the second term J_2, which is responsible for the satisfaction of the boundary conditions, we used

$$\sum_{j=1}^{M_1} \left(u(x''_j, 0)\right)^2 + \sum_{j=1}^{M_1} \left(u(0, y''_j)\right)^2 + \sum_{j=1}^{M_1} \left(u\left(x''_j, 1\right) - 1\right)^2 + \sum_{j=1}^{M_1} \left(u\left(1, y''_j\right) - 1\right)^2$$

We choose the trial points, where the values of the function are calculated, at the square boundaries.

Optimization was carried out using the RProp method. The experiment was conducted at two values of the number of points (M = 10,100) at which the operator is calculated.

The results are compared with an approximate solution

$$v(x,y) = \frac{4}{\pi}\sum_{i=1}^{100} \frac{sh[\pi(2i-1)x]\sin[\pi(2i-1)y]}{sh[\pi(2i-1)](2i-1)} + \frac{4}{\pi}\sum_{i=1}^{100} \frac{sh[\pi(2i-1)y]\sin[\pi(2i-1)x]}{sh[\pi(2i-1)](2i-1)},$$

based on the application of the Fourier method.

Quality assessment of neural network model was carried out in two ways. The first one is estimation for the sum of squares of errors of satisfaction to the equation inside the square, with 10 000 random test points (in notations J_1). The second one – estimation of the misalignment sum for all four parts of the border, with the points distributed at equal distance on each of the border part (in notations J_2). The process was run 10 times for each set of parameters. The results were used to calculate the average value, expectation, and variance of the summands J_1 and J_2.

3 Results of Computational Experiments

In the first series of computational experiments, the number of neurons is constant, and the entire network is trained.

In the second series, the number of neurons is variable, neurons are added one by one, and after the addition, the entire network is trained.

In the third series of computational experiments, there is also the addition of neurons one by one, but the learning process (selection of weights) is done only for the last added neuron. The maximum number of neurons is 10 in all series.

Let us compare neural network solutions for different system parameters.

With a small number of test points ($M = 10, M_1 = 10$), according to Tables 1 and 2, the third computational series (in which the last added neuron is trained) is the most optimal in time.

Table 1. Quality assessment of neural network model $M = 10, M_1 = 10, d_t = 0$

	Laplacian		Addition, all		Addition, last	
	$10^{-4}J_1$	$10^{-4}J_2/4$	$10^{-4}J_1$	$10^{-4}J_2/4$	$10^{-4}J_1$	$10^{-4}J_2/4$
Minimum	1,5	0,01	1,12	0,0190	1,25	0,0031
Average	5,19	0,0206	7,54	0,0249	1,94	0,0454
Variance	18,5	3,47E–05	52,9	1,36E–05	1,02	8,9E–05
Time	93,8		103,5		81,9	

According to Table 1, for all three series of experiments, a more accurate result is observed for points taken at the boundary of the square (in J_2 terms). With given system parameters, the most stable solution (the variance is minimal) for points within

Table 2. Quality assessment of neural network model $M = 10, M_1 = 10, d_t = 1$

	Laplacian		Addition, all		Addition, last	
	$10^{-4}J_1$	$10^{-4}J_2/4$	$10^{-4}J_1$	$10^{-4}J_2/4$	$10^{-4}J_1$	$10^{-4}J_2/4$
Minimum	8,91	0,0136	8,91	0,0170	0,04	0,0228
Average	136,9	0,0211	141	0,0222	37,4	0,0436
Variance	28668	1,77E–05	46300	1,16E–05	1444	0,0002
Time	93,8		112,1		38,4	

a square (in J_1 notations) belongs to the third series of experiments (the last added neuron is trained). But with the same parameters for the points on the border of the square, the most accurate solution is observed in the second series of experiments (all neurons are trained).

Let us refer to Table 2. With given parameters of the system, there is no regeneration of test points, but according to the table, the conclusions are similar, as for Table 1.

Comparing the data given in Tables 1 and 2, we can conclude that the average approximation error decreases tenfold with the regeneration of test points. With a larger number of test points ($M = 100, M_1 = 100$), the third series remains optimal in time (the last added neuron is trained).

According to data from Tables 3 and 4 for the points taken at the square boundary, the second series of experiments is the most accurate. For the points taken inside the square, provided that all test points be re-generated ($d_t = 0$), the third series is the most stable. But in the case where the points are not re-generated, the most stable solution is observed in the second series of experiments.

Table 3. Quality assessment of neural network model $M = 100, M_1 = 100, d_t = 0$

	Laplacian		Addition, all		Addition, last	
	$10^{-4}J_1$	$10^{-4}J_2/4$	$10^{-4}J_1$	$10^{-4}J_2/4$	$10^{-4}J_1$	$10^{-4}J_2/4$
Minimum	0,54	0,01	0,64	0,0124	0,71	0,0269
Average	0,85	0,0173	0,87	0,0181	1,05	0,0439
Variance	0,395	5,73E–05	0,065	2,16E–05	0,033	7,61E–05
Time	918,8		974		375	

Comparing the data given in Tables 3 and 4, we conclude that the error reduction at the regeneration of test points. At that, the program operation time is reduced.

Table 4. Quality assessment of neural network model $M = 100, M_1 = 100, d_t = 1$

	Laplacian		Addition, all		Addition, last	
	$10^{-4}J_1$	$10^{-4}J_2/4$	$10^{-4}J_1$	$10^{-4}J_2/4$	$10^{-4}J_1$	$10^{-4}J_2/4$
Minimum	0,05	0,0135	0,66	0,0129	0,69	0,0235
Average	0,86	0,0191	1,43	0,0185	3,73	0,0423
Variance	0,309	4,07E–05	0,212	1,38E–05	55,67	7,78E–05
Time	928,2		1007,2		389,1	

We illustrate the results with graphs of the neural network approximate solution on the diagonal of the square (Figs. 1, 2, 3 and 4).

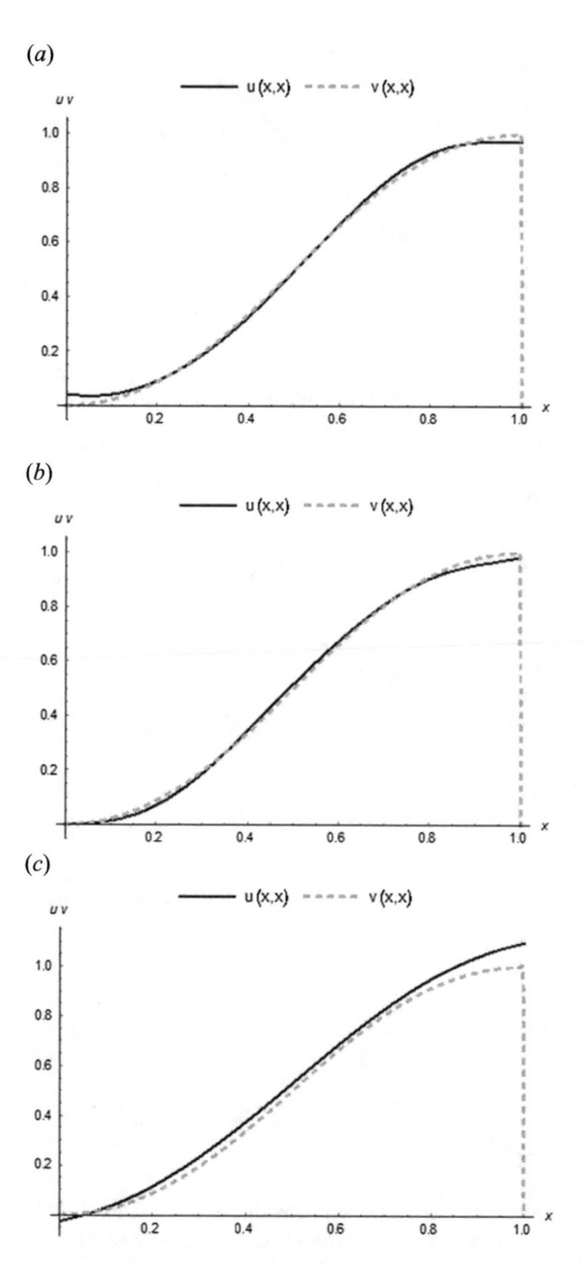

Fig. 1. Comparison of neural network solutions $u(x, x)$ for the parameters $M = 10, M_1 = 10$, constructed at $d_t = 0$, and approximate solution $v(x, x)$ for three cases: (*a*) the constant number of neurons, (*b*) adding one neuron and additional training of the entire network, (*c*) adding one neuron and additional training of the latter

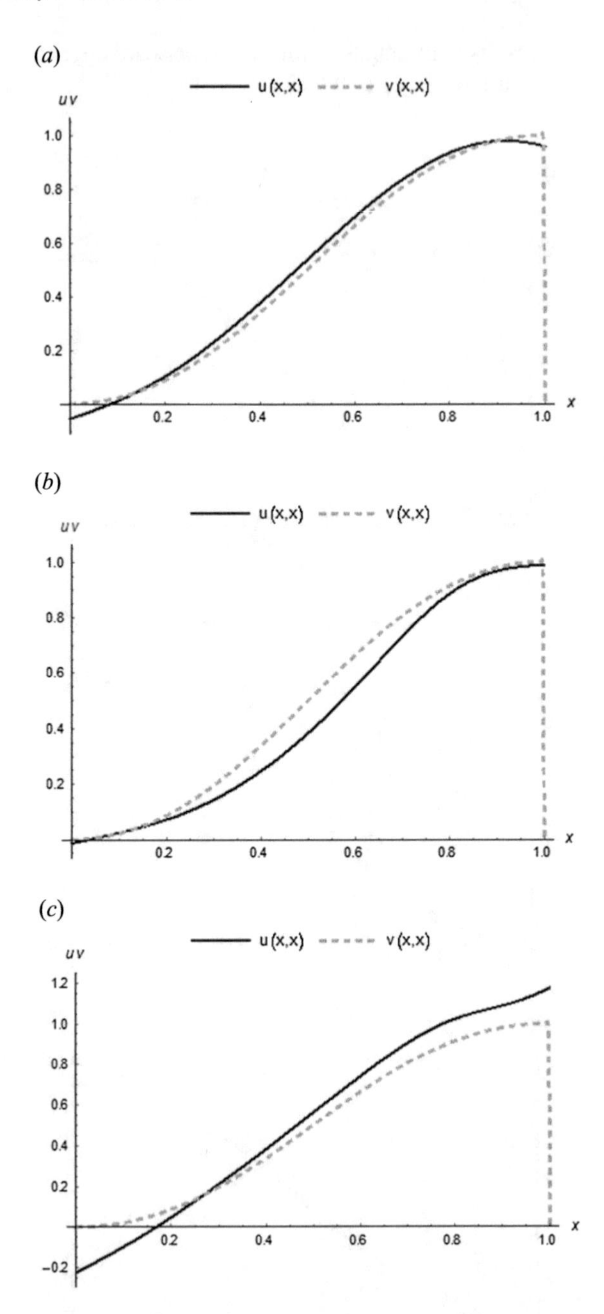

Fig. 2. Comparison of neural network solutions $u(x,x)$ for the parameters $M = 10, M_1 = 10$, constructed at $d_t = 1$, and approximate solution $v(x,x)$ for three cases: (*a*) the constant number of neurons, (*b*) adding one neuron and additional training of the entire network, (*c*) adding one neuron and additional training of the latter

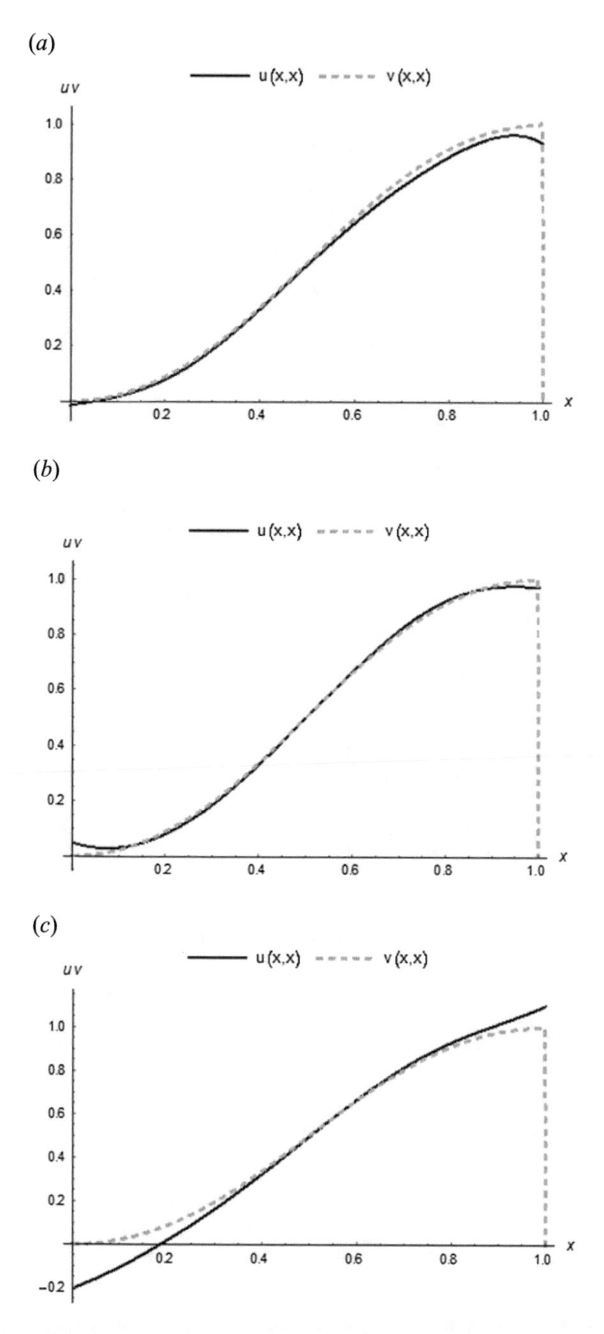

Fig. 3. Comparison of neural network solutions $u(x, x)$ for the parameters $M = 100, M_1 = 100$, constructed at $d_t = 0$, and approximate solution $v(x, x)$ for three cases: (*a*) the constant number of neurons, (*b*) adding one neuron and additional training of the entire network, (*c*) adding one neuron and additional training of the latter

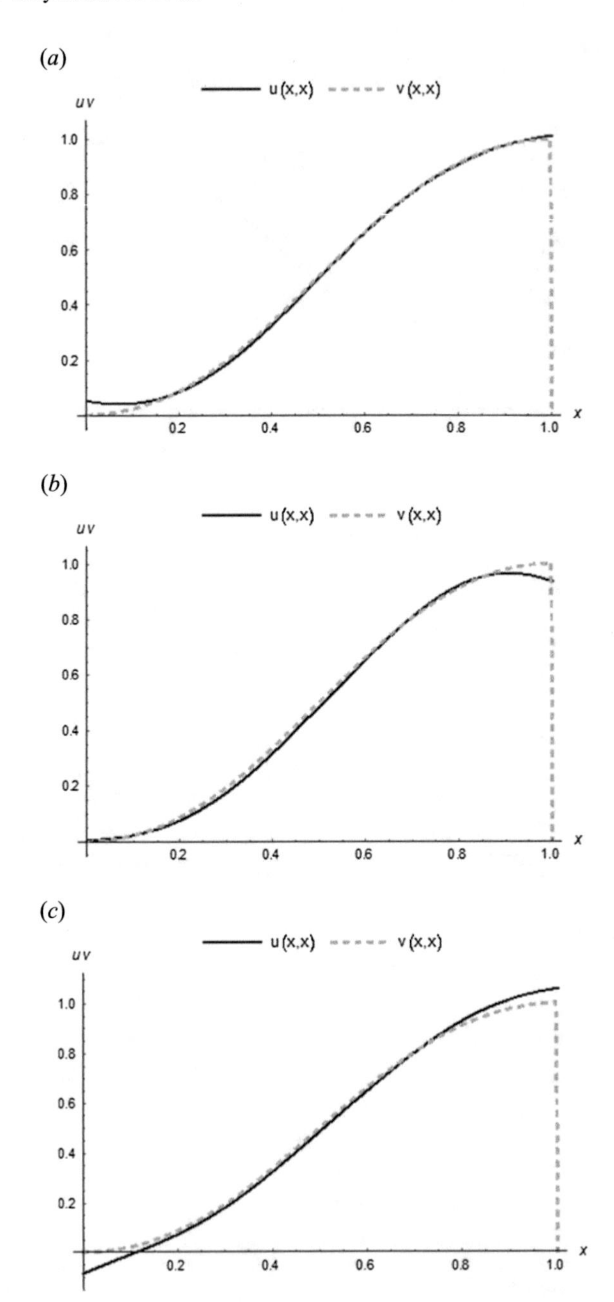

Fig. 4. Comparison of neural network solutions $u(x, x)$ for the parameters $M = 100, M_1 = 100$, constructed at $d_t = 1$, and approximate solution $v(x, x)$ for three cases: (*a*) the constant number of neurons, (*b*) adding one neuron and additional training of the entire network, (*c*) adding one neuron and additional training of the latter

Let us analyze the graphs given for different system parameters.

With a small number of test points, provided they be re-generated $(M = 10, M_1 = 10, d_t = 0)$, we can conclude that there is the growth of the error for the case with the addition of one neuron and additional training of the latter. However, with the same number of test points, but without taking into account their re-generation, the approximation error is maximal at a constant number of neurons.

With a larger number of test points $(M = 100, M_1 = 100)$ under the condition of their re-generation $(d_t = 0)$, similarly to the case with a small number of points, the maximum error occurs in the case of the addition of one neuron and additional training of the latter. In the case of non-regeneration of test points, the growth of the error is observed when adding neurons one by one and training the last neuron.

Comparing the graphs at different numbers of test points, we notice a decrease in the approximation error with an increase in the number of test points.

4 Conclusions

From computational experiments, we conclude that the algorithm with the addition of neurons one by one and additional training of the last neuron at the same other parameters gives a smaller approximation error than with the same addition of neurons and additional training of the entire network, especially in cases where the number of test points is large. Both in one and in the other case, when the test points are re-generated, the approximation error falls, at a small number of test points in tens times. We expect this result to be enhanced for more complex problems – at increasing dimension and complexity of the area geometry.

Acknowledgments. This paper is based on research carried out with the financial support of the grant of the Russian Science Foundation (project no. 18-19-00474).

References

1. Lagaris, I.E., Likas, A., Fotiadis, D.I.: Artificial neural networks for solving ordinary and partial differential equations. IEEE Trans. Neural Netw. **9**(5), 987–1000 (1998)
2. Dissanayake, M.W.M.G., Phan-Thien, N.: Neural-network-based approximations for solving partial differential equations. Commun. Numer. Methods Eng. **10**(3), 195–201 (1994)
3. Fasshauer, G.E.: Solving differential equations with radial basis functions: multilevel methods and smoothing. Adv. Comput. Math. **11**, 139–159 (1999)
4. Förnberg, B., Larsson, E.: A numerical study of some radial basis function based solution methods for elliptic PDEs. Comput. Math. Appl. **46**, 891–902 (2003)
5. Gorbachenko, V.I., Zhukov, M.V.: Solving boundary value problems of mathematical physics using radial basis function networks. Comput. Math. Math. Phys. **57**(1), 145–155 (2017)
6. Tarkhov, D.A., Vasilyev, A.N.: New neural network technique to the numerical solution of mathematical physics problems. I: simple problems. Opt. Mem. Neural Netw. (Inf. Opt.) **14**(1), 59–72 (2005)

7. Lazovskaya, T.V., Tarkhov, D.A., Vasilyev, A.N.: Parametric neural network modeling in engineering. Recent. Pat. Eng. **11**(1), 10–15 (2017)
8. Lazovskaya, T., Tarkhov, D.: Multilayer neural network models based on grid methods. IOP Conference Series: Materials Science and Engineering, vol. 158 (2016). http://iopscience.iop.org/article/10.1088/1757-899X/158/1/01206
9. Gorbachenko, V.I., Lazovskaya, T.V., Tarkhov, D.A., Vasilyev, A.N., Zhukov, M.V.: Neural network technique in some inverse problems of mathematical physics. In: Cheng, L., et al. (eds.) ISNN 2016. LNCS, vol. 9719, pp. 310–316. Springer, Cham (2016)
10. Budkina, E.M., Kuznetsov, E.B., Lazovskaya, T.V., Tarkhov, D.A., Shemyakina, T.A., Vasilyev, A.N.: Neural network approach to intricate problems solving for ordinary differential equations. Opt. Mem. Neural Netw. **26**(2), 96–109 (2017)
11. Vasilyev, A.N., Tarkhov, D.A., Tereshin, V.A., Berminova, M.S., Galyautdinova, A.R.: Semi-empirical neural network model of real thread sagging. In: Studies in Computational Intelligence, vol. 736, pp. 138–146. Springer (2018)

Spectral Density of 1D Ising Model in n-Vicinity Method

Leonid Litinskii(✉) and Inna Kaganowa

Scientific Research Institute for System Analysis RAS, Moscow, Russia
litin@mail.ru

Abstract. For a 1D-Ising system we obtain an exact combinatorial expression for its spectral density in the n-vicinity of the ground state. We show that the energies of the states from each n-vicinity take n different values and obtain a formula for a degeneracy of each energy level. We find out that in any n-vicinity there is N states whose energies differ from the energy of the ground state by an infinitesimal small value, where N is the number of spins. The obtained expressions are generalized to the case $N \to \infty$ when the variables of the problem become continuous. We compare the obtained expression with the normal approximation of the spectral density that is usually used in the framework of the n-vicinity method and discuss the reasons why it does not work in the case of the 1D-Ising system. The normal distribution provides an accurate approximation of the spectral density at the center of the energy interval but at it boundaries the behavior of the functions differs significantly. It is reasonable to say that this discrepancy leads to incorrect results when we apply the n-vicinity method to the 1D-Ising system.

Keywords: 1D-Ising system · n-Vicinity method · Spectral density

1 Introduction

The n-vicinity method is a method that claims to be universal when we calculate the free energy of a system. Its main idea is in the possibility to use the normal distribution in place of the exact spectral density of the spin system if we do not know it. In the case of high-dimensional cubic Ising systems ($d \geq 3$), for which there are no exact solutions, we obtained previously an analytical expression for the critical temperature for [1]. The calculated results are in good agreement with computer simulations [2].

In the same time, low-dimensional cubic Ising systems ($d = 1, 2$) are well examined theoretically [3]. However in these cases the n-vicinity method either does not applicable ($d = 1$) or predicts the type of the phase transition incorrectly ($d = 2$). It is important to try to improve the n-vicinity method and to extend the boundaries of its applicability.

In the present paper, we analyze the 1D-Ising system for which we succeeded in deriving the exact expression for the spectral density of the states from the n-vicinity: We obtained an analytical expression for the frequency of each value of the energy (the energy degeneration). This allows us to compare the exact spectral density with its approximation by the normal distribution.

B. Kryzhanovsky et al. (Eds.): NEUROINFORMATICS 2018, SCI 799, pp. 77–83, 2019.
https://doi.org/10.1007/978-3-030-01328-8_6

2 Basic Expressions

1. The principle idea of the n-vicinity method of calculation of the free energy is as follows [1, 4, 5]. Let us fix the initial configuration $\mathbf{s}_0 \in \mathbf{R}^{\mathbf{N}}$. Other configurations we distribute over its n-vicinities Ω_n; for given n the vicinity Ω_n contains all configurations that differ from $\mathbf{s}_0$ by n spins with opposite signs

$$\Omega_n = \{\mathbf{s} : (\mathbf{s}, \mathbf{s}_0) = N - 2n\},\ 0 \le n \le N.\ \ |\Omega_n| = C_N^n.$$

We denote by $D_n(E)$ the true distribution of the energy states from Ω_n. As a rule it is not known. However, for the distribution $D_n(E)$ we know the exact expressions for the mean E_n and the variance σ_n^2 in terms of the connection matrix and the initial configuration $\mathbf{s}_0$ [5]. In the framework of the n-vicinity method, we replace the unknown distribution $D_n(E)$ by the normal distribution with the mean E_n and the variance σ_n^2:

$$G_n(E) \sim \exp\left(-\frac{1}{2}\left(\frac{E - E_n}{\sigma_n}\right)^2\right).$$

The justification of the n-vicinity method we presented in [5] and the boundaries of its applicability for Ising models we analyzed in [1].

2. Let us analyze a finite one-dimensional chain of spins with cyclic boundary conditions in the presence of the magnetic field H. Then the energy of the state per one spin has the form [3]:

$$E(\mathbf{s}, H) = -\frac{(\mathbf{Js}, \mathbf{s})}{2N} - H\frac{\sum_{i=1}^{N} s_i}{N} = E(\mathbf{s}) - H \cdot \left(1 - 2\frac{n}{N}\right),$$

where $\mathbf{J}$ is the connection matrix of the 1D-Ising model.

In Appendix, we show that for the states from the n-vicinity the energy $E(\mathbf{s})$ has exactly n different values

$$E(k) = -\left(1 - 4\frac{k}{N}\right),\ k = 1, 2, \ldots n, \tag{1}$$

and the degeneration of the k-th energy is defined by the expression

$$D(n, k) = \frac{N \cdot k}{(N - n)n} C_{N-n}^k C_n^k,\ \ k = 1, 2, .., n;\ \ n = 1, 2, .., \frac{N}{2}. \tag{2}$$

In the case of the 1D-Ising system the expressions (1) and (2) define the spectral density of the states in the n-vicinity. In what follows we will need a continuous analogue of these expressions when $N \to \infty$. Let us introduce two parameters $x = n/N$ and $y = k/N$. Using these parameters, we rewrite the formulae (1) and (2) in the form

$$E(y) = -(1-4y),\ D(x,y) = \frac{\exp\left[-N\cdot\left(xS\left(\frac{y}{x}\right)+(1-x)S\left(\frac{y}{1-x}\right)\right)\right]}{2\pi N\sqrt{x(1-x)(x-y)(1-x-y)}}. \tag{3}$$

Here $S(a) = a\ln a + (1-a)\ln(1-a)$ is the Shannon function; the parameter y, that defines the value of the energy, belongs to the interval $y \in [0,x]$. The parameter x defines the magnetization $m = 1-2x$ and x changes inside the interval $x \in [0,1/2]$. Note that if we use Stirling's formula, the value of $n = xN$ has to be sufficiently large. Let us say that n has to be larger than 100. This means that $x \geq x_{\min} = 100/N$. Since finally we suppose that N tends to infinity, this restriction is rather formal.

We also need asymptotic expressions for the averages of the energy and the variance over the n-vicinities. They are

$$E_x = \lim_{N\to\infty} E_n = -(1-2x)^2 \text{ and } \sigma_x^2 = \lim_{N\to\infty} \sigma_n^2 = 16x^2(1-x)^2/N.$$

Then the asymptotic expression for the normal approximation of the exact distribution density takes the form

$$G(x,y) = \frac{e^{-\frac{1}{2}\left(\frac{E-E_x}{\sigma_x}\right)^2}}{\sqrt{2\pi}\sigma_x} = \frac{\exp\left[-N\left(S(x)+\frac{1}{2}\left(1-\frac{y}{x(1-x)}\right)^2\right)\right]}{2\pi N(x(1-x))^{3/2}}. \tag{4}$$

We normalize the functions (3) and (4) so that the equalities $\iint D(x,y)dxdy = 2^N$ and $\iint G(x,y)dxdy = 2^N$ are fulfilled.

3 Comparison of Distributions

Let us compare the distributions $D(x,y)$ and $G(x,y)$ for different values of the parameter x. When we take into account the exponential forms of the functions (3) and (4) it is evident that we have to compare not the functions $D(x,y)$ and $G(x,y)$ themselves but the functions in the exponents divided by N:

$$\begin{aligned} d(x,y) &= \lim_{N\to\infty}\frac{D(x,y)}{N} = -xS\left(\frac{y}{x}\right) - (1-x)S\left(\frac{y}{1-x}\right) \\ g(x,y) &= \lim_{N\to\infty}\frac{G(x,y)}{N} = -S(x) - \frac{1}{2}\left(1-\frac{y}{x(1-x)}\right)^2. \end{aligned} \tag{5}$$

The pre-exponential factors can be neglected.

In Fig. 1 we present the graphs of the functions $d(x,y)$ and $g(x,y)$ when the parameter x takes on four values from the interval $[0.05, 0.5]$. We see (this is also easily seen from Eq. (5)) that the maxima of these functions are in the same point $y_{\max} = x(1-x)$ and their maximal values are the same too: $d(x,y_{\max}) = g(x,y_{\max}) = -S(x)$. This is the reason why the normal distribution approximates the central part of the true distribution correctly. At the same time, the functions $d(x, y)$ and $g(x, y)$ can differ substantially at the “tails”.

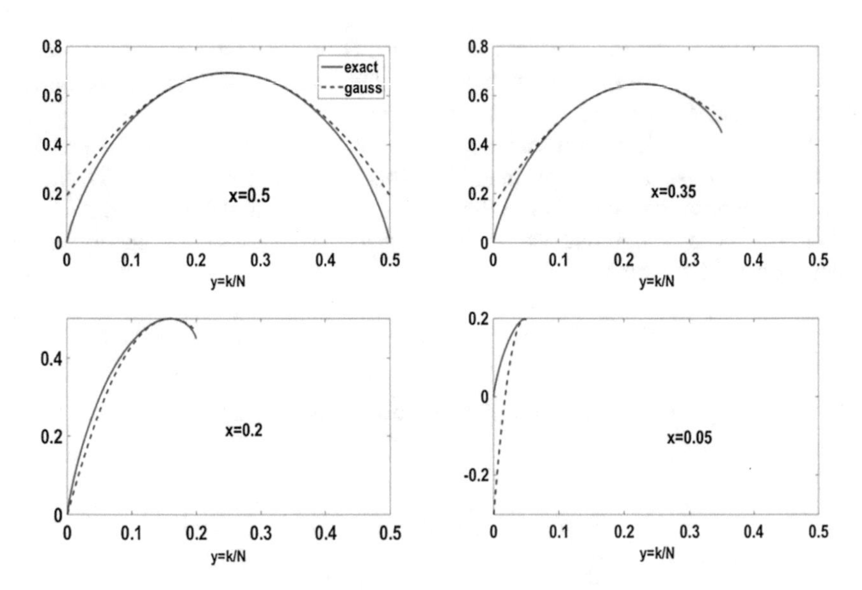

Fig. 1. Functions $d(x, y)$(solid line) and $g(x, y)$ (dashed line) for $x = 0.5,\ 0.35,\ 0.2,\ 0.05$.

The smaller the value of x the narrower the variation interval of y and, consequently, the contribution of this vicinity to the total distribution. The vicinity corresponding to $x = 0.5$, gives the maximal contribution to the total distribution. Here the number of states exceeds exponentially the numbers of the states in other vicinities. On the other hand, we see that at $x = 0.5$ the difference between $d(x, y)$ and $g(x, y)$ becomes maximal: at the tails the distributions $d(0.5, y)$ and $g(0.5, y)$ are substantially different. Probably that is the reason why the normal approximation does not work in the case of the 1D-Ising system.

4 Conclusions

Since the spectral density defines completely the free energy of the system, it also defines all macroscopic characteristics of the model. However, in general it is very difficult to derive an exact expression for the spectral density. In the present paper, for the 1D-Ising system we were able to obtain the exact combinatorial expression for the spectral density in the n-vicinity of the ground state. This allows us to compare the exact spectral density with its normal approximation, which is usually used in the n-vicinity method. We suggest the explanation why our n-vicinity method does not work in the case of the one-dimensional Ising model. On the other hand, it proves to be good for systems of larger dimensions [1]. The obtained results are preliminary and we plan to continue our studies.

The authors age grateful to Prof. B.V. Kryzhanovsky for helpful discussions and support in the course of our work.

The work is done in the framework of the Government Program 14, Project 0065-2018-0001.

Appendix

1. Let us write the connection matrix $\mathbf{J}$ of the 1D-Ising model as a sum of a one-step shift matrix $\mathbf{T}$ and the transpose of this matrix $\mathbf{T}^+$:

$$\mathbf{J}=\begin{pmatrix} 0 & 1 & 0 & 1 \\ 1 & 0 & \ddots & 0 \\ 0 & \ddots & \ddots & 1 \\ 1 & 0 & 1 & 0 \end{pmatrix}=\mathbf{T}+\mathbf{T}^+,\ \text{where}\ \mathbf{T}=\begin{pmatrix} 0 & 0 & 0 & 1 \\ 1 & 0 & \ddots & 0 \\ 0 & \ddots & \ddots & 0 \\ 0 & 0 & 1 & 0 \end{pmatrix}.$$

We set the interaction constant equal to one: $J_{ii+1}=1$.

Let $\mathbf{e}_i$ be the i-th unit vector in the space $\mathbf{R}^\mathbf{N}$: $(\mathbf{e}_i,\mathbf{e}_j)=\delta_{ij}$, where δ_{ij} is the Kronecker symbol and $i=1,\ldots,N$. The matrix $\mathbf{T}$ transforms the i-th unit vector into the $(i+1)$-th unit vector: $\mathbf{T}\mathbf{e}_i=\mathbf{e}_{i+1}$. Since we imposed the cyclic boundary conditions, we have $\mathbf{T}\mathbf{e}_N=\mathbf{e}_1$. By $F(\mathbf{s})$ we denote the quadratic form $F(\mathbf{s})=(\mathbf{T}\mathbf{s},\mathbf{s})$. Then the energy of the state $\mathbf{s}$ is equal to

$$E(\mathbf{s})=-\frac{(\mathbf{J}\mathbf{s},\mathbf{s})}{2N}=-\frac{F(\mathbf{s})}{N}. \tag{A1}$$

It is more convenient to work with the function $F(\mathbf{s})$ than with the energy $E(\mathbf{s})$.

As the initial configuration we choose the ground state $\mathbf{s}_0=(1,1,\ldots,1)=\sum_{i=1}^N\mathbf{e}_i$. By $\mathbf{s}_{i_1i_2..i_n}$ we denote a configuration from the n-vicinity of $\mathbf{s}_0$ in which at i_1,i_2,.., and i_n the values of the spins are equal to -1: $\mathbf{s}_{i_1i_2..i_n}=\mathbf{s}_0-2(\mathbf{e}_{i_1}+\mathbf{e}_{i_2}\ldots+\mathbf{e}_{i_n})$. It is easy to see that

$$F(\mathbf{s}_{i_1i_2..i_n})=N-4\cdot n+4\cdot(\mathbf{e}_{i_1+1}+\mathbf{e}_{i_2+1}+\ldots+\mathbf{e}_{i_n+1},\mathbf{e}_{i_1}+\mathbf{e}_{i_2}+\ldots+\mathbf{e}_{i_n}). \tag{A2}$$

By $\Delta_{i_1i_2..i_n}$ we denote the scalar product

$$\Delta_{i_1i_2..i_n}=(\mathbf{e}_{i_1+1}+\mathbf{e}_{i_2+1}+\ldots+\mathbf{e}_{i_n+1},\mathbf{e}_{i_1}+\mathbf{e}_{i_2}+\ldots+\mathbf{e}_{i_n}). \tag{A3}$$

The energy spectrum of the states from the n-vicinity is defined by a set of different $\Delta_{i_1i_2..i_n}$. It is easy to understand the structure of these scalar products. At first, we suppose that the indices come in sequence: $i_2=i_1+1,\ i_3=i_2+1,..,i_n=i_{n-1}+1$. It is straightforward to see that in this case the value of $\Delta_{i_1i_2..i_n}$ is equal to $n-1$.

Next, let a set of indices falls into two isolated groups and inside each of the group the indices come in sequence. That is the first group the indices are $i,i+1,..,i+a$ and in the second group they are $k, k+1, .., k+b$. Evidently, the equality $a+b+2=n$ is fulfilled and there is a gap between the last index of the first group and the first index of the second group: $k > i+a+1$. It is easy to see that independent of the content of the groups the value of $\Delta_{i_1 i_2 .. i_n}$ is equal to $n-2$. Similarly, when the set of the indices $i_1, i_2, .., i_n$ falls into three isolated groups inside each of which the indices i_j come in sequence, independent of the content of the groups the value of $\Delta_{i_1 i_2 .. i_n}$ is equal to $n-3$ and so on. When the set of the indices $i_1, i_2, .., i_n$ falls into k isolated groups inside each of which the indices come in sequence, independent of the content and the size of the groups the value of $\Delta_{i_1 i_2 .. i_n}$ is equal to $n-k$. When $n \leq N/2$ the maximal number of the isolated groups is $k_{\max} = n$. Using Eqs. (A1), (A2), (A3) we obtain the energies of the states from the n-vicinity of the configuration $\mathbf{s}_0$. These energies take on exactly n values

$$E_k = -(1 - 4k/N),\ k = 1, 2, \ldots n;\ n = 1, 2, .., N/2.$$

2. Now, let us find out the degeneration of an energy E_k for the states from the n-vicinity. (As a rule, in different n-vicinities the degenerations of the same energy E_k are different.) By $D(n,k)$ we denote the number of states in the n-vicinity whose energy is equal to E_k.

The 0-vicinity consists of the configuration $\mathbf{s}_0$. Here $F(\mathbf{s}_0) = (\mathbf{s}_0, \mathbf{s}_0) = N$ and we can write that $E_0 = -1$ and $D(0,0) = 1$.

The 1-vicinity of $\mathbf{s}_0$ consists of N configurations that differ from $\mathbf{s}_0$ in the opposite value of one spin only. Consequently, all N configurations from the 1-vicinity are characterized by the same energy E_1 and $D(1,1) = N$.

It is easy to see that $\binom{N}{2}$ configurations of the 2-vicinity of the $\mathbf{s}_0$ fall into two groups. They are, first, N different configurations where two "-1" are in succession. They are characterized by the energy E_1. The second group consists of all the other configurations from the 2-vicinity, which are characterized by the energy E_2. Consequently,

$$D(2,1) = N,\ D(2,2) = \frac{N(N-3)}{2}.$$

Next, in the 3-vicinity there are exactly N configurations characterized by the energy E_1, $N(N-4)$ configurations, which are characterized by the energy E_2, and the energy of all other configurations is E_3. Then we obtain

$$D(3,1) = N,\ D(3,2) = N(N-4),\ D(3,3) = \frac{N(N-4)(N-5)}{3!}.$$

When n increases, the analysis is somewhat more difficult, but with the aid of standard methods it is not a problem to obtain the following expressions for the 4-th and 5-th vicinities

$$D(4,1) = N,\ \ D(4,2) = \tfrac{3}{2}N(N-5),\ \ D(4,3) = \tfrac{N(N-5)(N-6)}{2}$$
$$D(4,4) = \tfrac{N(N-5)(N-6)(N-7)}{4!},$$

and

$$D(5,1) = N, \quad D(5,2) = 2N(N-6), \quad D(5,3) = N(N-6)(N-7),$$
$$D(5,4) = \tfrac{N(N-6)(N-7)(N-8)}{3!}, \ D(5,5) = \tfrac{N(N-6)(N-7)(N-8)(N-9)}{5!},$$

respectively. The analysis of the obtained expressions for the frequencies of occurrence of the energy E_k for the five n-vicinities allowed us to write down the following combinatorial formula,

$$D(n,k) = \frac{Nk}{(N-n)n} C_{N-n}^{k} C_{n}^{k}, \quad k = 1, 2, \ldots n; \quad n = 1, 2, \ldots N/2.$$

We confirmed these formulae by means of a lot computer simulations performed for different values of N, n and k.

References

1. Kryzhanovsky, B., Litinskii, L.: Applicability of n-vicinity method for calculation of free energy of Ising model. Phys. A **468**, 493–507 (2017)
2. Lundow, P.H., Markstrom, K.: The discontinuity of the specific heat for the 5D Ising model. Nucl. Phys. B **895**, 305–318 (2015)
3. Baxter, R.J.: Exactly Solved Models in Statistical Mechanics. Academic Press, London (1982)
4. Kryzhanovsky, B., Litinskii, L.: Generalized Bragg-Williams equation for system with an arbitrary long range interaction. Dokl. Math. **90**, 784–787 (2014)
5. Kryzhanovsky, B., Litinskii, L.: Generalized approach to description of energy distribution of spin system. Opt. Mem. Neural Netw. (Inf. Opt.) **24**(3), 165–185 (2015)

Match-Mismatch Detection Neural Circuit Based on Multistable Neurons

Yury S. Prostov(✉) and Yury V. Tiumentsev

Moscow Aviation Institute (National Research University), Moscow, Russia
prostov.yury@yandex.ru

Abstract. An ability to detect matches and mismatches between the real world and its representation are both crucial for intelligent autonomous systems to work efficiently. Here we recall our previously proposed model of the neuron which activation characteristic (dependence between an input pattern and output signal) depends on the value of the modulation parameter and varies from a smooth sigmoid-like function to the form of a quasi-rectangular hysteresis loop. Then we propose the neural circuit based on this model of multistable hysteresis neuron. Such a neural circuit can compare expectations represented by downward signals and reality represented by upward signals. In case of matching between them the neural circuit transfers both up and downward signals. In another case, internal inhibitory neurons are activated, and transferring becomes blocked until the expected conditions are met. Besides, the changes in the modulation parameter allow fine-tuning the behavior of this neural circuit. In the end, the results of the numerical simulation are presented.

Keywords: Neurodynamic model · Hystersis · Mismatch detection

1 Introduction

One of the most critical tasks in the field of neural networks modeling is to search and develop new data processing models in the sense of solving the problem of intelligent autonomous systems [1] invention. It is crucial for such models to have the properties of autonomy and adaptability to apply them as a core part of the control and information processing systems in various unmanned vehicles. One possible way to obtain these properties is to have a mechanism for detecting matches and mismatches between the world and its representation stored in memory and use this mechanism in the processes of recognition and learning on-the-fly.

There are several approaches to implement match-mismatch detection mechanism. One of them is based on usage of the oscillatory neural networks [2] where a model of the oscillator represents the neuron while effects of inter-element synchronization and desynchronization are used to detect matches and mismatches. Another approach is based on usage of a simpler neural model but with particular networks topology as in Hierarchical Temporal Memory [3]. Also, there is

B. Kryzhanovsky et al. (Eds.): NEUROINFORMATICS 2018, SCI 799, pp. 84–90, 2019.
https://doi.org/10.1007/978-3-030-01328-8_7

an approach called Dynamic Field Theory (DFT) [4] in which a model of the multi-stable recurrent neuron is used, and some match-mismatch detection neural circuit are proposed [5]. Nevertheless, no comprehensive model has all the necessary properties.

Previously we proposed the model of hysteresis neuron [6,7] which has special properties and as a result could be used to solve the problem outlined above. Our model is similar to one which is used in the DFT-models, but it has specific modulation parameter to adjust the behavior of the neurons. More precisely, we can change on-the-fly behavior of the neuron from mono-stability to multi-stability and also control the bifurcation points. In this article, we propose the neural circuit based on hysteresis neurons with match-mismatch detection properties and show how these properties depend on the value of the modulation parameter.

2 Neuron Model

The model of a neuron is defined as follows:

$$\begin{cases} \dot{u} &= \alpha y + i(\mathbf{w}, \mathbf{x}) - \mu u, \\ y &= f(h(u, \theta)), \end{cases} \tag{1}$$

where $u \in \Re$ is a potential variable, $y \in [0; 1]$ is an output variable, $\mathbf{w} \in \Re^M$ is a weights vector, $\mathbf{x} \in [0; 1]^N$ is an input vector, $\alpha \in [0; +\infty)$ is a recurrent connection weight, $\mu \in (0; 1)$ is a potential dissipation parameter, $\theta \in (0; +\infty]$ is a modulation parameter, $i(\mathbf{w}, \mathbf{x})$ is an external excitation function (in the following we will omit the arguments for brevity) which can be specified as a scalar product or as a Gaussian radial basis function or as any other distance measure function, $h(u, \theta) = u/\theta$ is a potential modulation function, $f(z) = \sigma((z - \Delta)/\lambda)$ is a sigmoidal activation function with $\Delta = 3.0$ and $\lambda = 1.0$. Also, we assume that the values of the parameters α and μ are fixed and selected in advance while the value of the modulation parameter θ is changeable during model operating.

This model partially has been analyzed in [6–9]. Here we briefly recall the essential properties that are necessary for the construction of neural ensemble with required features. First of all, the Eq. (1) has no analytical solution due to non-linearities, but we can represent the solution graphically using the equation:

$$i(y^\star) = -\alpha y^\star + \mu\theta g(y^\star), \tag{2}$$

which defines the dependence of the external excitation value i on the equilibrium point value $y^\star$ and where $g(z) = \Delta + \lambda \log(z/(1 - z))$ is the function inverse to the function f. Furthermore, any equilibrium point $y^\star$ is stable if the function $i(y^\star)$ increases in its neighborhood and unstable otherwise. The form of this function under various parameter values depicted in Fig. 1a, b.

Also it can be shown that model (1) has cusp catastrophe [10] with pitchfork bifurcation defined by condition $\alpha = 4\mu\theta$. Thus, in the case of $\alpha \leq 4\mu\theta$ only one

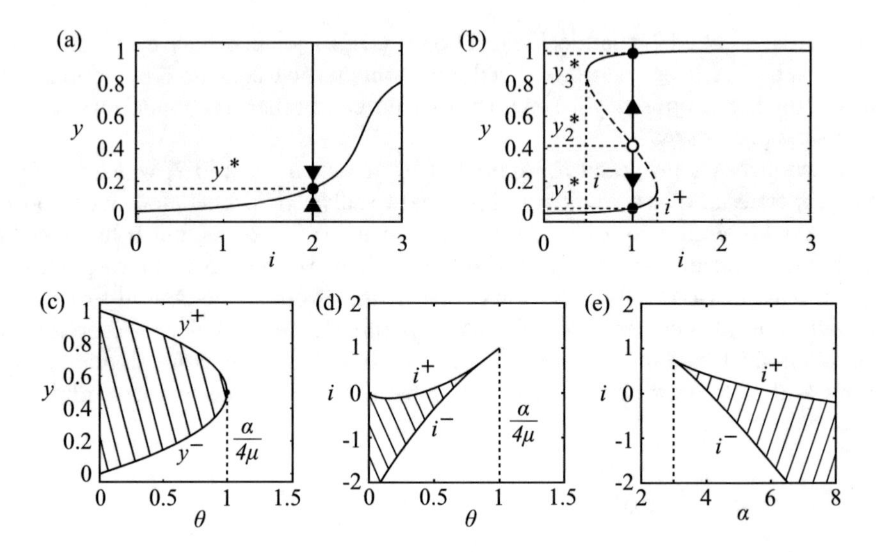

Fig. 1. Graphical solutions of a model (1): (a) the stable point (black circle) in the case of mono-stability; (b) the stable points (black circle) and unstable points (white circle) in the case of bi-stability; (c) the boundary between areas of stable (not shaded region) and unstable (shaded region) points; (d), (e) the dependences of the thresholds $i^{\pm}$ on the parameters α and θ (shaded region corresponds to a bi-stability area)

stable equilibrium point exists as shown in Fig. 1a. But in the case of $\alpha > 4\mu\theta$ a hysteresis loop arises in the form of function $i(y^{\star})$ as shown in Fig. 1b. As a result a bi-stability region arises with boundaries defined by condition $i \in (i^{-}; i^{+})$ where $i^{\pm} = -\alpha y^{\pm} + \mu\theta g(y^{\pm})$ and $y^{\pm} = 0.5 \mp \sqrt{0.25 - \mu\theta/\alpha}$ as shown in Fig. 1c–e. Essential feature here is that to get transition of an equilibrium point $y^{\star}$ from low stable region $(0; y^{+})$ to high stable region $(y^{-}; 1)$ the value of external excitation i must exceed threshold value i^{+}. And vice versa, to get transition from high stable region to low stable region the value of external excitation i must drop lower than threshold value i^{-}.

3 Ensemble Model

Architecture of the proposed neural ensemble demonstrated at Fig. 2a where neuron 1 provides features recognition in the upward data stream while neuron 3 do the same in the downward data stream. The purpose of the neuron 2 is to block upward signals (the activity of the neuron 1) if there are no downward signals (the activity of the neuron 3) and neuron 4 does the same block for downward signals. We call the neurons 1 and 3 as recognizing neurons and the neurons 2 and 4 as inhibitory neurons. This circuit can be defined as follows:

$$\begin{cases} \dot{\mathbf{u}}\tau &= \mathbf{W}\mathbf{y} + \mathbf{i} - \mu\mathbf{u}, \\ \mathbf{y} &= f(h(\mathbf{u}, \theta)), \end{cases} \tag{3}$$

where the corresponding scalar variables from Eq. (2) become vectors, $\tau = [1, 2, 1, 2]^{\mathrm{T}}$ is a vector that provides greater inertness of the inhibitory neurons, $\mathbf{i} = [i_{up}, 0, i_{down}, 0]^{\mathrm{T}}$ is a vector of external excitations with the upward stream excitation i_{up} and the downward stream excitation i_{down}, $\mathbf{W} = \left[\begin{smallmatrix} \mathbf{W}^{in} & \mathbf{W}^{em} \\ \mathbf{W}^{em} & \mathbf{W}^{in} \end{smallmatrix}\right]$ is a weight matrix with blocks $\mathbf{W}^{in} = \left[\begin{smallmatrix} \alpha_1 & \gamma_1 \\ -\beta_1 & \alpha_2 \end{smallmatrix}\right]$ and $\mathbf{W}^{em} = \left[\begin{smallmatrix} 0 & -\beta_2 \\ 0 & 0 \end{smallmatrix}\right]$.

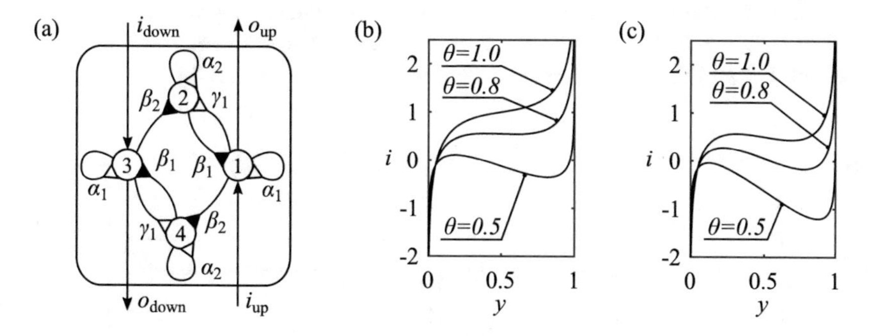

Fig. 2. (a) Architecture of the mismatch detection neural circuit. (b) Solutions curve $i(y)$ for the recognizing neurons 1 and 3. (b) Solutions curve $i(y)$ for the inhibitory neurons 2 and 4.

Note that in the case of model bi-stability an output variable y takes values from the neighborhood of 0 (inactive state) or from the neighborhood of 1 (active state) as shown in Fig. 1c, i.e. $y \in O^{+}_{\theta,\alpha}(0)$ or $y \in O^{-}_{\theta,\alpha}(1)$ where the subscript emphasizes the dependence of the neighborhoods width on the values of the parameters θ and α. Neuron 2 should become active if neuron 1 is active while neuron 1 should become inactive if neuron 2 is active to get locked the upward data stream. This means that the following conditions must be satisfied: $\gamma_1 \cdot \min O^{-}_{\theta,\alpha_1}(1) > i^{+}_{2}$ and $-\beta_1 \cdot \min O^{-}_{\theta,\alpha_2}(1) < i^{-}_{1}$. At the same time when neuron 3 is active which means the presence of downward signals neuron 2 should not be active to pass upward signals. As a result, another condition must be satisfied too: $-\beta_2 \cdot \min O^{-}_{\theta,\alpha_1}(1) < i^{-}_{2}$. Also, note that we did not introduce separate parameters of β and γ for the neurons 3 and 4 due to the logical symmetry of the downward data stream.

As a result, we choose the next values for the model parameters: $\mu = 0.75$, $\alpha_1 = 2.5$, $\alpha_2 = 3.5$, $\gamma_1 = 1.65$, $\beta_1 = 2.5$ and $\beta_2 = 2.5$. The solution curve $i(y)$ for the neurons 1 and 3 which is defined by Eq. (2) takes the form shown in Fig. 2b while for the inhibitory neurons 2 and 4 it takes the form shown in Fig. 2c. As we see, in the case of $\theta = 1.0$ the recognizing neurons are mono-stable and their activation characteristic has the form of a sigmoid-like function while the inhibitory neurons are bi-stable but have a very tiny hysteresis loop in activation characteristic. In case of $\theta = 0.5$ the former ones become bi-stable with the low value of the threshold parameter i^{+}_{l} $(l = 1, 3)$ while the last ones stay bi-stable but the value of difference between thresholds $|i^{+}_{k} - i^{-}_{k}|$ $(k = 2, 4)$ becomes significantly greater. Such model parameters configuration allows to obtain a set of behavioral strategies which are controlled through the modulation parameter.

4 Experimental Results

The results of numerical simulation of the proposed neural circuit are presented in Fig. 3 where the following values of the modulation parameter were applied: (b) $\theta = 0.5$, (c) $\theta = 0.8$ and (d) $\theta = 1.0$. In all these cases the values of the variables i_{up} and i_{down} were varied equally as shown in Fig. 3a.

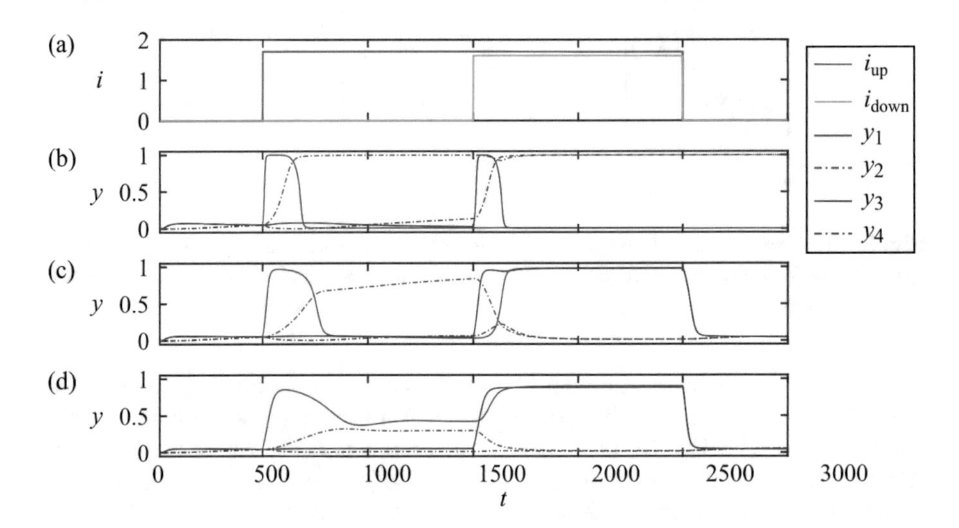

Fig. 3. (a) The values of the external excitation variables used during simulations. Simulation results in case of $\theta = 0.5$ (b), $\theta = 0.8$ (c) and $\theta = 1.0$ (d).

As we can see, in the first case (see Fig. 3b) after a short time after the appearance of the upward signal neuron 1 becomes inactive. As well neuron 3 becomes inactive a short time after the arrival of the downward signal. In this case, timing mismatch occurred, and as a result, there are no signals passed through the neural circuit.

In the second case (see Fig. 3c) the presence of the downward signal leads to a re-activation of neuron 1. In other words, both upward and downward signals have passed through the neural circuit successfully. So, the requirement of matching by time disappeared, and there is the only requirement of signals matching.

In the last case (see Fig. 3d) recognizing neurons are mono-stable. As a result, there is no full inhibition of neuron 1 and the upward signal is passed through the neural circuit with no presence of the downward signal. However, the output value of this neuron is scaled due to inhibitory influence. But in the presence of the downward signal, this influence disappears.

5 Conclusions

We briefly recalled the previously proposed neurodynamic model of a neuron which can be monostable as well as bistable. More precisely, activation characteristic of such a neuron can vary from a smooth sigmoid-like function to a function with a quasi-rectangular hysteresis loop. The final form of this characteristic is controlled by the value of the specific modulation parameter.

We demonstrated how a neural circuit with useful properties could be build using the describer neuron model. General topology of this circuit was proposed, and also some heuristics for choosing the values of constant parameters are introduced. As a result, we got a neural element which compares the signals from two data streams and the logic of this comparison changes depending on the value of the modulation parameter.

In case of signals matching the recognizing neurons which have adjustable weights become active while the inhibitory neurons become inactive. Otherwise, the opposite situation occurs. As a result, the activity of the recognizing neurons can be interpreted as a proper matching between the real world (upward signals) and its representation (downward signals). At the same time, the activity of the inhibitory neurons indicates on mismatches between the reality and predictions. Both of these facts can be used to train and adapt the model, and this will be the subject of our further research.

Acknowledgements. This research is supported by the Ministry of Education and Science of the Russian Federation under Contract No. 9.9124.2017/VU.

References

1. Tiumentsev, Y.V.: Intelligent autonomous systems — information technology challenge. In: Proceedings of 8th National Conference "Conference on Artificial intelligence — 2002", vol. 1, pp. 33–38. Fizmatlit, Moscow (2002). (in Russian)
2. Borisyuk, R., Denham, M., Hoppensteadt, F., Kazanovich, Y., Vinogradova, O.: An oscillatory neural network model of sparse distributed memory and novelty detection. Biosystems **58**(1), 265–272 (2000)
3. George, D., Hawkins, J.: Towards a mathematical theory of cortical micro-circuits. PLoS Comput. Biol. **5**(10), e1000532 (2009)
4. Schöner, G., Spencer, J.: Dynamic Thinking: A Primer on Dynamic Field Theory. Oxford University Press, New York (2015)
5. Barrero, E.P.: Mismatch detection neural circuit applied to navigation. Dissertation, University of Zurich and ETH Zurich, Institute of Neuroinformatics (2017)
6. Prostov, Y.S., Tiumentsev, Y.V.: A study of neural network model composed of hysteresis microensembles. In: Proceedings of XVII All-Russian Scientific Engineering and Technical Conference "Neuroinformatics-2015", vol. 1, pp. 116–126. NRNU MEPhI, Moscow (2015). (In Russian)
7. Prostov, Y.S., Tiumentsev, Y.V.: A hysteresis micro ensemble as a basic element of an adaptive neural net. Opt. Mem. Neural Netw. **24**(2), 116–122 (2015)
8. Prostov, Y.S., Tiumentsev, Y.V.: Adaptive gateway element based on a recurrent neurodynamical model. In: International Conference on Neuroinformatics, pp. 33–38. Springer, Cham (2017)

9. Prostov, Y.S., Tiumentsev, Y.V.: Functional plasticity in a recurrent neurodynamic model: from gradual to trigger behavior. Procedia Comput. Sci. **123**, 366–372 (2018)
10. Zeeman, E.C.: Catastrophe Theory: Selected Papers, 1972–1977. Addison-Wesley, Reading (1977)

Hopfield Associative Memory with Quantized Weights

Mikhail S. Tarkov([✉])

Rzhanov Institute of Semiconductor Physics SB RAS, Novosibirsk, Russia
tarkov@isp.nsc.ru

Abstract. The use of binary and multilevel memristors in the hardware neural networks implementation necessitates their weight coefficients quantization. In this paper we investigate the Hopfield network weights quantization influence on its information capacity and resistance to input data distortions. It is shown that, for a weight level number of the order of tens, the quantized weights Hopfield-Hebb network capacitance approximates its continuous weights version capacity. For a Hopfield projection network, similar result can be achieved only for a weight levels number of the order of hundreds. Experiments have shown that: (1) binary memristors should be used in Hopfield-Hebb networks, reduced by zeroing all weights in a given row which moduli are strictly less than the maximum weight in the row; (2) in the Hopfield projection networks with quantized weights, multilevel memristors with a weight levels number significantly more than two should be used, with a specific levels number depending on the stored reference vectors dimension, their particular set and the permissible input data noise level.

Keywords: Associative memory · Hopfield network
Binary and multilevel memristors · Quantized weight coefficients

1 Introduction

In the Hopfield network [1, 2] functioning process, two modes can be distinguished, i.e. learning and classification. The network weights are calculated in the learning mode based on the known vectors. In the classification mode for the fixed weight values and a specific initial neurons state input, a transient process appears, terminating in one of the local network energy minima. When entering training vectors $x^k, k = 1, \ldots, p$, weights w_{ij} are calculated according to the generalized Hebb rule

$$w_{ij} = \frac{1}{N} \sum_{k=1}^{p} x_i^k x_j^k. \tag{1}$$

An important associative memory parameter is its capacity. A capacity is the maximum stored patterns number. It is shown [1] that when using the Hebb rule for training, the memory capacity estimate is $p_{\max} = \frac{N}{2 \ln N}$. Such network will be referred to below as the Hopfield-Hebb network.

B. Kryzhanovsky et al. (Eds.): NEUROINFORMATICS 2018, SCI 799, pp. 91–97, 2019.
https://doi.org/10.1007/978-3-030-01328-8_8

The Hopfield network projection learning method [3, 4] has the iterative weight matrix W dependence on the learning vector sequence $x^k, k = 1,\ldots,p$:

$$y^k = (W^{k-1} - E) \cdot x^k$$
$$W^k = W^{k-1} + \frac{y^k \cdot y^{kT}}{y^{kT} \cdot y^k} \qquad (2)$$

under the initial conditions $W^0 = 0$ (E is the identity matrix). As the vectors presentation result, the network weight matrix takes on a value $W = W^p$. This method application increases the Hopfield network maximum capacity up to $p_{\max} = N - 1$. Further, such network will be called the Hopfield projection network.

2 Binary and Multilevel Memristors

The neural network hardware implementation requires a lot of memory to store the neurons layer weight matrix and it is expansive. The solution of this problem is simplified by using a device called memristor (a resistor with a memory) as a memory cell. The memristor was predicted theoretically in 1971 by Chua [5]. The first physical realization of the memristor was demonstrated in 2008 by the Hewlett Packard laboratory as a thin-film TiO_2 structure [6]. The memristor behaves like a synapse: it "remembers" the total electrical charge that has passed through it. The memory based on the memristors can reach the integration degree of 100 Gbits/cm^2, several times higher than that based on the flash memory technology. These unique properties make the memristor a promising device for creating massively parallel neuromorphic systems.

Binary memristors realize two conductivity values. Multilevel memristors realize a set of discrete conductivity levels (the levels number can reach tens and hundreds). Binary and multilevel memristors [7–9] are based on the filament switching mechanism and are more widespread than analog memristors, which conductivities can be changed continuously. The analog memristor materials are encountered much less often and they require a more complex making process. Multilevel memristors are more stable to statistical fluctuations than the analog memristors. The use of binary and multilevel memristors for setting Hopfield networks weight coefficients makes it topical to investigate the weight quantization influence on the network information capacity and its resistance to the network input data noise level. In this case, the weights quantization problem can be considered as a reduction problem [10] generalization.

3 Estimating Hopfield Network Capacity

To estimate the Hopfield network capacity [11], a set M of random vectors with components from set $\{-1, +1\}$ is generated. Set M is used to construct the Hopfield network according to rules (1) or (2). Each of the set M vectors is fed to the constructed Hopfield network input. If, after the network operation step, its output coincides with its input for all the set M vectors, then a new random vector is added to M and the process is repeated. Otherwise, the network capacity is assumed equal to $|M| - 1$.

This algorithm approbation results are given in Tables 1 and 2. Here C denotes the capacity estimate averaged over 100 tests, $N/(2\ln N)$ is an asymptotic estimate of the Hopfield-Hebb network maximum capacity and $N - 1$ is an estimate of the Hopfield projection network maximum capacity.

Table 1. Hebb method

N	32	64	128	256	512
C	5.14	7.51	11.22	18.07	29.88
$N/(2\ln N)$	4.62	7.69	13.19	23.08	41.04

Table 2. Projection method

N	32	64	128	256	512
C	30.83	62.39	125.97	253.52	508.52
$N-1$	31	63	127	255	511

It follows from Table 2 that the algorithm [11] results are very close to the estimates [4] for the Hopfield projection network (the difference is less 1%).

4 Reduced Hopfield Network Capacity

The Hopfield network weighs reduction is carried out as follows [10]:

1. Network weights are normalized:

$$w_{ij} \leftarrow \frac{w_{ij}}{\max\limits_{j=1,\ldots,N} |w_{ij}|}$$

2. Normalized weights are reduced:

$$w'_{ij} = \begin{cases} +1, w_{ij} = +1 \\ -1, w_{ij} = -1 \\ 0, -1 < w_{ij} < +1 \end{cases} \tag{3}$$

In order to preserve the weight matrix symmetry, the Hopfield network reduction rule (3) takes the following form: weight w_{ij} is zeroed simultaneously with weight w_{ji} with a fulfillment of the conditions $|w_{ij}| < |w_{i,\max}|$ and $|w_{ji}| < |w_{j,\max}|$, where $w_{i,\max}$ and $w_{j,\max}$ are the maximum weights in rows i and j, respectively.

As a result of reduction, we obtain a network with weights from the set $\{-1, 0, +1\}$ that can be realized using binary memristors. The reduced Hopfield-Hebb network capacity values averaged over 100 experiments are shown in Table 3. Their values are close to $\log_2 N$.

Table 3. The reduced Hopfield-Hebb network capacity

N	32	64	128	256	512
C	4,62	5,43	6,52	7,85	8,9
$\log_2 N$	5	6	7	8	9

5 Quantized Weights Hopfield Network Capacity

We propose the Hopfield network weights quantization algorithm:

1. Find the maximum modulus $W_{\max}$ of the Hopfield network weights.
2. Find the magnitude of the weight quantum (jump) $\Delta = W_{\max}/(L-1)$, where $L \geq 2$ is the quantization levels number.
3. Convert all weights $w_{ij}, i, j = 1, 2, \ldots, N$ of the network according to the principle: if $(k-1)\Delta \leq |w_{ij}| < k\Delta$ then w_{ij} gets the value $(k-1)\Delta \cdot \mathrm{sgn}(w_{ij})$, $k = 1, \ldots, L$, where $\mathrm{sgn}(a) = \begin{cases} 1, a > 0 \\ -1, a \leq 0 \end{cases}$.

The Hopfield network capacity graphs (averaged over 100 tests) as a function of the quantization levels number $L \in \{2, 4, 8, 16, 32, 64, 128, 256\}$ of weight coefficients are presented in Fig. 1. These graphs show that, for a small L, the quantized weight networks capacity values are small, but, for the Hopfield-Hebb network starting from $L = 16$, it becomes close to the continuous weights network capacity (Table 1). For the Hopfield projection network, the capacitance dependence on L is much stronger. The quantized weights projection network approaches the continuous weights projection network by the capacitance only when L approaches $128 \div 256$ (Fig. 1b).

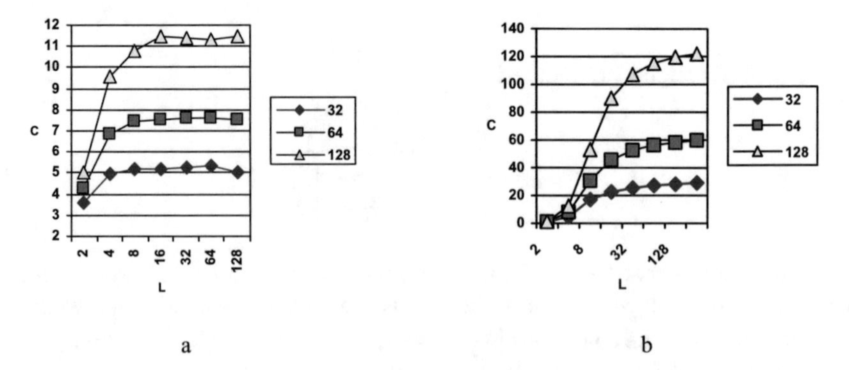

Fig. 1. The Hopfield network capacity C dependence on the weights quantization levels L number ($N \in \{32, 64, 128\}$): (a) Hopfield-Hebb network, (b) Hopfield projection network

6 Weight Quantization Levels Selection for Noise Filtering

The choice of minimum weight levels number for a noise filtering depends on the Hopfield network type (Hopfield-Hebb network or Hopfield projection network) and a set of particular reference vectors.

Our experiments show that the reduced Hopfield-Hebb network [10] is better than that of the quantized weights. So, the optimal weights levels number for the Hopfield-Hebb network is equal to two, regardless of the reference vector dimension. Such network weights can be realized on binary memristors. With a sufficiently large vector dimension, these networks are resistant to noise, with a significant reduction in the connections number (more than 90%) [10].

The experimentally obtained minimum amounts of the Hopfield projection network weight quantization levels as a function of the distorted pixels fraction ξ for a set of 10 reference vectors with dimension $n \times n$, $n \in \{8, 16, 32, 64\}$, corresponding to the digits images in Fig. 2, are shown in Fig. 3 and Table 4. The experimental results show that to provide acceptable noise levels corresponding to the Hopfield projection network [10] with continuous weights, it is required to specify tens of weight levels (see Fig. 3 and Table 4).

Fig. 2. Reference patterns

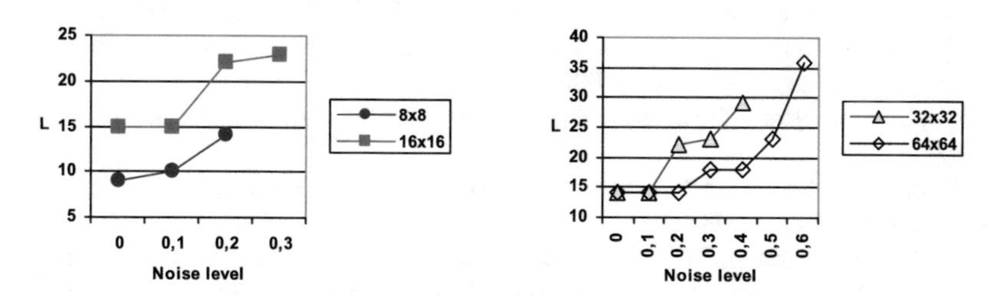

Fig. 3. Dependence of quantization levels number L on noise level ξ

Table 4. The weights quantization levels number L of the Hopfield projection network for the maximum permissible percentages ξ of distorted pixels [10]

$n \times n$	8×8	16×16	32×32	64×64
ξ	0,14	0,33	0,48	0,64
L	14	26	29	36

7 Conclusion

At present, the task of constructing neural networks based on the use of multilevel (including two-level - binary) memristors for the hardware synapses implementation is topical. Since the distinguishable memristor conductivity levels number is limited, the effect of Hopfield network weights quantization number on its information capacity is investigated. It is shown that when the weight levels number reaches approximately $2 \div 3$ tens, the quantized weights Hopfield-Hebb network capacity approaches its continuous weights network capacity. For the Hopfield projection network, this can be reached with a samples number of the order of hundreds.

Experiments have shown that: (1) Binary memristors should be used in Hopfield-Hebb networks, reduced by zeroing all weights in a given row which moduli are strictly less than the maximum weight in the row. (2) In Hopfield projection networks with discrete weights, multi-level memristors with a number of levels significantly greater than two (tens and hundreds) should be used, with a specific number of levels (tens and hundreds) depending on the dimension of the stored reference vectors, their specific set and the permissible noise level.

References

1. Haykin, S.: Neural Networks: A Comprehensive Foundation. Prentice Hall, Inc., Upper Saddle River (1999)
2. Hopfield, J.: Neural networks and physical systems with emergent collective computational abilities. Proc. Natl. Acad. Sci. USA **79**, 2554–2558 (1982)
3. Personnaz, L., Guyon, I., Dreyfus, G.: Collective computational properties of neural networks: new learning mechanisms. Phys. Rev. A **34**(5), 4217–4227 (1986)
4. Michel, A.N., Liu, D.: Qualitative Analysis and Synthesis of Recurrent Neural Networks. Marcel Dekker Inc., New York (2002)
5. Chua, L.: Memristor – the missing circuit element. IEEE Trans. Circuit Theory **18**, 507–519 (1971)
6. Strukov, D.B., Snider, G.S., Stewart, D.R., Williams, R.S.: The missing memristor found. Nature **453**, 80–83 (2008)
7. He, W., Sun, H., Zhou, Y., Lu, K., Xue, K., Miao, X.: Customized binary and multi-level HfO2 − x-based memristors tuned by oxidation conditions. Sci. Rep. **7**, 10070 (2017)
8. Yu, S., Gao, B., Fang, Z., Yu, H., Kang, J., Wong, H.-S.P.: A low energy oxide-based electronic synaptic device for neuromorphic visual systems with tolerance to device variation. Adv. Mater. **25**, 1774–1779 (2013)
9. Tarkov, M.S.: Crossbar-based hamming associative memory with binary memristors. In: Huang, T., Lv, J., Sun, C., Tuzikov, A.V. (eds.) ISNN 2018. LNCS, vol. 10878, pp. 380–387. Springer, Cham (2018). https://link.springer.com/chapter/10.1007/978-3-319-92537-0_44. Accessed 12 July 2018

10. Tarkov, M.S.: Synapses reduction in autoassociative Hopfield network. In: 2017 International Multi-conference on Engineering, Computer and Information Sciences (SIBIRCON). Novosibirsk, Russia, 18–22 September, pp. 158–160. IEEE (2017). https://ieeexplore.ieee.org/document/8109860/. Accessed 12 July 2018
11. Folli, V., Leonetti, M., Ruoccol, G.: On the maximum storage capacity of the Hopfield model. Front. Comput. Neurosci. **10**, Article 144 (2017). https://www.frontiersin.org/articles/10.3389/fncom.2016.00144/full. Accessed 12 July 2018

Applications of Neural Networks

Linear Filtering Based on a Pulsed Neuron Model with an Orthogonal Filter Bank

Vladimir Bondarev(✉)

Sevastopol State University, Sevastopol, Russian Federation
bondarev@sevsu.ru

Abstract. This paper deals with the model and learning rules of a pulsed neuron, which provide the linear filtering of signals represented by pulse trains. To reduce the number of training parameters of a pulsed neuron we propose using a bank of orthogonal filters as a model of synaptic connections. For this model of pulsed neuron, we derive a supervised learning rule in a general form that can include various orthogonal basis functions. The rules minimize the mean square error between the desired and the actual output signal of a linear filter realized on the base of the pulsed neuron model. We derive two special learning rules: with set of exponential complex orthogonal functions and set of block-pulse orthogonal functions. For both set of these functions, we demonstrate rule's properties by computer simulation of linear filters that implement high-pass filtering and double integration of the input signal transformed to pulse train. We show the impulse and frequency responses of the filters as well as the dependencies of the normalized mean square error on the number of training iterations.

Keywords: Pulse neuron · Pulse train · Supervised learning · Linear filtering

1 Introduction

Pulsed neural networks use the mechanism of pulse coding of signals (stimuli), according to which neurons perceive, process and transmit information in form of pulse sequences [1, 2]. The representation and processing of signals in the pulse domain has a number of advantages: low power consumption, noise immunity, simplicity of conversion to digital form, etc. [3, 4]. The papers [4–12] consider the features of the pulse representation of signals and the conditions for their reconstruction by pulse sequences generated by integrate-and-fire (IAF) and leakage integrate-and-fire (LIF) neuron models, as well as the implementation of various signal processing operators in the pulse domain.

In [12], we proposed the rules for adapting the impulse responses of synaptic connections to realize various linear operators on the basis of a pulsed neuron (PN). However, since impulse responses are represented by their direct samples, this requires the calculation of a large number of adaptable parameters of the PN model to provide an acceptable approximation error.

B. Kryzhanovsky et al. (Eds.): NEUROINFORMATICS 2018, SCI 799, pp. 101–108, 2019.
https://doi.org/10.1007/978-3-030-01328-8_9

In this paper, we consider the possibility to reduce the number of PN model adaptable parameters by using a bank of orthogonal filters to model the dynamics of synaptic connections.

2 Description of the PN Model

Let us consider the approach to the synthesis of linear filters with the help of adaptive modeling method (Fig. 1) [13]. A special feature of the case under consideration is the realization of an adaptive filter in the form of a serial connection of the coding input IAF-neuron and the adaptive multi-input PN [9]. The coding neuron converts the input signal $z(t)$ into a bipolar pulse train $u(t)$, which is processed by the PN. Such a scheme of an adaptive filter constructed on the basis of pulsed neurons combines the discretization of signals with their processing and is alternative to the circuit that digitizes analog signals using an ADC [8].

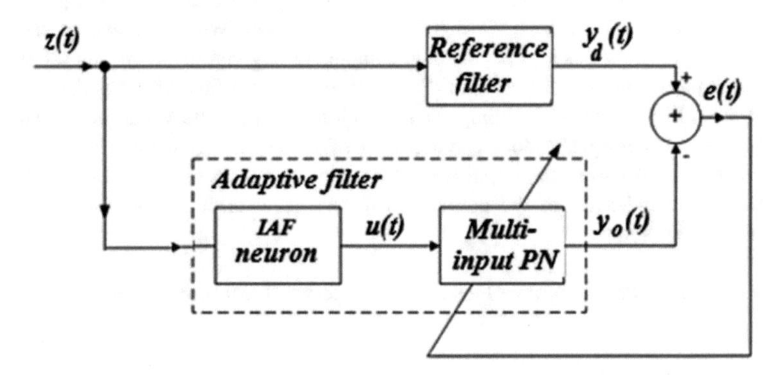

Fig. 1. The scheme of adaptive modeling

For our study, we will use the model of the multi-input PN described in [10, 11]. The model of the multi-input PN (Fig. 2) consists of input linear filters with impulse responses $h_i(t)$ that simulate the dynamic properties of synaptic connections, multipliers by synaptic weighting coefficients w_i and an adder. Bipolar input pulse trains $u(t)$ generated by the encoding presynaptic neuron arrive at inputs of the PN linear filters with impulse responses $h_i(t)$. Filter reactions $x_i(t)$ are weighted with synaptic weights w_i and summed to form the total postsynaptic potential $y_o(t)$ of the PN. If the integral of the module of $y_o(t)$ exceeds a threshold Θ then PN emits an output pulse with a sign corresponding to the sign of $y_o(t)$ and integrator is reset. Generally speaking, this chain of the PN transformations corresponds to the pulsed LIF-neuron. Here we will consider the multi-input PN model without a reset operation.

As a result of the adaptation process, the parameters of the PN model are adjusted in such a way that the specified requirements for the dynamic characteristics of the adaptive filter are met. These requirements are set by a reference filter (Fig. 1) having the desired frequency response or impulse response.

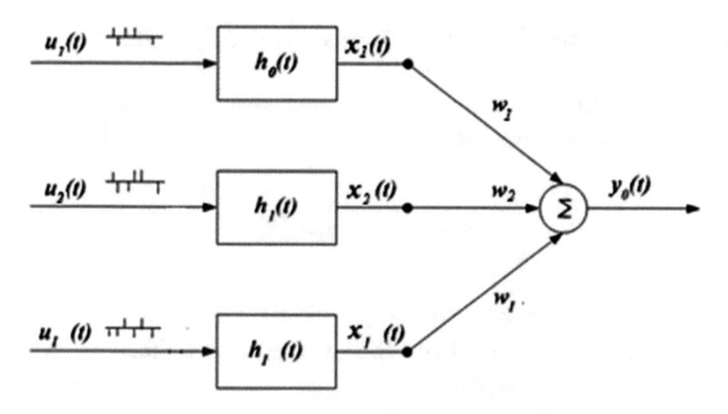

Fig. 2. Model of a multi-input pulsed neuron

Let the learning problem of the PN consists in the determining of the impulse responses $h_i(t)$ that provide a minimum of the mean square error $e(t) = y_d(t) - y_o(t)$, where $y_d(t)$ is the output (desired) signal of the reference filter with the given frequency response or impulse response. We will assume here that $w_i = 1, \forall i$.

If we calculate the values of $y_o(t)$ at discrete time $t_n = n\Delta t$, where Δt is the time sampling step, then as it shown in [8]

$$y_o(n) = \sum_{i=1}^{I} x_i(n), \quad x_i(n) = \mathbf{b}^{\mathrm{T}}(n)\mathbf{h}_i, \tag{1}$$

where $\mathbf{b}^{\mathrm{T}}(n)$ is the sliding binary vector with the elements that equal to the signs of the input pulses, $\mathbf{h}_i$ denotes the impulse response vector $\mathbf{h}_i = (h_i(0), h_i(1), \ldots, h_i(K-1))^{\mathrm{T}}$. Using (1) you can get the following supervised learning rule to adapt the impulse responses of each input filter of the PN [12]

$$\mathbf{h}_i(n) = \mathbf{h}_i(n-1) + \mu(n)e(n)\mathbf{b}\,(n), \tag{2}$$

where $\mu(n)$ is the learning rate.

In expression (2), the impulse response of filters in individual channels is represented by their direct samples at discrete instants of time. This requires computing a large number of elements $h_i(0), h_i(1), \ldots, h_i(K-1)$ to provide a given approximation error.

3 The Model of the PN with Orthogonal Filter Bank

To obtain a limited number of adaptable PN parameters, we approximate $\mathbf{h}_i$ with the help of a generalized finite Fourier series:

$$h_i(k) \approx \sum_{q=1}^{Q} c_{iq}\varphi_q(k), \tag{3}$$

where $\{\varphi_q(k)\}$ is the set of known orthogonal basis functions.

Substituting (3) into (1), we obtain (taking into account the fact that all the inputs of PN receive the same pulse sequence)

$$x_i(n) = \sum_{k=0}^{K-1} b(n-k)h_i(k) = \sum_{q=1}^{Q} c_{iq}(k)\sum_{k=0}^{K-1} b(n-k)\varphi_q(k). \tag{4}$$

The internal sum in (4) determines the output signal of a filter with a finite impulse response which corresponds to basis function $\varphi_q(k)$:

$$x_q(n) = \sum_{k=0}^{K-1} b(n-k)\varphi_q(k) = \mathbf{b}^{\mathrm{T}}(n)\boldsymbol{\varphi}_q. \tag{5}$$

Since the vector $\mathbf{b}^{\mathrm{T}}(n)$ is binary, the signal $x_q(n)$ is computed by summing the samples of the q-th basic orthogonal function. In this case, the individual channels of the PN become orthogonal to each other and form a bank of orthogonal filters.

Taking into account (4) and (5), we derive from (1) the output signal of the PN

$$y_o(n) = \sum_{i=1}^{I}\sum_{q=1}^{Q} c_{iq}x_q(n) = \sum_{q=1}^{Q} x_q(n)\sum_{i=1}^{I} c_{iq} = \sum_{q=1}^{Q} c_q x_q(n) = \mathbf{c}^{\mathrm{T}}\mathbf{x}(n), \tag{6}$$

where $c_q = \sum_{i=1}^{I} c_{iq}$, and $\mathbf{x}(n) = (x_1(n), x_2(n), \ldots, x_Q(n))$ is the vector of the outputs of the bank of orthogonal filters.

Using (6) and the objective function in the form of mean square error, we obtain the learning rule for the vector $\mathbf{c}$:

$$\mathbf{c}(n) = \mathbf{c}(n-1) + \mu(n)e(n)\mathbf{x}(n), \tag{7}$$

where $\mathbf{x}(n)$ is calculated in accordance with (5).

The rule (7) is generalized and allows us to determine the values of the vector $\mathbf{c}$ in the course of the adaptive process for various types of basis functions. The element values of the vector $\mathbf{c}$ can be interpreted as the coefficients of the orthogonal decomposition of the impulse response (3) of a non-recursive filter constructed on the basis of the PN.

Let us consider specific types of learning rules for two classes of basis functions:

1. *The complex exponential function set.*

In this case $\{\varphi_q(k)\} = \{\exp(j2\pi qk/K)\}$, $K = \lceil T/\Delta t \rceil$, T is the period of the fundamental harmonic. Then

$$y_o(n) = \sum_{q=1}^{Q} [c_q^{re} x_q^{re}(n) - c_q^{im} x_q^{im}(n)], \tag{8}$$

where $x_q^{re}(n) = \mathbf{b}^{\mathrm{T}}(n)\boldsymbol{\varphi}_q^{re}$ and $x_q^{im}(n) = \mathbf{b}^{\mathrm{T}}(n)\boldsymbol{\varphi}_q^{im}$ are the real and imaginary components of the output signals of the bank of orthogonal filters. In accordance with rule (7)

$$c_q^{re}(n) = c_q^{re}(n-1) + \mu(n)e(n)x_q^{re}(n), \tag{9}$$

$$c_q^{im}(n) = c_q^{im}(n-1) + \mu(n)e(n)x_q^{im}(n), \tag{10}$$

Rules (9) and (10) determine the complex coefficient $c_q(n)$ at the frequency $2\pi q/T$.

2. *The block-pulse function set.*

These functions are defined over a time interval $k\Delta t \in [0, T]$ [14]:

$$1_q(T_s, k\Delta t) = \begin{cases} 1, & k\Delta t \in [(q-1)T_s, qT_s] \\ 0, & elsewhere \end{cases}, \quad T_s = T/Q, \tag{11}$$

where $q = 1, 2, \ldots, Q$. Then the output signal of the orthogonal filter will be equal to

$$S_q(n) = x_q(n) = \sum_{k=0}^{K-1} b(n-k) 1_q(T_s, k\Delta t), \tag{12}$$

where $S_q(n)$ is the sum of elements of a binary vector $\mathbf{b}(n)$ with indices k satisfying the condition $k\Delta t \in [(q-1)T_s, qT_s)$. The learning rule will accordingly be written in the form

$$c_q(n) = c_q(n-1) + \mu(n)e(n)S_q(n). \tag{13}$$

In this case, the coefficient c_q is an estimate of the average value of the desired impulse response at the interval $[(q-1)T_s, qT_s)$. If only one input pulse falls into the interval T_s, then $S_q(n) = b_q(n)$ and rule (13) will coincide with rule (2).

4 Computer Simulation

During the simulation, the simple model of bipolar IAF-neuron with single input was used as the model of encoding neuron. The encoding neuron converts an input signal $z(t)$ to a bipolar pulse train. This pulse train simultaneously goes to all inputs of the multi-input PN.

In simulation, we investigate the PN model, which implements high-pass filtering and double integration of the input signal $z(t)$. The similar problem arises in the case of accelerometer signal processing [12].

During the training, the input signal $z(t)$ equal to the sum of sine signals with the multiple frequencies $f = f_0 l$ was applied to the input of the encoding neuron. The desired output signal $y_d(t)$ is defined as follows

$$y_d(t) = \sum_{l=1}^{L} H(f_0 l) \sin(2\pi f_0 l t + P_l), \tag{14}$$

where $f_0 = 1/T_0$ is the fundamental harmonic, T_0 is the duration of the impulse response of the reference filter, $H(f_0 l)$ are the normalized values of frequency response of the reference filter (double integrator):

$$|H(f_0 l)| = \begin{cases} 0, & f_0 l < F_c \\ F_c^2 \big/ (f_0 l)^2, & f_0 l \geq Fc \end{cases}, \tag{15}$$

where F_c is the cut-off frequency. Let the reference filter have a linear phase characteristic. Then all harmonics that form the signal $y_d(t)$ should be shifted in phase by the angle $P_l = -\pi l - \pi$.

The pulse and frequency responses, as well as normalized values of mean square errors (NMSE) of the PN filters adapted with the help of rules (9–10) and (13) are shown in the Fig. 3. In all our simulations, the frequency response of the reference filter (15) was set in 30 points uniformly distributed in the frequency range indicated in Fig. 3. The cut-off frequency was F_c = 0.25 Hz. We used the time sampling step Δt = 6.7 ms and the constant learning rate $\mu = 0.005$.

Figure 3 shows that the NMSE is decreasing with growth of n. The frequency responses of the synthesized filters based on the PN model with the orthogonal filter bank are close to the frequency response of the reference filter (15). Note that the NMSE for the filter that adapts using (9–10) is less than the NMSE for the filter that adapts based on (13). You can also use fewer orthogonal filters for rules (9–10). However, due to simplicity of the block-pulse functions the PN filter based on (13) is more efficient from a computational point of view.

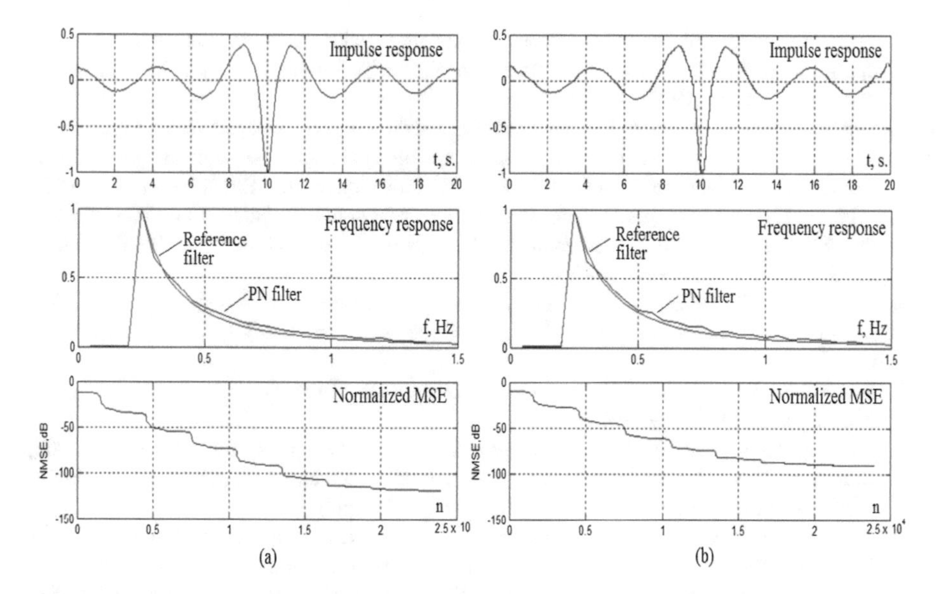

Fig. 3. Simulation results: (a) – for the complex exponential function set; (b) – for the pulse-block function set

5 Conclusions

The developed approach to adaptive synthesis of linear filters, based on the use of the PN model with the bank of orthogonal filters, is quite general and allows synthesizing arbitrary linear filters with a given frequency response or impulse response. Using a bank of orthogonal filters can significantly reduce the number of adaptable PN parameters.

We presented generalized and special learning rules for two classes of orthogonal basis functions. Learning rules that use complex exponential orthogonal functions allow direct control of the frequency response of the filter. Learning rules with block-pulse orthogonal functions allow direct monitoring of the filter impulse response. In addition, filters based on orthogonal block-pulse functions are very effective in software and hardware implementation.

References

1. Grüning, A., Bohte, S.M.: Spiking neural networks: principles and challenges. In: Proceedings of the 22nd European Symposium on Artificial Neural Networks, Computational Intelligence and Machine Learning (ESANN 2014), Bruges (Belgium), 23–25 April 2014. i6doc.com publ, Louvain-La-Neuve (2014)
2. Ponulak, F., Kasiński, A.: Introduction to spiking neural networks: information processing, learning and applications. Acta Neurobiol. Exp. **71**, 409–433 (2011)

3. Lee, H.C.: Integral pulse frequency modulation with technological and biological applications. Ph.D. thesis, Department of Electrical Engineering, McGill University, Montreal, Quebec (1969)
4. Feichtinger, H.G., Príncipe, J.C., Romero, J.L., Alvarado, A.S., Velasco, G.A.: Approximate reconstruction of bandlimited functions for the integrate and fire sampler. Advances in Computational Mathematics **36**(1), 67–78 (2012). https://doi.org/10.1007/s10444-011-9180-9
5. Wei, D., Harris, J.G.: Signal reconstruction from spiking neuron models. In: Proceedings of the 2004 International Symposium on Circuits and Systems (ISCAS 2004), vol. 5, pp. 353–356. IEEE Press (2004)
6. Zeevy, Y.Y., Bruckstein, A.M. A note on single signed integral pulse frequency modulation. In: IEEE Transactions on Systems, Man, and Cybernetics, vol. SMC-7, No. 12, pp. 875–877 (1977)
7. Bruckstein, A.M., Zeevi, Y.Y.: Analysis of "Integrate-to-Threshold" neural coding schemes. Biol. Cybern. **34**, 63–79 (1979)
8. Lazar, A.A.: A simple model of spike processing. Neurocomputing **69**, 1081–1085 (2006)
9. Bondarev, V.N.: Adaptive pulse-frequency modeling aimed at digital signal processing problems. Vestnik SevGTU **18**, 46–51 (1999). (in Russian)
10. Bondarev, V.: Vector-matrix models of pulse neuron for digital signal processing. In: Cheng, L., Liu, Q., Ronzhin, A. (eds.) Advances in Neural Networks—ISNN 2016. Lecture Notes in Computer Science, vol. 9719, pp. 647–656. Springer, Cham (2016). https://doi.org/10.1007/978-3-319-40663-3_74
11. Bondarev, V.: Pulse neuron learning rules for processing of dynamical variables encoded by pulse trains. In: Kryzhanovsky, B., Dunin-Barkowski, W., Redko, V. (eds.) Neuroinformatics 2017. Studies in Computational Intelligence, vol. 736, pp. 53–58. Springer, Cham (2018). https://doi.org/10.1007/978-3-319-66604-4_8
12. Bondarev, V.: Pulse neuron supervised learning rules for adapting the dynamics of synaptic connections. In: Huang, T., et al. (eds.) Advances in Neural Networks—ISNN 2018. Lecture Notes in Computer Science, vol. 10878, pp. 183–191, Springer, Cham (2018). https://doi.org/10.1007/978-3-319-92537-0_22
13. Widrow, B., Stearns, S.D.: Adaptive Signal Processing. Prentice-Hall, Englewood Cliffs (1985)
14. Beregovenko, G.Y., Puhov, G.E.: Stupenchatye izobrazheniya i ih primenenie (Step representations and their applications). Naukova Dumka, Kiev (1983). (in Russian)

Comparative Testing of the Neural Network and Semi-empirical Method on the Stabilization Problem of Inverted Pendulum

Egor A. Degilevich, Aleksandra M. Kobicheva, Valery A. Kozhin, Anastasia D. Subbota, Ilya Y. Surikov, Dmitry A. Tarkhov(✉), and Valery A. Tereshin

Peter the Great St. Petersburg Polytechnic University, Saint Petersburg, Russia
dtarkhov@gmail.com

Abstract. In the article we have presented the results of comparative testing of non-standard methods of dynamic systems' control. We've implemented the testing by the example of solving the problem of bringing the pendulum to the neighborhood of unstable equilibrium position in the shortest time under conditions of limited control. We've compared four approaches - a one-step approach based on the exact solution of the pendulum equation and three two-step approaches. The first two-step approach we've based on the method of restarts. The second two-step approach we've built on the neural network training. In the third two-step approach, we've used our modifications of the algorithm for constructing approximate solutions of ordinary differential equations. Based on computational experiments, we've concluded that the one-step approach is significantly less effective than the two-step one. All three two-step approaches have approximately the same rate of convergence. At the same time, the approach based on the method of restarts requires much more time, although it is more resistant to changes in the starting point. Approaches based on the use of a pre-trained neural network and our modification of the implicit Euler method require significantly less computing resources for their work. The disadvantage of these methods is the process's elongation of bringing the pendulum into the neighborhood of an unstable equilibrium position for some initial conditions.

Keywords: Inverted pendulum · Stabilization · Control law · Euler method
The Pontryagin maximum principle

1 Introduction

The control's study of the inverted pendulum (Furuta pendulum) is a good illustration of the possibilities of various algorithms to stabilize the motion of nonlinear systems near the position of unstable equilibrium [1, 2]. The scientific papers devoted to this problem usually describe the results of experiments performed on two types of installations. It can be a horizon-tally and rectilinearly moving trolley with an inverted pendulum [1, 2, 4, 7] or a vertical axis motor with an inverted pendulum mounted on its flange [3, 5].

B. Kryzhanovsky et al. (Eds.): NEUROINFORMATICS 2018, SCI 799, pp. 109–114, 2019.
https://doi.org/10.1007/978-3-030-01328-8_10

The axis of the pendulum support crosses the motor axis at a right angle. To stabilize the motion of the pendulum near the vertical position for small deviations, it is efficient to use the linearized mathematical model and linear control [1, 5, 6]. At the same time it is completely justified to use all well-developed algorithms for optimal control. A special and, in our opinion, the most important interest is the stabilization of the motion of the pendulum near the vertical point for significant initial deviations [1, 2, 3, 7]. To solve this problem, various control principles are used, including fuzzy logic [4], relay control (bang-bang) [3, 7], two-loop control of separately slow and fast processes [1, 2]. Some optimal algorithms create at the beginning of the movement an increasing buildup of the pendulum and after its rise provide damped oscillations around zero [6]. Other algorithms immediately lift the pendulum up with-out a build-up [2]. In this paper we research the automated search for optimal control of nonlinear systems by training neural networks. It is shown that the control laws obtained in this way for large initial deviations in a number of cases translate the pendulum into the required small neighbor-hood of the equilibrium position faster than the control laws obtained on the basis of exact solutions.

2 Materials and Methods

The behavior of the pendulum is modeled by differential equations:

$$\ddot{\varphi} = a \sin \varphi + bu, \tag{1}$$

where φ is deviation's angle of the pendulum from the vertical, u (force moment) and b are coefficients that depend on the parameters of the object. The task is to choose a control u, so that $\varphi, \dot{\varphi} \underset{t \to +\infty}{\longrightarrow} 0$. Consider the model case with $a = b = 1$. Let us pass in Eq. (1) to the coordinates in the phase plane: $x = \varphi, y = \dot{\varphi}$:

$$\begin{cases} \dot{x} = y \\ \dot{y} = \sin x + u. \end{cases} \tag{2}$$

We will solve the problem of stabilizing the pendulum in the upper equilibrium position under the condition $|u| \leq u_0$ in a minimum time. In accordance with the Pontryagin maximum principle [8], the control is selected by maximizing the Hamiltonian function, as a result, we obtain $u = \pm 1$. Exact research and analytical construction of control is difficult, therefore we apply approximate methods. In what follows we assume that $u_0 = 1$.

The first approach is one-step. We construct approximate solutions on the interval Δt with $u = \pm 1$ and choose the control sign, proceeding from the minimum $x^2 + y^2$ at the end of the specified time interval. Next, we make a transition to a new point in accordance with the solution of system (2) on a given interval, and the choice of the control sign is repeated. Computational experiments have shown that this method works slowly for any choice Δt.

The next three approaches are two-step, where we select two consecutive time intervals on which the control has different signs. The second approach is based on an exact solution. We make two steps with alternating +1 and −1 control's sign or conversely (random selection) and random duration from 0 to T_0, the repetition number is 10000. Using the built-in Mathematics function we construct the solution in the Wolfram Mathematica system. We choose a control option based on the minimum $x^2 + y^2$ at the end of the time interval of the two steps.

The third approach is to repeat 50 times for different starting points for $0 < x_0 < 2; -1 < y < 1$ procedure of the second approach for 1000 repetitions with controls of different signs. Further, six radially-basic neural networks of the form are trained: $c_0 + \sum_{i=1}^{n} c_i e^{-\alpha_i\left((x-x_i)^2 + (y-y_i)^2\right)}$ (three for each set of control characters). Two networks are for the lengths of the optimal time intervals, one is for the minimum $x^2 + y^2$ at the end of the time interval from the two steps indicated. At the next stage of control, the choice of the control sign is determined on the basis of the minimum of the outputs of the third networks, and the times are set by the appropriate pair of the first two networks.

The fourth approach is to use the modification [9] of the one-step implicit Euler method to find the time for control of each sign. At the same time we select a pair of time slots to control various characters so that the approximate solution came to an end at the origin. As a result, we obtain $u = -sign(x)$. In addition, for the lengths of the intervals we get: $\begin{cases} t = 0.5\left(\sqrt{y_0^2 + 4|x_0|} - y_0 u\right) \\ \tau = 0.5\left(\sqrt{y_0^2 + 4|x_0|} - y_0 u\right). \end{cases}$

After the transition is made in accordance with the solution (2) over the length interval $t + \tau$, the control selection is repeated.

3 Results

We give the results of calculations for the point. $x_0 = 2.5, y_0 = 0.5$ (Figs. 1 and 2).

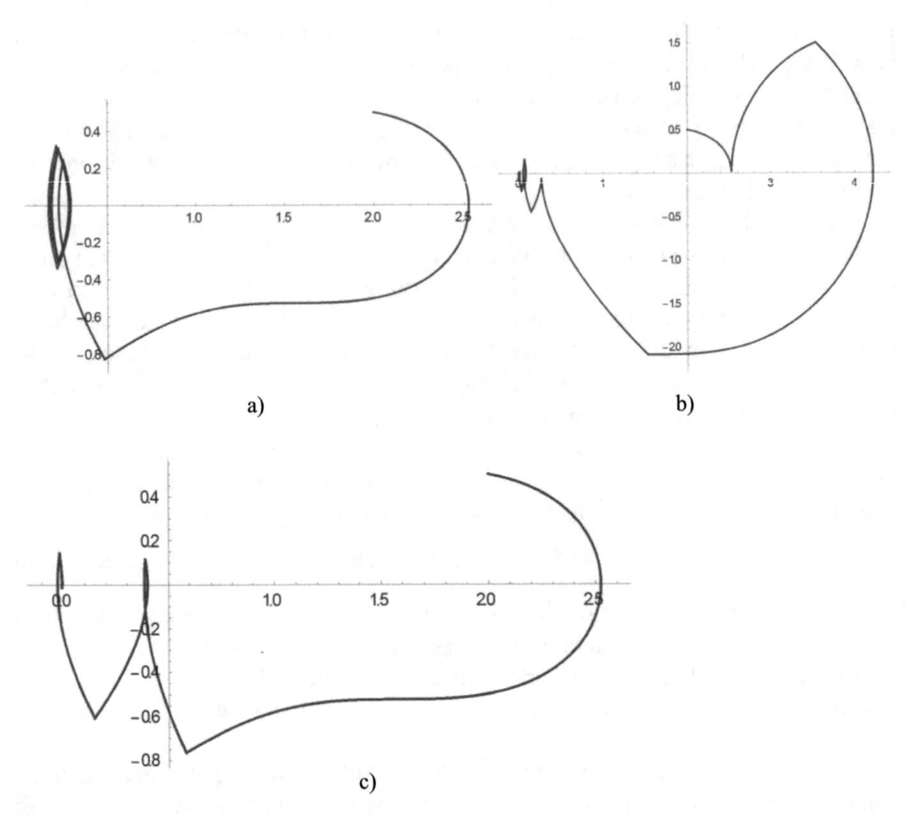

Fig. 1. Phase trajectories formed with: (a) a neural network, (b) an implicit Euler method, (c) an exact solution. From these graphs we can see that control leads to a neighborhood of the equilibrium position.

We can see from these graphs that the solution obtained with the help of the neural network and the implicit Euler method is no worse than an exact solution in the sense of the time of bringing the target point (unstable equilibrium position) to the neighborhood. To have control in the neighborhood of origin we need a separately trained network.

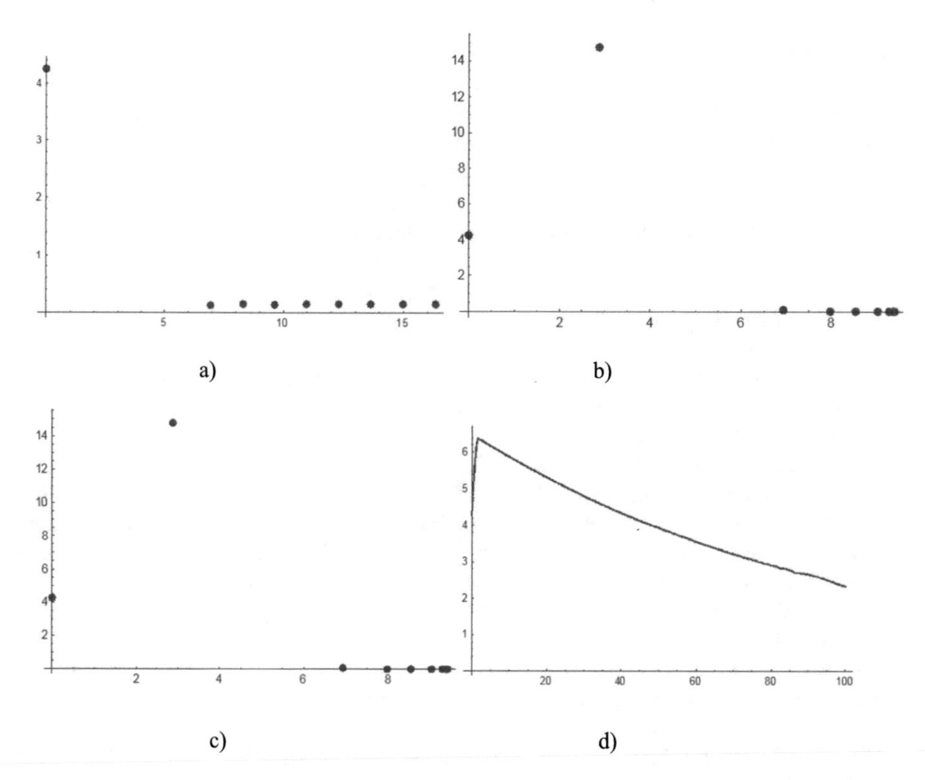

Fig. 2. Graphs of the dependence of the distance's square to the origin of coordinates on the phase plane on time, formed with the help of: (a) a neural network, (b) the implicit Euler method, (c) an exact solution, (d) a one-step method.

4 Conclusions

The computational experiments carried out have showed that the modifications of classical numerical methods proposed in [9] can be successfully applied to control problems, including those related to control under instability conditions. These methods can be particularly useful in situations where the mathematical model of the controlled object is inaccurate and can be refined in the process of controlling it as data on the processes in the modeled system are accumulated.

Acknowledgement. This paper is based on research carried out with the financial support of the grant of the Russian Scientific Foundation (project №18-19-00474).

References

1. Wiboonjaroen, W., Sujitjorn, S.: Stabilization of an inverted pendulum system via state-PI feedback. Int. Math. Model. Methods Appl. Sci. **5**(4), 763–772 (2011)
2. Subudhi, B.: On stabilization of cart-inverted pendulum system: an experimental study. A thesis submitted in partial fulfillment of the requirements for the award of the award of degree Master of Technology by Research in Electrical Engineering by T. Rakesh Krishnan, National Institute of Technology Rourkela, India, p. 98 (2012)
3. Furuta, K., Yamakita, M., Kobayashi, S.: Swing up control of inverted pendulum. CH2976-9/91/0000-2193 $1.00 0, pp. 2195–2198. IEEE (1991)
4. Nurfakhirah, B.I.: Fuzzy logic controller design for inverted pendulum system. A project report submitted in partial fulfillment of the requirement for the award of the degree Master of Electrical Engineering, Faculty of Electrical and Electronic Engineering University, Tun Hussein Onn Malaysia, p. 87 (2013)
5. Mandic, P.D., Lazarevic, M.P., Sekara, T.B.: D-decomposition technique for stabilization of Furuta pendulum: fractional approach. Bull. Pol. Acad. Sci. Teh. Sci. **64**(1) 189–196 (2016)
6. Hangos, K.M., Bokor, J., Szederkenyi, G.: Analysis and control of nonlinear process systems. Springer – Verlag London Berlin Heidelberg a member of Bertelsmann Springer Science +Business Media GmbH springeronline.com, p. 335 (2004)
7. Quintero, S.A.P.: Controlling the Inverted Pendulum. Department of Electrical and Computer Engineering, University of California, Santa Barbara, p. 5 (2014)
8. Pontryagin, L.S., Boltyanskij, V.G., Gamkrelidze, R.V., Mishchenko, E.F.: Matematicheskaya teoriya optimal'nyh processov, M.: Nauka, 384s (1969)
9. Lazovskaya, T., Tarkhov, D.: Multilayer neural network models based on grid methods. IOP Conf. Ser. Mater. Sci. Eng. **158** (2016). http://iopscience.iop.org/article/10.1088/1757-899X/158/1/01206

Homotopy Continuation Training Method for Semi-Empirical Continuous-Time State-Space Neural Network Models

Mikhail Egorchev(✉) and Yury Tiumentsev

Moscow Aviation Institute, National Research University, Moscow, Russia
mihail.egorchev@gmail.com

Abstract. Recurrent neural networks present a powerful class of nonlinear dynamical system models that hold great potential for real-world applications. Semi-empirical state-space neural network based models are of particular interest since they utilize both the theoretical domain-specific knowledge as well as experimental data of system behavior in order to improve the generalization performance. A continuous-time form of the model grants additional flexibility in the choice of suitable numerical integration methods. However, recurrent neural network training for multi-step-ahead prediction is known to be a very difficult optimization problem. Traditional gradient-based methods often fail to find a sufficiently good solution unless the initial guess for parameter values lies very close to it. We propose a training method based on the homotopy continuation approach with variable prediction horizon which allows circumventing this issue. The efficiency of this training method is confirmed by computational experiments for the maneuverable aircraft angular motion simulation and identification problem.

Keywords: Nonlinear dynamical system · Semi-empirical model
Neural network · Homotopy continuation

1 Introduction

Semi-empirical neural network based approach to simulation and identification of nonlinear controlled dynamical systems proposed in [1,2] allows incorporating theoretical domain-specific knowledge into the neural network model structure to reduce the number of unknown parameters required to achieve the desired model accuracy while retaining its flexibility. This approach may also be viewed as a form of regularization that improves the model generalization performance. Semi-empirical models were shown to significantly outperform purely empirical neural network based nonlinear autoregressive models with exogenous inputs.

A continuous-time form of semi-empirical models provides additional flexibility in the choice of suitable numerical methods to solve the corresponding initial value problems. However, this formulation also requires the derivation of

B. Kryzhanovsky et al. (Eds.): NEUROINFORMATICS 2018, SCI 799, pp. 115–120, 2019.
https://doi.org/10.1007/978-3-030-01328-8_11

continuous-time versions of algorithms for estimation of the error function derivatives with respect to the semi-empirical model parameters. In this paper, we present a continuous-time version of the Real-Time Recurrent Learning (RTRL) algorithm [3] for computation of the gradient and the Hessian of the error function. A continuous-time version of the BackPropagation Through Time (BPTT) algorithm [4] for gradient computation in case of a purely empirical neural network model is given in [5].

Recurrent neural network training for multi-step-ahead prediction is known to be a severe problem due to several issues: vanishing or exploding gradient effects [6]; possible bifurcations of the recurrent neural network model [7]; the presence of spurious valleys in the error function landscape [8]. Thus, traditional gradient-based optimization methods fail to locate a good minimum unless the initial guess for parameter values lies very close to the solution.

In this paper, we propose a homotopy continuation training method designed to overcome these issues and provide for efficient learning of semi-empirical state-space neural network models. We confirm the efficiency of the approach by computational experiments for the maneuverable aircraft angular motion simulation and identification problem.

2 Semi-Empirical Continuous-Time State-Space Neural Network Models

Semi-empirical continuous-time state-space neural network models of controlled lumped dynamical systems have the following form:

$$\begin{aligned} \frac{d\hat{\mathbf{x}}}{dt}(t) &= \hat{\mathbf{f}}(\hat{\mathbf{x}}(t), \mathbf{u}(t); \mathbf{wf}) \\ \hat{\mathbf{y}}(t) &= \hat{\mathbf{g}}(\hat{\mathbf{x}}(t); \mathbf{wg}) \end{aligned}, \tag{1}$$

where $\hat{\mathbf{x}}\colon [0,T] \to \mathbb{R}^{n_x}$ is an estimate of a state-space trajectory, $\hat{\mathbf{y}}\colon [0,T] \to \mathbb{R}^{n_y}$ is an estimate of observable outputs, $\hat{\mathbf{f}}\colon \mathbb{R}^{n_x+n_u} \times \mathbb{R}^{n_{\mathbf{wf}}} \to \mathbb{R}^{n_x}$ and $\hat{\mathbf{g}}\colon \mathbb{R}^{n_x} \times \mathbb{R}^{n_{\mathbf{wg}}} \to \mathbb{R}^{n_y}$ are feedforward semi-empirical neural networks with parameters $\mathbf{wf}$ and $\mathbf{wg}$. It is assumed that the measured outputs $\tilde{\mathbf{y}}(t)$ correspond to the true outputs $\mathbf{y}(t)$ corrupted by an additive white Gaussian noise $\eta(t)$ with zero mean.

To train the model, we require a set of experimental data that describes the behavior of an unknown dynamical system. In contrast to purely empirical models, the theoretical knowledge allows us to select a meaningful set of state variables interpretable in domain-specific terms. Therefore, the initial values of the state variables may be measured or estimated and included in the training set. Also, in contrast to discrete-time models which assume uniform timesteps, the continuous-time models allow us to specify non-uniform timesteps in the training set. Thus, the training set has the following form:

$$\left\{\left\langle\left\langle{}^{p}\tilde{\mathbf{x}}^{0}, {}^{p}\tilde{\mathbf{y}}^{0}\right\rangle, \left\{\left\langle{}^{p}\Delta t^{k}, {}^{p}\mathbf{u}^{k}, {}^{p}\tilde{\mathbf{y}}^{k}\right\rangle\right\}_{k=0}^{{}^{p}K}\right\rangle\right\}_{p=1}^{P},$$

where P is the total number of trajectories, ${}^p\tilde{\mathbf{x}}^0$ is the initial state estimate for p-th trajectory, ${}^pK$ is the number of timesteps for p-th trajectory, ${}^p\Delta t^k$ are the values of timesteps, ${}^p\mathbf{u}^k$ are the known controls and ${}^p\tilde{\mathbf{y}}^k$ are the measured outputs. We also denote the trajectory time duration by ${}^pT = \sum_{k=0}^{{}^pK} {}^p\Delta t^k$.

One of the critical factors contributing to generalization performance of neural network based models is the representativeness of training data set. To obtain such a training set, we need to design the suitable experiments by selecting appropriate initial state values as well as the controls. Handcrafted experiment designs tend to be suboptimal; hence we propose an automatic procedure for this purpose. We decompose the control signals for each trajectory into low-frequency high-amplitude reference maneuvers and high-frequency low-amplitude excitations. We represent reference maneuvers by a solution of the optimal control problem for the initial theoretical model of the system to maximize the differential entropy of predicted states and controls over the feasible compact set. In turn, the excitations are designed as multisine signals with a wideband frequency content and minimum peak-factor value. Description of a version of this procedure can be found in [9].

3 Homotopy Continuation Training Method

One of the techniques that allow circumventing the requirement of a good initial guess for a nonlinear optimization problem is a family of globally convergent probability-one homotopy continuation methods. These methods have been previously applied to feedforward neural network training problem. In particular, [10] applied a convex homotopy for the sum of squared errors objective function. Also, [11] proposed a natural homotopy that transforms neuron activation functions from linear to nonlinear ones. However, these homotopies are less efficient for a recurrent neural network training problem because the error function parameter sensitivity grows exponentially over time. Thus, even for moderate prediction horizons, the error function landscape becomes quite complicated. To overcome this issue, we propose the following homotopy for the error function which controls the prediction horizon value:

$$E(\mathbf{a}, \mathbf{w}, \tau) = (1-\tau)\frac{\|\mathbf{w}-\mathbf{a}\|^2}{2} + \frac{1}{2}\sum_{p=1}^{P} \int_0^{\tau\, {}^pT} \sum_{j=1}^{n_y} \mathbf{EW}_j \left({}^p\tilde{\mathbf{y}}_j(t) - {}^p\hat{\mathbf{y}}_j(t)\right)^2 dt, \quad (2)$$

where $\tau \in [0, 1]$ is the homotopy parameter, $\mathbf{w} \in \mathbb{R}^{n_w}$ is the total set of neural network model parameters which includes both $\mathbf{wf}$ and $\mathbf{wg}$, $\mathbf{a} \in \mathbb{R}^{n_w}$ are the initial neural network model parameter values and $\mathbf{EW}_j$ are the error weights usually taken to be inversely proportional to the measurement noise variances. Note that for $\tau = 0$ this error function has unique minimum at $\mathbf{w} = \mathbf{a}$, while for $\tau = 1$ the error function is equivalent to the usual difficult problem with maximum prediction horizon. We can obtain estimates of desired outputs $\tilde{\mathbf{y}}(t)$

at arbitrary time instants $t \in [0, T]$ by interpolation of measured outputs with respect to time.

Error function gradient is given by:

$$\frac{\partial E(\mathbf{a}, \mathbf{w}, \tau)}{\partial \mathbf{w}_i} = (1 - \tau)(\mathbf{w} - \mathbf{a}) - \sum_{p=1}^{P} \int_0^{\tau\, {}^pT} \sum_{j=1}^{n_y} \mathbf{EW}_j \left({}^p\tilde{\mathbf{y}}_j(t) - {}^p\hat{\mathbf{y}}_j(t)\right) \frac{\partial\, {}^p\hat{\mathbf{y}}_j(t)}{\partial \mathbf{w}_i}\, dt.$$

In the following we will denote the error function gradient homotopy as $\mathbf{H}(\mathbf{a}, \mathbf{w}, \tau) = \frac{\partial E(\mathbf{a}, \mathbf{w}, \tau)}{\partial \mathbf{w}}$. In order to obtain the gradient, we first compute the Jacobians by solving the following initial value problem:

$$\begin{aligned} &\frac{\partial \hat{\mathbf{x}}_j(0)}{\partial \mathbf{wf}_i} = 0, \\ &\frac{d}{dt}\frac{\partial \hat{\mathbf{x}}_j(t)}{\partial \mathbf{wf}_i} = \frac{\partial \hat{\mathbf{f}}_j(\hat{\mathbf{x}}(t), \mathbf{u}(t); \mathbf{wf})}{\partial \mathbf{wf}_i} + \sum_{l=1}^{n_x} \frac{\partial \hat{\mathbf{f}}_j(\hat{\mathbf{x}}(t), \mathbf{u}(t); \mathbf{wf})}{\partial \hat{\mathbf{x}}_l(t)} \frac{\partial \hat{\mathbf{x}}_l(t)}{\partial \mathbf{wf}_i}, \end{aligned} \tag{3}$$

which we can consider as a continuous-time version of the RTRL algorithm. Then we use the solution of this initial value problem to compute:

$$\begin{aligned} \frac{\partial \hat{\mathbf{y}}_j(t)}{\partial \mathbf{wf}_i} &= \sum_{l=1}^{n_x} \frac{\partial \hat{\mathbf{g}}_j(\hat{\mathbf{x}}(t); \mathbf{wg})}{\partial \hat{\mathbf{x}}_l(t)} \frac{\partial \hat{\mathbf{x}}_l(t)}{\partial \mathbf{wf}_i}, \\ \frac{\partial \hat{\mathbf{y}}_j(t)}{\partial \mathbf{wg}_i} &= \frac{\partial \hat{\mathbf{g}}_j(\hat{\mathbf{x}}(t); \mathbf{wg})}{\partial \mathbf{wg}_i}. \end{aligned} \tag{4}$$

According to the parameterized Sard's theorem [11], for almost all initial parameter values $\mathbf{a}$ there is a smooth curve $\gamma \in \mathbb{R}^{n_w} \times [0, 1)$ emanating from $(\mathbf{a}, 0)$ and satisfying the necessary conditions for a local extremum. Moreover, if this curve is bounded then it has an accumulation point at $(\bar{\mathbf{w}}, 1)$. Finally, if the Hessian of the error function has full rank at this point, then the curve γ has finite arc length. Hence, to find the local extremum of the difficult error function with maximum prediction horizon, we need to follow the curve γ numerically starting at $(\mathbf{a}, 0)$ until we reach $\tau = 1$. In the special case when the Hessian $\frac{\partial^2 E(\mathbf{a}, \mathbf{w}, \tau)}{\partial \mathbf{w}^2}$ has full rank along the curve γ, the following additional properties hold: the Hessian is positive definite, therefore the curve represents local minima of the error functions; the homotopy parameter τ monotonically increases along the curve γ. In this case, we can follow the curve by solving the initial value problem for Davidenko's system of equations:

$$\frac{d\mathbf{w}}{d\tau} = -\frac{\partial \mathbf{H}(\mathbf{a}, \mathbf{w}, \tau)}{\partial \mathbf{w}}^{-1} \frac{\partial \mathbf{H}(\mathbf{a}, \mathbf{w}, \tau)}{\partial \tau}, \tag{5}$$

with initial condition $\mathbf{w}(0) = \mathbf{a}$. To avoid the accumulation of truncation error for the numerical solution of the initial value problem, we use the Levenberg-Marquardt method as the corrector, i. e. we solve the minimization problem for the error function with a fixed value of τ (we denote this procedure by $\mathbf{LM}(E, \mathbf{a}, \mathbf{w}, \tau)$). We can obtain the required Gauss-Newton approximation for the Hessian as follows:

$$\frac{\partial \mathbf{H}_i(\mathbf{a}, \mathbf{w}, \tau)}{\partial \mathbf{w}_k} = \frac{\partial^2 E(\mathbf{a}, \mathbf{w}, \tau)}{\partial \mathbf{w}_i \partial \mathbf{w}_k} \approx (1-\tau)\delta_{ik} + \sum_{p=1}^{P} \int_0^{\tau\, {}^pT} \sum_{j=1}^{n_y} \mathbf{EW}_j \frac{\partial\, {}^p\hat{\mathbf{y}}_j(t)}{\partial \mathbf{w}_i} \frac{\partial\, {}^p\hat{\mathbf{y}}_j(t)}{\partial \mathbf{w}_k}\, dt,$$

where δ_{ik} is the Kronecker delta. The total homotopy continuation training algorithm also involves a simple form of steplength adaptation and is summarized below. If the norm of model parameters change exceeds $\bar{\delta}$, then the predictor steplength $\Delta\tau$ is decreased and the corrector step is re-evaluated. Conversely, if the norm of model parameters change does not exceed $\underline{\delta}$, the step length is increased. Initial model parameter values $\mathbf{a}$ are picked at random from a set $\mathbf{W} \subset \mathbb{R}^{n_w}$. Articles [12,13] proposed similar ideas involving a gradually increasing prediction horizon.

Algorithm 1. Homotopy continuation training algorithm for models of the form (1)

Require: $\underline{\delta}$, $\bar{\delta}$, $\Delta\tau^{\min}$, $\Delta\tau$
1: $\mathbf{a} \sim U(\mathbf{W})$
2: $\mathbf{w} \leftarrow \mathbf{a}$
3: $\tau \leftarrow 0$
4: **while** $\tau < 1$ **and** $\Delta\tau > \Delta\tau^{\min}$ **do**
5: $\quad\tilde{\tau} \leftarrow \min\{\tau + \Delta\tau, 1\}$
6: $\quad\tilde{\mathbf{w}} \leftarrow \mathbf{LM}(E, \mathbf{a}, \mathbf{w}, \tau)$
7: $\quad$**if** $\|\tilde{\mathbf{w}} - \mathbf{w}\| < \bar{\delta}$ **then**
8: $\qquad\mathbf{w} \leftarrow \tilde{\mathbf{w}}$
9: $\qquad\tau \leftarrow \tilde{\tau}$
10: $\qquad$**if** $\|\tilde{\mathbf{w}} - \mathbf{w}\| < \underline{\delta}$ **then**
11: $\qquad\quad\Delta\tau \leftarrow 2\Delta\tau$
12: $\qquad$**end if**
13: $\quad$**else**
14: $\qquad\Delta\tau \leftarrow \frac{1}{2}\Delta\tau$
15: $\quad$**end if**
16: **end while**

We can also implement more sophisticated Euler predictor. To compute tangents to the curve γ we also need the derivative of error function gradient homotopy with respect to the parameter τ:

$$\frac{\partial \mathbf{H}_i(\mathbf{a}, \mathbf{w}, \tau)}{\partial \tau} = \frac{\partial^2 E(\mathbf{a}, \mathbf{w}, \tau)}{\partial \mathbf{w}_i \partial \tau} = -(\mathbf{w} - \mathbf{a}) - \sum_{p=1}^{P} {}^pT \sum_{j=1}^{n_y} \mathbf{EW}_j \left({}^p\tilde{\mathbf{y}}_j(t) - {}^p\hat{\mathbf{y}}_j(t)\right) \frac{\partial\, {}^p\hat{\mathbf{y}}_j(t)}{\partial \mathbf{w}_i}.$$

We have successfully applied the proposed algorithm to the training problem for a semi-empirical continuous-time state-space model of a maneuverable F-16 aircraft angular motion. Training set included trajectories with a prediction horizon of 20 s, while the test set included trajectories with a 40 s prediction horizon. Root-mean-squared prediction error on the test set was: 0.017 deg for angle of attack, 0.008 deg for sideslip angle, 0.097 deg/sec for roll rate, 0.04 deg/sec for pitch rate and 0.019 deg/sec for yaw rate.

4 Conclusions

Presented results of computational experiments evidence that the proposed homotopy continuation algorithm with variable prediction horizon is an efficient training method that allows overcoming problems inherent to traditional gradient-based optimization methods. We can also conclude that semi-empirical continuous-time state-space neural network based model class is a suitable representation for nonlinear controlled dynamical systems.

This research is supported by the Ministry of Education and Science of the Russian Federation under Contract No. 9.9124.2017/VU.

References

1. Oussar, Y., Dreyfus, G.: How to be a gray box: dynamic semi-physical modeling. Neural Networks **14**(9), 1161–1172 (2001)
2. Egorchev, M.V., Tiumentsev, Y.V.: Semi-empirical neural network based approach to modelling and simulation of controlled dynamical systems. Procedia Comput. Sci. **123**, 134–139 (2018)
3. Williams, R.J., Zipser, D.: A learning algorithm for continually running fully recurrent neural networks. Neural Comput. **1**(2), 270–280 (1989)
4. Werbos, P.J.: Backpropagation through time: what it does and how to do it. Proc. IEEE **78**, 1550–1560 (1990)
5. Sato, M.: A real time learning algorithm for recurrent analog neural networks. Biol. Cybern. **62**(3), 237–241 (1990)
6. Bengio, Y., Simard, P., Frasconi, P.: Learning long-term dependencies with gradient descent is difficult. Trans. Neur. Netw. **5**(2), 157–166 (1994)
7. Doya, K.: Bifurcations in the learning of recurrent neural networks. IEEE Int. Symp. Circuits Syst. **6**, 2777–2780 (1992)
8. De Jesus, O., Horn, J.M., Hagan, M.T.: Analysis of recurrent network training and suggestions for improvements. Proc. IJCNN **4**, 2632–2637 (2001)
9. Egorchev, M.V., Tiumentsev, Yu.V.: Neural network semi-empirical modeling of the longitudinal motion for maneuverable aircraft and identification of its aerodynamic characteristics. In: Advances in Neural Computation, Machine Learning, and Cognitive Research. SCI, vol. 736, pp. 65–71 (2018)
10. Chow, J., Udpa, L., Udpa, S.S.: Homotopy continuation methods for neural networks. IEEE Int. Symp. Circuits Syst. **5**, 2483–2486 (1991)
11. Coetzee, F.M.: Homotopy approaches for the analysis and solution of neural network and other nonlinear systems of equations. Ph.D. thesis. Carnegie Mellon University, Pittsburgh (1995)
12. Elman, J.L.: Learning and development in neural networks: the importance of starting small. Cognition **48**(1), 71–99 (1993)
13. Egorchev, M.V., Tiumentsev, Y.V.: Learning of semi-empirical neural network model of aircraft three-axis rotational motion. Opt. Mem. Neural Netw. **24**(3), 201–208 (2015)

Temperature Forecasting Based on the Multi-agent Adaptive Fuzzy Neuronet

E. A. Engel(✉) and N. E. Engel

Katanov State University of Khakassia,
90 Lenina Prospekt, Abakan 655000, Russia
ekaterina.en@gmail.com

Abstract. This article presents a multi-agent adaptive fuzzy neuronet for average monthly ambient temperature forecasting. We fulfilled the agents of the MAFN based on neural networks. The automatic generation of the of the neuronet parameters of the optimal architecture is the most complex task. Due to train the optimum multi-agent adaptive fuzzy neuronet we modified the Ant Lion Optimizer and combined it with the Levenberg-Marquardt algorithm. We first applied the modified ALO to globally optimize the multi-agent adaptive fuzzy network's structure in multi-dimensional space, and then we elaborated the Levenberg-Marquardt algorithm to speed up the convergence process. We generated an optimum multi-agent adaptive fuzzy neuronet architecture from the obtained global optimum which represented the MAFN optimum architecture's parameters. The simulation results show that the proposed training algorithm outperforms the modified Ant Lion Optimizer and Levenberg-Marquardt algorithms in training the optimum multi-agent adaptive fuzzy neuronet for average monthly ambient temperature forecasting.

Keywords: Temperature forecasting · Multi-agent adaptive fuzzy neuronet
Ant Lion Optimizer · Levenberg-Marquardt algorithm

1 Introduction

The average monthly ambient temperature forecasting is important for different reasons in multiple areas, including renewable energy and transport systems. We present a multi-agent adaptive fuzzy neuronet (MAFN) for average monthly ambient temperature forecasting. We fulfilled the agents of the MAFN based on neural networks [1]. An automatic definition of the optimal architecture's parameters of a neuronet is very complex task which requires an extensive analysis of the system and the trial-error process. This process is demanding because it is difficult to anticipate all conditions of optimal neuronet architecture. This forms the motivation to modify evolutionary optimization techniques such as the Ant Lion Optimizer (ALO) [2] for detection of an optimum multi-agent adaptive fuzzy neuronet architecture. We generated the MAFN architecture's parameters (an agent's number, a number of nodes in hidden layer,

B. Kryzhanovsky et al. (Eds.): NEUROINFORMATICS 2018, SCI 799, pp. 121–128, 2019.
https://doi.org/10.1007/978-3-030-01328-8_12

corresponded weights and biases) from the global optimum. Due to train the optimum multi-agent adaptive fuzzy neuronet we elaborate algorithm, in which we modified the ALO and combined it with the Levenberg-Marquardt algorithm. We first applied the modified ALO to globally optimize the multi-agent adaptive fuzzy network's structure in multi-dimensional space, and then we elaborated the Levenberg-Marquardt algorithm to speed up the convergence process. The simulation results show that proposed training algorithm outperforms modified ALO and Levenberg-Marquardt algorithm in training the optimum MAFN for average monthly ambient temperature forecasting.

2 The Multi-agent Adaptive Fuzzy Neuronet for Average Monthly Ambient Temperature Forecasting

The multi-agent adaptive fuzzy neuronets are fulfilled based on the data

$$\begin{aligned} Z_h =&(x_h^0 = (d_{h-1},\, d_{h-2}, d_{h-3},\, d_{h-4},\, d_{h-5},\, d_{h-6}), \\ x_h^1 =&(p_h^1, p_h^2),\, x_h^2 = (d_{h-1},\, d_{h-6}),\, d_h), \end{aligned} \tag{1}$$

where $d_{h\text{-}i}$ is the historical data of average monthly ambient temperature; p_h^1, ${p^{2t}}_h$ are the first two principal components of the x_h^0 (the first three principal component variances are 49.3, 48.9 and 0.6, respectively), $h = \overline{1..1626}$, $i = \overline{0..6}$. This database was collected at the site of Irkutsk from 01/1882 through 12/2017.

The agents of the MAFN $F_{jq}(x_h^z)$ are fulfilled as two-layered networks, $z = \overline{1..2}$, $j = \overline{1..2}$. We coded the MAFN architecture's parameters (an agent's number – q, number of nodes in hidden layer, corresponded weights and biases) into particles X. For modified ALO these particles X represented the swarm of ants and ant lions. The fitness function evaluated as follows:

$$f(X, x^z) = (1/H)\sum_{l=1}^{H} \left|d_l - F_{jq}(X, x_l^z)\right|, \tag{2}$$

where H is number of evaluated samples. We used function (2) as a fitness function for the modified ALO.

3 The Modified ALO Algorithm for Training the MAFN

We represented the modified ALO process at $t - th$ iteration by the following characteristics:

$xy_{X,j}^{xd_X(t)}(t)$: j^{th} component (dimension) of the personal best (*pbest*) position of ant lion X, in dimension $xd_X(t)$;

$*x_{X,j}^{xd_X(t)}(t)$: j^{th} component (dimension) of the position of ant lion $(* = x)$/ant $(* = v)$ X (represents the $j - th$ MAFN architecture's parameters), in dimension $xd_X(t)$;

$gbest(d)$: Global Best position of the elite ant lion index in dimension d;

$x\hat{y}_j^d(t)$: j^{th} component (dimension) of the global best position of the elite ant lion, in dimension d;

$xd_X(t)$: current dimension of ant lion X position;

$vd_X(t)$: dimensional of ant position X;

$x\underset{X}{\tilde{d}}(t)$: personal best dimension of ant lion position X

$best(xd_X(t))$: best ant lion.

The function **Randomize** generates random MAFN architecture's parameters values (all walks of ants) based on the following equation.

$$vx_{X,j}^{xd_X(t)}(t) = X = [0, cumsum(2r(t_1)-1), \ldots, cumsum(2r(t_T) - 1)], \qquad (3)$$

where *cumsum* is the cumulative sum, t shows the step of the random walk, T is the maximum number of iterations, $r(t)$ is a stochastic function defined as follows:

$$r(t) = \begin{cases} 1 & if \quad rand > 0.5 \\ 0 & if \quad rand \leq 0.5 \end{cases}. \qquad (4)$$

In order to preserve random walks of ants inside the search space, they are normalized in the following way:

$$vx_X^{xd_X(t)}(t) = c + \left(vx_X^{xd_X(t)}(t) - a\right) \times (b - c^t)/(d^t - a), \qquad (5)$$

where for this research $min(a) = -1$ is the minimum of the random walk of variable, $max(a) = 1$ is the maximum of the random walk of variable, c^t is the minimum of the variable at the $t - th$ iteration, and d^t is the maximum of the variable at the $t - th$ iteration. Mathematical modeling of ants trapping in the ant lion's pits is given as follows:

$$c_i^t = xx_j^{xd_X(t)}(t) + c^t, \; d_i^t = xx_j^{xd_X(t)}(t) + d^t, \qquad (6)$$

where c^t is the minimum of variable at the $t - th$ iteration, d^t is the maximum of variable at the $t - th$ iteration, and $vx_j^{xd_X(t)}(t)$ is the position of the selected

$j-th$ ant lion at the *t-th* iteration. The ant lion's hunting capability is modeled by fitness proportional roulette wheel selection. The mathematical model that describes the way the trapped ant slides down towards the ant lion is given as follows:

$$c^t = c^t \cdot T/(10^W \cdot t),\ d^t = d^t \cdot T/(10^W \cdot t), \tag{7}$$

where t is the current iteration and w is the constant that depends on the current iteration.

We expressed the modified ALO algorithm for training the MAFN (the termination criteria is $\{\mathrm{T}, \varepsilon_C, \ldots\}$; S is the number of ants and ant lions) as follows:

1 For $\forall X \in \{1,S\}$ **do:**

1.1 Randomize $xd_X(1)$, $vd_X(1)$.

1.2 Initialize $\tilde{xd}_X(0) = xd_X(1)$.

1.3 For $\forall$d=4*h+3 $\in \{D_{\min}, D_{\max}\}$ **do:**

1.3.1 Randomize $xx_X^d(1)$, $vx_X^d(1)$. The function **Randomize** generated the position of the ants $vx_X^d(1)$ by usage (3)-(7).

1.3.2 Initialize $xy^d(t) = xx_X^d(1)$, $\hat{xy}^d(t) = xx_X^d(1)$.

1.4 End For.

2 End For.

3 For $\forall$t $\in \{1,T\}$ **do:**

3.1 For $\forall X \in \{1,S\}$ **do:**

3.1.1 If $f(xx_X^{xd_X(t)}(t)) > f(xy_X^{xd_X(t-1)}(t-1))$

3.1.1.1 then Do: $xy_X^{xd_X(t)}(t) = xx_X^{xd_X(t)}(t)$

3.1.1.2 If $f(xx_X^{xd_X(t)}(t)) < f(xy_X^{\tilde{xd}_X(t-1)}(t-1))$

3.1.1.2.1 then $xd_X(t) = xd_X(t-1)$

3.1.1.3 End If

3.1.1.4 else $xd_X(t) = xd_X(t)$

3.1.2 else $xy_X^{xd_X(t)}(t) = xy_X^{xd_X(t)}(t-1)$

3.1.3 End If

3.1.4 If $(f(xx_X^{xd_X(t)}(t)) < \max(f(xy_X^{xd_X(t)}(t-1)), \max\limits_{1 \le p < X}(f(xx_p^{xd_X(t)}(t))))$

3.1.4.1 then Do: $gbest(xd_X(t)) = X$. For the $t-th$ iteration the best ant lion $best(xd_X(t)) = X$ is considered as elite. This rule is fulfilled elitism. The elitism indicates that every ant randomly walks near selected the ant lion and has a position $vx_{X_i}^{xd_X(t)}(t) = (R_A^t + R_E^t)/2$, where R_A^t is the random walk around the ant lion selected by the roulette wheel at the $t-th$ iteration and R_E^t is the random walk around the elite ant lion at the *t-th* iteration.

3.1.4.2 If $f(xx_X^{xd_X(t)}(t)) > f(x\hat{y}^{dbest}(t-1))$

3.1.4.2.1 then $dbest = xd_X(t)$

3.1.4.3 End If

3.1.5 End If

3.1.6 In other dimensions $\forall d \in [D_{\min}, D_{\max}]\text{-}\{xd_X(t)\}$ **do** updates $xy_{X,j}^d(t) = xy_{X,j}^d(t-1)$, $xy_j^d(t) = xy_j^d(t-1)$.

3.2 End For

3.3 If the termination criteria are met

3.3.1 then Stop.

3.4 End If

3.5 For $\forall X \in \{1, S\}$ **do:**

3.5.1 For $\forall \mathrm{j} \in \{1, xd_X(t)\}$ **do:**

3.5.1.1 Compute

$$vx_{X,j}^{xd_X(t)}(t+1) = w(t)vx_{X,j}^{xd_X(t)}(t) + c_1 r_{1,j}(t)(xy_{X,j}^{xd_X(t)}(t) - xx_{X,j}^{xd_X(t)}(t)) + c_2 r_{2,j}(t)(\hat{xy}_j^{xd_X(t)}(t) - xx_{X,j}^{xd_X(t)}(t)),$$

$$xx_{X,j}^{xd_X(t)}(t+1) = \begin{Bmatrix} xx_{X,j}^{xd_X(t)}(t) + vx_{X,j}^{xd_X(t)}(t+1) & if\ X_{\min} \le vx_{X,j}^{xd_X(t)}(t+1) \le X_{\max} \\ U(X_{\min}, X_{\max}) + xx_{X,j}^{xd_X(t)}(t+1) & else \end{Bmatrix},$$

$$xx_{X,j}^{xd_X(t)}(t+1) \leftarrow \begin{Bmatrix} xx_{X,j}^{xd_X(t)}(t+1) & if\ X_{\min} \le xx_{X,j}^{xd_X(t)}(t+1) \le X_{\max} \\ U(X_{\min}, X_{\max}) & else \end{Bmatrix}$$

3.5.1.2 In other dimensions $\forall d \in [D_{\min}, D_{\max}]\text{-}\{xd_X(t)\}$ **do** updates locations of ants and ant lions $vx_{X,j}^d(t+1) = vx_{X,j}^d(t)$, $xx_{X,j}^d(t+1) = xx_{X,j}^d(t)$.

3.5.2 End For

3.5.3 Compute

$$vd_X(t+1) = \left\lfloor vd_X(t) + c_1 r_1(t)(x\tilde{d}_X(t) - xd_X(t)) + c_2 r_2(t)(dbest - xd_X(t)) \right\rfloor,$$

$$xd_X(t+1) = \begin{Bmatrix} xd_X(t) + vd_X(t+1) & if\ VD_{\min} \le vd_X(t+1) \le VD_{\max} \\ xd_X(t) + VD_{\min} & if\ vd_X(t+1) < VD_{\min} \\ xd_X(t) + VD_{\max} & if\ vd_X(t+1) > VD_{\max} \end{Bmatrix},$$

$$xd_X(t+1) \leftarrow \begin{Bmatrix} xd_X(t) & if\ P_d(t+1) \ge \max(15, xd_X(t+1)) \\ xd_X(t) & if\ xd_X(t+1) < D_{\min} \\ xd_X(t) & if\ xd_X(t+1) > D_{\max} \\ xd_X(t+1) & else \end{Bmatrix}$$

3.6 End For

4 End For

To illustrate the benefits of the modified ALO in finding the optimal architecture's parameters of a MAFN, we revisited the numerical examples.

4 Fulfillment of the MAFN

In order to train the optimum agents of the MAFN for average monthly ambient temperature forecasting we applied the modified ALO to globally optimize the network's structure (the modified ALO will stop after a global solution is localized within small region), and then we used Levenberg-Marquardt to speed up convergence process. The algorithm of the agent's interaction elaborates a fuzzy-possibilistic method and includes three steps:

Step 0: **for each** *agent*$_q$ **in** *subculture* S_k **do**

$F_{kq}(x_h^z) \leftarrow$ **GetResponse**($agent_q$; x_h^z);

$v_q \leftarrow$ **TakeAction**($F_{kq}(x_h^z)$):

Evaluate $v_q = 1 - f(w, x_H^z)$, where f is the objective function (2), x_H^z is a sample which we chose from data (3) according condition: $\min\limits_{i \in 1..N} \left| x_i^z - x_H^z \right|$.

end for.

$w = [F_{k1}(x_h^z), \ldots, F_{kq}(x_h^z)]$.

We **compute** $I_h = Fes_{kh}(F_{kq}(x_h^z))$ based on $(w, [v_1, \ldots, v_q])$ as fuzzy expected solution (Fes) in 2 steps [1]

Step 1: **Solve equation** $\left[\prod\limits_{i=1}^{q} (1 + \lambda w_i) - 1 \right] / \lambda = 1, \quad -1 < \lambda < \infty.$

Step 2: We **compute** $s = \int h \circ w_\lambda \ \mathrm{Sup}_{\alpha \in [0,1]} \min\{\alpha, W_\lambda(F_\alpha(v_j))\}$, where $F_\alpha(v_j) = \{F_i | F_i, v_j \geq \alpha\}, v_j \in v, W_\lambda(F_\alpha(v_j)) = \left[\prod\limits_{F_i \in F_\alpha(v_j)}^{k} (1 + \lambda w_i) - 1 \right] / \lambda$; $\mathrm{I}_\mathrm{h}^\mathrm{t} = \max\limits_{v_i \in V} s(w_j)$.

We briefly described the fulfillment of the MAFN as follows

Step 1. We classified all samples of the data (1) (x_h^z, t_h) into two groups: A_1 is year with normal temperatures ($C_h^t = 1$), A_2 is year with abnormal temperatures ($C_h^t = -1$). This classification provides vector with elements C_h^t.

Step 2. We trained two two-layer networks: $Y(x_h^z)$ (number of hidden neurons and delays are 7 and 2, respectively). The vector x_h^z was network's input. The vector C_h^t was network's target. We defined fuzzy sets A_j, (A_1 is year with normal temperatures, A_2 is year with abnormal temperatures) with membership function $\mu_j(s)$ based on aforementioned two-layer networks $Y(x_h^z)$, $j = \overline{1..2}$.

Step 3. We trained MAFN $s(o, z)$ based on the data (1). We elaborated optimization algorithm o (if $o = 1$ then optimization algorithm is modified ALO; if $o = 2$ then optimization algorithm is Levenberg-Marquardt; if $o = 3$ then optimization algorithm is proposed algorithm, in which the modified ALO combined with the Levenberg-Marquardt algorithm). This step provides neural networks $F_{jq}(x_h^z)$ which forecast average monthly ambient temperature $Ir_{jq} = F_{jq}(x_h^z)$. We defined two agent's subcultures s_j based on aforementioned two-layer networks $F_{jq}(x_h^z)$.

If-then rules are defined as:

$$\Pi_j : \mathbf{IF} x_h^z \text{ is } A_j \text{ } \mathbf{THEN} \text{ } I_h = Fes\left(F_{jq}\left(x_h^z\right)\right), \tag{8}$$

We briefly described simulation of the trained MAFN $\forall c \in [1215, 1226]$ as follows.

Step 1. Aggregation antecedents of the rules (8) maps input data x_c^z into their membership functions and matches data with conditions of rules. These mappings are then activate the k rule, which indicates the k year's temperature mode and k agent's subcultures – s_k, $k = \overline{1..2}$.

Step 2. According the k mode the multi-agent adaptive fuzzy neuronet (generated base on the data x_d^z , where $d = \overline{1..c-1}$) forecasts average monthly ambient temperature $I_h = Fes(F_{kq}(x_c^z))$ as a result of multi-agent interaction of subculture s_k.

The fuzzy-possibilistic algorithm allows for the forecasting of average monthly ambient temperature in a intelligent manner, so as to take into account the responses of all agents based on fuzzy measures and the fuzzy integral.

5 Results

To show the advantages of the MAFN in forecasting of the average monthly ambient temperature, the numerical examples from the previous sections are revisited using the software "The multi-agent adaptive fuzzy neuronet" [3]. There the three MAFN $s(o, z)$ were fulfilled based on the training set of the data (1), $t = \overline{1\ldots1214}$. We trained the first MAFN $s(1, z)$ using modified ALO ($o = 1$). In order to obtain statistical results, we perform 120 modified ALO runs with following parameters: $S = 50$ (we use 50 ants and ant lions), $T = 110$ (we terminate at the end of 110 iterations), the dimension is $d_h = 4 * h + 3 \in \{D_{\min} = d_1 = 7, D_{\max} = d_{10} = 43\}$. The vector $f(best(d_h)) = (2.5, 2.1, 1.5, 1.1, 0.8, 1.2, 1.7, 2.4, 2.7, 3.1)$ shows that only one set of MAFN architecture with $d_5 = 23$ can achieves the fitness function (2) above 0.9 over data (1), $t = \overline{1215\ldots1226}$. The MAFN $s(1, z)$ have three agents of each subculture s_k. We generated the aforementioned agents as the two-layered network with five hidden neurons. All the MAFN $s(o, z)$ have the same architecture. Forecast accuracies of the MAFN $s(o, z)$ we evaluated as the fitness function (2). Table 1 shows comparison of the results between the MAFN $s(o, z)$. Comparisons between temperature forecasting models MAFN $s(o, z)$ show that the $s(3, 1)$ is definitely more accurate. Figure 1(a) shows that MAFN $s(3, 1)$ has definitely more convergence speed than MAFN $s(1, 1)$ or $s(1, 2)$ in the average monthly ambient temperature forecasting. In Fig. 1(b) the ineffectiveness of the $s(2, z)$ can be seen.

Table 1. The results of MAFN at an average monthly ambient temperature forecasting

The MAFN $s(o, z)$	$s(1, 1)$	$s(1, 2)$	$s(2, 1)$	$s(2, 2)$	$s(3, 1)$	$s(3, 2)$
The function (2) over the data (1), $t = \overline{1\ldots1226}$	0.8	1.7	2.8	3.4	0.7	1.5

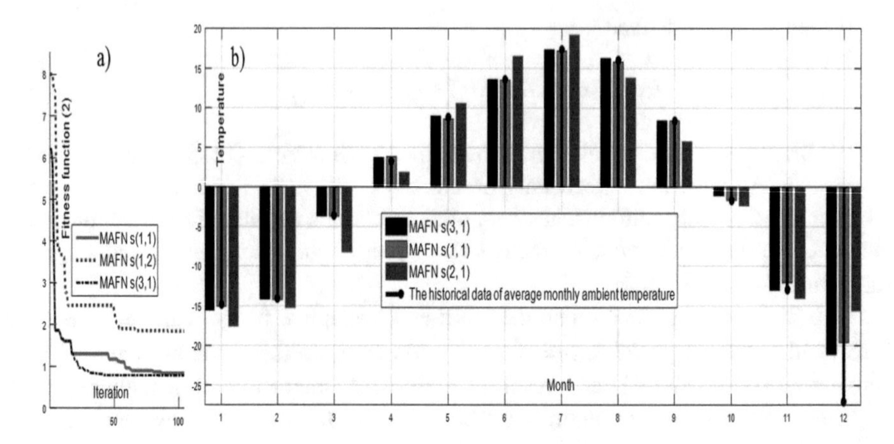

Fig. 1. (a) The mean convergence curves (b) The results of MAFN

The performance of the multi-agent adaptive fuzzy neuronet generated by proposed algorithm is superior to the same one trained by modified ALO or Levenberg-Marquardt algorithm (Table 1, Fig. 1).

6 Conclusions

This paper presents a MAFN for average monthly ambient temperature forecasting. Simulation comparison results for average monthly ambient temperature forecasting demonstrate the effectiveness of the MAFN generated by proposed algorithm, in which we combined the modified ALO with the Levenberg-Marquardt algorithm as compared with the same ones generated by modified ALO or Levenberg-Marquardt algorithm.

References

1. Engel, E.A.: Dump truck fault's short-term forecasting based on the multi-agent adaptive fuzzy neuronet. In: Kryzhanovsky, B., Dunin-Barkowski, W., Redko, V. (eds.) Advances in Neural Computation, Machine Learning, and Cognitive Research, pp. 72–77. Springer (2018)
2. Mirjalili, S.: The ant lion optimizer. Adv. Eng. Softw. **83**, 80–98 (2015)
3. Engel, E.A.: The multi-agent adaptive fuzzy neuronet. Certificate about State registration of computer programs. Federal Service for Intellectual Property (Rospatent). No 2016662951 (2016)

The Neural Network with Automatic Feature Selection for Solving Problems with Categorical Variables

Yuriy S. Fedorenko(✉) and Yuriy E. Gapanyuk

Bauman Moscow State Technical University, Moscow, Russia
fedyural1235@mail.ru, gapyu@bmstu.ru

Abstract. The article deals with the problem of feature selection in data analysis. It is critically important for any tasks which are solved by machine learning algorithms. The difficulty is that features depend on the domain area, so the feature selection requires a lot of manual work. In this paper, the regression task with categorical input data is considered. The problem is that when using simple methods such as logistic regression, it is necessary to combine features into separate levels to obtain an acceptable result. This requires expert work or computationally expensive full search of all variants. In the present paper, the neural network for solving this task is proposed. Its architecture and details of learning are discussed. The advantage of this architecture is that it doesn't require selection of complicated levels. For its operation, it is sufficient to list all input features. Experiments have shown that such neural network with raw features works better than logistic regression with handcrafted chosen levels. Also in this article, the metagraph representation of proposed neural networks is discussed. This makes it more convenient to work with this neural net.

Keywords: Feature selection · Sparse neural network · Categorical variables Logistic regression · Logistic loss · ROC curve · Metagraph · Metagraph agent

1 Introduction

The solving of data analysis tasks consists of two parts: feature extraction and prediction [1]. The performance metrics values of machine learning algorithm depend on proper input features. The process of feature extraction is often difficult and requires handcraft of domain experts. So methods which allow automatically feature engineering are very perspective. The advantage of neural networks especially deep is that they can learn representations in data. In other words, they form features automatically during work. The last layer of such a neural net can be considered as a linear classifier [2]. The other layers serve for obtaining representation for this classifier. During training, each hidden layer makes the classification task simpler. For example, linearly nonseparable classes in input data may come linearly separable in hidden layers. This essentially lightens the researcher's work.

B. Kryzhanovsky et al. (Eds.): NEUROINFORMATICS 2018, SCI 799, pp. 129–135, 2019.
https://doi.org/10.1007/978-3-030-01328-8_13

2 Problem Definition

Consider a set of examples at the input of the model. Sample is described by m categorical features, each of that has n values (one-hot encoding is used). Some samples lead to the event. It is required building the model for event probability prediction for each sample. For solving this task logistic regression algorithm may be used, but in this case, the handcrafted preparing of features is necessary. The proposed neural network automatically finds needed transformations.

3 Model Description

The proposed neural network has one hidden layer. Unlike usual perceptron [3] its architecture is sparse: neurons of the hidden layer are connected only with some input layer neurons. In particular, each hidden layer neuron is connected with one or two inputs responsible for the concrete feature (all possible pairs are participated). So a number of neurons in the hidden layer is $m \cdot (m-1)/2 + m = m \cdot (m+1)/2$, where m is the number of features. Each neuron of the hidden layer has ReLU activation. Output neuron is connected with all hidden layer neurons and has sigmoid activation. The schema of the neural net is presented on Fig. 1.

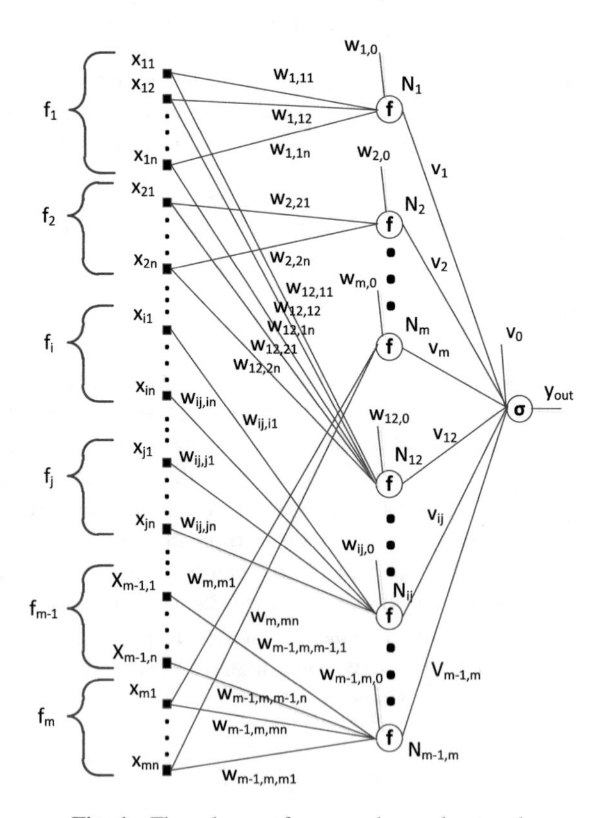

Fig. 1. The schema of proposed neural network

The signals of neural network are calculated by following expressions:

$$N_i = \omega_{i0} + \sum_{k=1}^{n} x_{ik} \cdot \omega_{i,ik} \tag{1}$$

$$N_{ij} = \omega_{ij,0} + \sum_{k=1}^{n} x_{ik} \cdot \omega_{ij,ik} + \sum_{k=1}^{n} x_{jk} \cdot \omega_{ij,jk}, \tag{2}$$

where N_i is neuron, connected with i-th feature on input and N_{ij} is neuron, connected with values of i and j features (i > j). In weight coeffs of hidden layer $\omega_{ij,ik}$ the first two indexes relate to a neuron of the hidden layer and the last two indexes relate to input. x_{ik} can take two values: 0 or 1. The signal of the output layer neuron is calculated as:

$$y = (\sum_{i=1}^{m} v_i \cdot f(N_i) + \sum_{i=1}^{m} \sum_{j=i+1}^{m} v_{ij} \cdot f(N_{ij}) + v_0), \quad y_{OUT} = \sigma(y),$$

where N_i, N_{ij} are calculated from Eqs. (1) and (2), v_i, v_{ij} are weights of output neuron, f is the activation function of hidden layer neurons.

4 Training of Neural Network

The neural network is trained by the stochastic gradient method. The logistic loss function [4] is used:

$$J(y_{OUT}) = -y_T \cdot \ln(y_{OUT}) - (1 - y_T) \cdot \ln(1 - y_{OUT}).$$

The calculation of gradients is below. The gradients of the output layer:

$$\frac{\partial J}{\partial v_i} = \frac{\partial J}{\partial y_{OUT}} \cdot \frac{\partial y_{OUT}}{\partial y} \cdot \frac{\partial y}{\partial v_i} = (y_{OUT} - y_T) \cdot f(N_i), \frac{\partial J}{\partial v_{ij}} = (y_{OUT} - y_T) \cdot f(N_{ij}).$$

The gradients of weights of hidden layer neurons:

$$\frac{\partial J}{\partial \omega_{ij,ik}} = \frac{\partial J}{\partial y_{OUT}} \cdot \frac{\partial y_{OUT}}{\partial y} \cdot \frac{\partial y}{\partial w_{ij,ik}} = (y_{OUT} - y_T) \cdot v_{ij} \cdot f'(N_i) \cdot x_{ik},$$

$$\frac{\partial J}{\partial \omega_{ik}} = (y_{OUT} - y_T) \cdot v_i \cdot f'(N_i) \cdot x_{ik}.$$

Updating of weights is made in the standard way on the basis of gradient descent method. Obviously weights are updated only for neurons with non zero inputs.

5 Experiments

In experimental analysis, two models were compared: logistic regression with handcrafted features and the proposed neural network. For training, the data for CTR prediction [5] were chosen. In online advertising click-through rate (CTR) is a very important metric for evaluating ad performance. Each sample has a set of features. On the one side, it is advertising banner features, such as ad campaign id, platform id, text, title, the image of banner, etc. On the other side, it is user features such as sex, age, interests, etc. For successful use of logistic regression model such primary features should be combined into pairs and triples to form levels of the model (for example, level "interests, campaign"). Searching such composed levels is a computationally expensive process. For example, for pairs of features one level selection requires searching C_n^2 combinations. For triples, the number of possible variants increases to C_n^3, which implies the cubic complexity of n. The proposed neural net enters a set of primary features and trains weights to extract relevant features pairs. So, this model is profitable even if its quality will be similar to the logistic regression model. The logloss function and area under ROC curve (AUC) are used as metrics. The first one characterizes the model calibration, and the second one describes ranking quality [6]. The results of the experiment are presented below (Fig. 2):

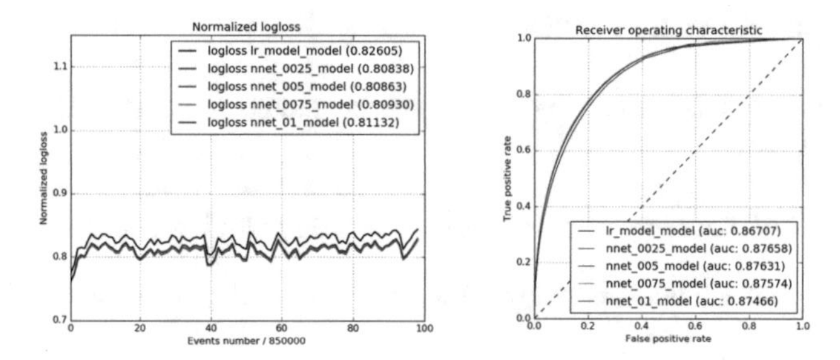

Fig. 2. Comparison of logloss and AUC for logistic regression and proposed neural network

So, proposed neural network with automatic feature selection has better quality than logistic regression with handcrafted features. Besides, the essential advantage of the neural net is that it doesn't require hard work for feature selection.

6 Special Aspects of Proposed Neural Network

An experiment was conducted in which neural net repeatedly had entered the training samples. The loss function value was measured after each pass. After 20–30 passes the loss function for neural net became lower than for logistic regression. But after random sorting of the train set the loss function value jumped up (Fig. 3).

The analysis shows that the problem is in "died" neurons [7]. If the weight of hidden layer neuron is negative, this neuron won't influence model, i.e., the

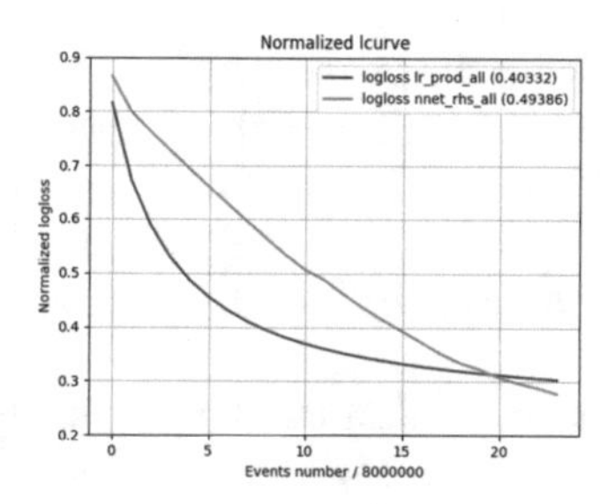

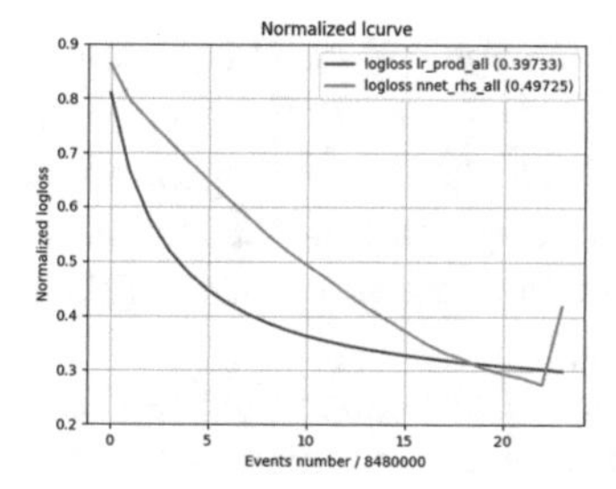

Fig. 3. The loss function value for logistic regression and neural net with multiple passes through train set

corresponding input dies. This input can again be positive if it will become part of a pair with positive weight with a large value, which often leads to an event. However, this is unlikely in practice, so about 80% of neurons were simply turned off in described above experiment. The other neurons, in fact, were trained to order of passed samples. After random sorting of data, the order of samples was changed. As a result, the error was increased, because "turned off" weights were restored too slow.

This problem can be solved by using different activation functions other from ReLU. The hyperbolic tangent is not a good choice because it decreases neural net quality. A leaky ReLU [8] which has a low slope for negative values can be used. Using such activation function reduces the problem of "died" neurons.

7 The Metagraph Representation of a Proposed Neural Network

In our paper [9] the metagraph representation of perceptron neural network was proposed. Metagraph is a kind of complex graph model: $MG = \langle V, MV, E \rangle$, where MG is metagraph; V is set of metagraph vertices; MV is set of metagraph metavertices; E is set of metagraph edges.

It is the metavertex that is the distinguishing feature of the model. The metavertex may include inner vertices, edges, and metavertices. From the general system theory point of view, metavertex is a special case of manifestation of emergence principle which means that metavertex with its private attributes, inner vertices, metavertices, and edges became whole that cannot be separated into its component parts.

For metagraph transformation, the metagraph agents are proposed. In this paper, we use two kinds of metagraph agents: the metagraph function agent (ag^F) and the metagraph rule agent (ag^R).

The metagraph function agent (ag^F) serves as a function with input and output parameter in the form of metagraph: $ag^F = \langle MG_{IN}, MG_{OUT}, AST \rangle$, where MG_{IN} is input parameter metagraph; MG_{OUT} is output parameter metagraph; AST is abstract syntax tree of metagraph function agent in the form of metagraph. The function agent is used for individual neurons description what is discussed in details in [9].

The metagraph rule agent (ag^R) uses a rule-based approach: $ag^R = \langle MG, R, AG^{ST} \rangle$, $R = \{r_i\}, r_i : MG_j \rightarrow OP^{MG}$, where MG is working metagraph, a metagraph on the basis of which the rules of agent are performed; R is set of rules r_i; AG^{ST} is start condition (metagraph fragment for start rule check or start rule); MG_j is a metagraph fragment on the basis of which the rule is performed; OP^{MG} is set of actions performed on metagraph.

The "active metagraph" means a combination of data metagraph with an attached metagraph agent. According to metagraph approach, the perceptron neural network may be represented as the hierarchical interconnection of active metagraphs. According to [9], each layer of perceptron neural network may be represented as an active metagraph which consists of the data metagraph part and the agent part. The data metagraph part represents the structure of the network: neurons are represented as low-level metavertices and connections between neurons are represented as edges. The agent part consists of "training agent" which is responsible for the training of the network and the "operational agent" which is responsible for the regular neural network operation. It should be noted two things. Firstly, both training and operational agents use the same data metagraph as data tier. Secondly, both training and operational agents are organized hierarchically. For example, the training agent of neural network layer interacts with training agents of individual neurons belonging to the layer.

Using the metagraph approach the problem of "died" neurons may be solved in a structural way. We can add new kind of agent "control agent" that structurally detects the "died" neuron as unbounded (not connected with edges) vertex. If a critical number of "died" neurons are detected then control agent send messages to the other kinds of agents in order to force a change in "died" neurons parameters.

Thus, the metagraph approach provides the unified technique for neural networks representation and helps to solve the problem of "died" neurons in a structural way.

8 Conclusions

The model proposed in this article (actually facilitated version of perceptron) allows reducing manual selection of features in different tasks. Experiments showed that the metric values in the suggested model are better than in logistic regression with handcrafted features. The problem of "died" neurons in the proposed model can be solved by using a leaky ReLU activation function.

References

Lecun, Y., et al.: Gradient-based learning applied to document recognition. Proc. IEEE **86**(11), 2278–2324 (1998)

Goodfellow, I., Bengio, Y., Courvile, A.: Deep Learning. MIT Press, Cambridge (2016)

Haykin, S.: Neural networks: a comprehensive foundation, 2nd edn. Prentice Hall, New Jersey (1999)

Bishop, C.: Pattern Recognition and Machine Learning. Springer, Heidelberg (2006)

He, X., Pan, J., Jin, O., et al.: Practical lessons from predicting clicks on ads at Facebook. In: Proceedings of International Workshop on Data Mining for Online Advertising (ADDKDD), pp. 1147–1156. ACM, New York (2014)

Bradley, A.: The use of the area under the ROC curve in the evaluation of machine learning algorithms. Pattern Recognit. **30**(7), 1145–1159 (1997)

Li, F., Johnson, J., Yeung, S.: Convolutional neural networks for visual recognition. http://cs231n.github.io/neural-networks-1. Accessed 5 May 2018

Avinash, S.: Understanding Activation Functions in Neural Networks. https://medium.com/the-theory-of-everything/understanding-activation-functions-in-neural-networks-9491262884e0. Accessed 5 May 2018

Fedorenko, Yu.S., Gapanyuk, Yu.E.: Multilevel neural net adaptive models using the metagraph approach. Opt. Mem. Neural Netw. **25**(4), 228–235 (2016)

Object Detection on Docking Images with Deep Convolutional Network

Ivan S. Fomin[1(✉)], Svetlana R. Orlova[2], Dmitrii A. Gromoshinskii[1], and Aleksandr V. Bakhshiev[2]

[1] The Russian State Scientific Center for Robotics and Technical Cybernetic (RTC), Saint Petersburg, Russia
i.fomin@rtc.ru

[2] Peter the Great St.Petersburg Polytechnic University, Saint Petersburg, Russia
bakhshiev_av@spbstu.ru

Abstract. The docking process requires evaluation of the relative position between space apparatus. The position can be determined using the apparatus based camera. The relative position determination algorithm requires information about camera calibration, precise 3D position of the visual landmarks and the landmarks position in frame received from camera. Using the data, it is possible to solve the PnP problem for determining the relative position with high accuracy. Precision of the detected landmarks position and number of the detected landmarks are the most important. This paper addresses the problem of the visual landmark detection on images of the space docking process. The source images retrieved with analog camera in hard conditions; all images are noisy, blurred and low quality, covered with special set of service symbols. We prepared special dataset of frames with positions of each visual landmark manually marked. After training on the dataset, the mAP metric of the selected deep convolutional network is up to 85,1%, it is better than result achieved with the previously studied approach. The image processing with several filter types does not significantly improve the detection result.

Keywords: Deep networks · Convolutional networks · Docking
Object detection · Visual landmarks

1 Introduction

There are a lot of known issues in the area of the autonomous apparatus. One of those problems is determination of the relative position between one apparatus and another or a special docking point (docking bay, or so called "node"), where apparatus should stay fixed for fueling, the electric batteries recharging and manned or unmanned maintenance. It is very important to determine precise relative position between two apparatus or apparatus and the docking node in the docking process. In the space docking process (which is used as an example of the docking process in this paper) the problem becomes more challenging because objects are not pinned to the flat surface but have all 6 degrees of freedom. All degrees must be determined on each camera frame to perform continuous control of the docking process.

B. Kryzhanovsky et al. (Eds.): NEUROINFORMATICS 2018, SCI 799, pp. 136–143, 2019.
https://doi.org/10.1007/978-3-030-01328-8_14

Nowadays, the spacecraft dockings are performed using special radio-electronic and optical systems. Components of such systems must be installed on the International Space Station and on a spacecraft that will be docked. Also every cargo and passenger spacecraft has been equipped with the television systems for already 45 years, it allows operator in the Mission Control Center to perform visual control of the docking process.

Earlier in articles [1], [2] we described the use of the spacecraft's television system for the detection of its position relative to the ISS. The determined relative position has precision even higher than results of the special set of analog measuring devices used by crew in the docking process. To approve these results we developed special software which can be installed on a notebook computer inside the ISS, or on a computer in the Mission Control Center. The video frames are received by the analog channel, then digitized and sent to the program input. Program performs simultaneous detection and tracking of the visual landmarks previously selected onboard of the ISS. Using positions of these features, known model of the camera and precise 3D model of the ISS, this program is able to calculate position of a spacecraft relative to the ISS with high precision by solving the PnP problem.

2 Problem Statement

One of the most important part of the computer program used to determine relative position between the spacecraft and the International Space Station is algorithm of the object detection. Quality and performance of this algorithm determine quality and performance of the overall system, because the object detection process takes more time than any other computational part of the algorithm, and methods of calculation of the relative position do not change between two different frames and use results of the object detection process (pixel coordinates of the detected visual landmarks) as input data. The object detection part of the algorithm depends on special conditions of each docking process, quality of the analog television signal and images that program receives in the docking process.

Special set of the analog television cameras used on the spacecraft and the ISS during the docking process has several features, which should be described more thoroughly. This system includes only analog TV cameras and many other analog components. In the Mission Control Center experiments the analog to digital signal conversion is performed at the moment when the television signal reaches the personal computer or laptop with the program installed. The analog components introduce many distortion effects while translating and receiving the signal. Examples of distortions in the docking process are shown on Fig. 1, images are extracted from video with open access [3]. These distortions are common for the docking process and they are included in the training dataset. Other distortions that may appear in the docking process are camera exposure failures due to sunrays (direct or reflected from the ISS surface), fast brightness changes on the scene and blurring of the image.

Current approach uses several algorithms with different complexity to compensate these distortions, from simple histogram equalization to complex noise reduction algorithm. Nowadays the object detection algorithms are mainly based on the deep

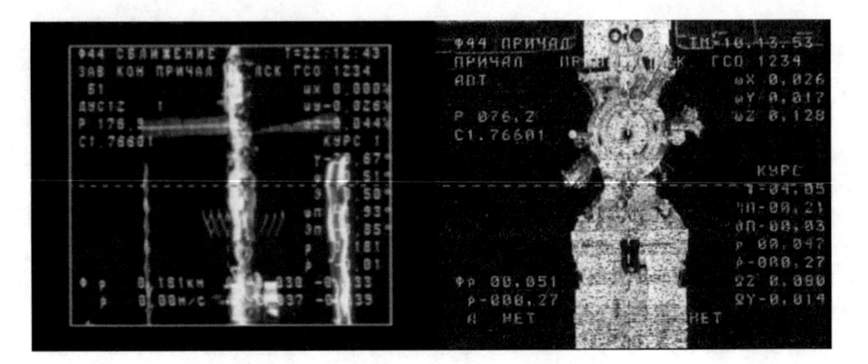

Fig. 1. Examples of the image distortions

convolutional networks. All convolutional networks have very good generalization ability, but deep convolutional nets are the best in generalization. In some situations, when other algorithms (like a cascade detector in the current program version) require complex image processing, for the convolutional network sometimes it is enough to include images with distortions in the training dataset.

Earlier in papers [4], [5] we discussed possibility of using object detection system based on the convolutional network and called Faster R-CNN [6]. The detection results are better than detection quality of the cascade detector but we need to increase performance and detection accuracy to allow deep convolutional network become our new detection pipeline. Eventually, we decided to train and test a new architecture of the object detection network.

3 Related Work

Earlier, when area of the object detection in computer vision only started to grow, such algorithms used both simple convolutional networks and traditional methods as the SVM or cascade detectors based on principle of the sliding window. Also sometimes cascade of the methods are used to make decision from the simplest to the most complicated. In the current solution the object detection is performed in a similar way.

The first famous and good enough object detection method is based on the R-CNN convolutional network and uses results of two different methods to detect some areas of interest (RoIs) on image. These methods used simple sliding window with some more complex algorithm to make decision, whether selected area can contain some object or not. The RoIs formed by those methods are passed to the input of the deep classification network that determines class of the object in the RoI. R-CNN combined these methods (detection and classification). This approach was improved by two reworks. The first variant processed all RoI for each image independently; network processed only the selected area and nothing more. In the improved version of R-CNN the convolutional layers of the classification network process the whole image first and then project all RoI on the output of the last convolutional layer, normalize resulting receptive fields and pass the normalized vectors to the input of the fully-connected classification part. This is how the Fast R-CNN architecture works. In the next improved version algorithm uses separate group of the convolutional and fully-connected layers to detect RoIs instead of external tool based on the "classic" image processing algorithms. In this version the whole algorithm is convolutional and works on GPU which significantly increases performance of the object detection. Special deep network for the RoI detection, that also trains its weights, and simultaneous usage of the convolutional layers also increase precision, recall and accuracy of the object detection. Final version of the algorithm was named Faster R-CNN.

Network with architecture named YOLO (or You Look Only Once) [7] does not use the sliding window approach at all. In the beginning it divides the input image by grid in rows and columns using preselected number of cells. Then those cells are projected to the output of the last convolutional layer and the special layers decide if cell contains any object of some type with the certain size (from the list of types and sizes that previously set at the training stage) or not. Before the network starts training user must determine the set of anchor boxes for each cell. There is a special tool to generate this set of the anchor boxes from the training dataset before the network starts training. Algorithm processes all variance of possible object sizes, then clusters it and uses probability of the certain sizes to determine the set of anchor boxes automatically. Then on the training/detection stage algorithm does not try to improve position of the box around every detected object in the special convolutional layers (as Faster R-CNN do) but uses only anchor boxes. Experiments show that this approach does not decrease recall or accuracy, but significantly increases the detection performance.

In addition to the automated anchor box selection authors made several significant improvements in the YOLO pipeline and presented another version, YOLO v2 [8], which is better than first basic and experimental version of YOLO. First, they began to use the batch normalization for training images, and increased the mAP value by 2%. Secondly, they increased resolution of the classification network. While the most of the "standard" classification networks uses resolution near 256x256 following the classic AlexNet architecture, the YOLO v2 uses the 448x448 resolution, which allows to increase the mAP value by 4%. Thirdly, they used the set of prebuilt anchor boxes for each cell allowing network to check more than one thousand boxes per image and increase detection recall, as well as the automated anchor boxes sizes selection. Fourthly, they used prediction of the object coordinates relatively to the current cell position, not to the left top corner of the image, it allows to simplify parametrization, make detector independent from the cell position on image and increase mAP. Extension of the feature map of the last layer with features from the previous one transformed in a special way allows system to detect smaller objects than the first version of YOLO. The random changes in size of the input image during the training process allow the trained network to work with images of very different sizes. When the input images size decreases, precision of the detector decreases as well, but it significantly increases the network performance. This provides user very simple and clear tradeoff between the system quality and performance.

4 Used YOLO Implementation

To achieve results represented in this paper, we used already implemented and shared on GitHub version of YOLO [9] developed for training and processing with couple of the convolutional network frameworks – Keras and TensorFlow. This implementation allows user to use several different architectures to train network and then detect objects. List of architectures includes the simplified model TinyYolo, basic YOLOv2, SqueezeNet, Inception3 and MobileNet. Unfortunately, all architectures except MobileNet require more than 4 Gb of video memory to train and operate. Therefore, MobileNet was only one accessible, so we chose it to perform the training and detection experiments.

5 Training Dataset

To train the chosen network we prepared special training dataset that contains images received in the docking process of two different spacecraft to four different docking nodes. Examples of the video data received in the docking process used to extract images to create a dataset are in open access [3]. For each frame in dataset we prepared a file with the bounding boxes of the visual landmarks (the set of visual landmarks was determined earlier when experiments with algorithm of determination of the relative position [1] were performed). For each of four docking nodes view of the ISS is unique and each node contains unique set of the visual landmarks. The dataset parameters for each docking node are represented in Table 1.

Table 1. Training and testing datasets

Parameters	Node 1	Node 2	Node 3	Node 4	All
Num of objects	20	33	38	51	96
Train dataset size (images)	2307	3443	5070	5146	18764
Test dataset size (images)	1709	904	980	1032	–

Because YOLO algorithm does not significantly decrease the detection performance when number of objects grows (the cascade object detector does), and increasing number of classes does not increase amount of time required for training, all images for four nodes were combined in one large dataset, where all identical objects have the same name. A lot of landmarks are on the different nodes and it is better to train network on all instances simultaneously. Network was trained on the desktop PC on the dataset described above. Full train time is ~ 120 h on the desktop PC with the GPU GeForce GTX960 (4 Gb) installed.

The main purpose of training and testing of this network is to compare results of the object detection with similar ones received in process of the Faster R-CNN testing in the same conditions. To make conditions identical we used dataset and hardware for training and testing the same to those used to train and test Faster R-CNN.

To compare quality of detection the mAP (Mean Average Precision) metric was evaluated for each test dataset for full set of test images without subdivision by objects. This metric is standard for the object detection quality comparison and described in rules for the result evaluation of PASCAL VOC CHALLENGE [11]. Precision-Recall Curve is required to evaluate metric, so it should be transformed to the special plot for mean average precision calculation (as described in [11]) and then mAP can be easily estimated.

Similar to Faster R-CNN, tests on testing datasets for each docking node were performed. Results of comparison are represented in Table 2 below.

Table 2. Results for each node

	Node 1	Node 2	Node 3	Node 4
VGG-1024, all objects	0,790	0,363	0,309	0,394
VGG-1024, augmented	0,765	0,345	0,297	0,508
VGG-1024, high contrast	0,779	0,377	0,250	0,454
ZF, all objects	0,708	0,316	0,219	0,345
ZF, high contrast	0,727	0,365	0,217	0,360
YOLO v2, all objects	0,851	0,716	0,455	0,406

In addition, research on depending of the object detection quality on different image pre-filtering for the first docking node was performed. It is very simple and fast to pose filter on image before sending it to the input of the network, especially if this will increase the detection quality. Results of this research are represented in Table 3.

Table 3. Quality related to image filtering

	F1	F2	F3	F4	F5
VGG-1024, all objects	0,846	0,849	0,843	0,851	0,857
YOLO v2, all objects	0,851	0,849	0,850	0,849	0,846

Filters are represented by headings F1-F5, where:

F1 – no filter; F2 – high contrast; F3 – histogram equalization; F4 – histogram equalization and high contrast; F5 – high contrast and high sharpness.

6 Conclusions

Training and testing of the deep convolutional network YOLO v2 with architecture named MobileNet were performed for this paper. Special dataset extracted from videos of the space docking process of different spacecraft to the International Space Station (exist in open access) was prepared. After training and testing we performed results evaluation and comparison of the object detection by the Faster R-CNN network with the VGG_CNN_M_1024 and ZF internal architectures, and the YOLO v2 with the MobileNet internal architecture on the same testing dataset.

Results of two different networks testing allow us to make a conclusion that quality of the YOLO + MobileNet convolutional network is slightly better than the object detection quality of the Faster R-CNN network. Performance of YOLO is significantly higher than Faster R-CNN. It is from 20 (for the highest allowed image resolution) to 60 (for the lowest resolution) images per second. The YOLO network training also requires significantly more time. It takes almost five days to train on the laptop PC with the GeForce GTX 960 with 4 Gb DDR5 video memory onboard, where Faster R-CNN converged in about 18 h.

The chosen deep convolutional network allows us to detect visual landmarks with performance up to 60 frames per second if it is installed on the desktop or laptop PC with the integrated graphic card that has enough computational capability to work with Keras and TensorFlow and has at least 4 Gb video memory.

For the future work we are planning to integrate previously tested algorithms for the relative position evaluation in the currently tested convolutional network system. Integration allows us to perform tests of the detection quality and the relative position calculation precision in the same conditions and decide whether we will use the convolutional network for the landmark detection or go on with the current cascade detector. Also, we are planning to evaluate more complex architectures inside YOLO and choose the best one in terms of quality and performance.

References

1. Stepanov, D., Bakhshiev, A., Gromoshinskii, D., Kirpan, N., Gundelakh, F.: Determination of the relative position of space vehicles by detection and tracking of natural visual features with the existing TV-cameras. In: International Conference on Analysis of Images, Social Networks and Texts, pp. 431–442. Springer, Cham, April 2015
2. Bakhshiev, A., Korban, P., Kirpan, N.: Software package for determining the spatial orientation of objects by tv picture in the problem space docking. Robotics and Technical Cybernetics **1**(1), 71–75 (2013)
3. InterSpace. The channel is hosting a record of all space launches in the world. https://www.youtube.com/channel/UC9Fu5Cry8552v6z8WimbXvQ, last accessed 2018/05/04
4. Fomin I., Gromoshinskii, D., Stepanov, D.: Visual features detection based on deep neural network in autonomous driving tasks. In: GraphiCon2016 The Conference Proceedings, pp. 430–434. IFTI, NNGASU, Nizny Novgorod (2016)
5. Fomin, I., Gromoshinskii, D., Bakhshiev, A.: Object detection on images in docking Tasks Using Deep Neural Networks. In: NEUROINFORMATICS 2017: Advances in Neural Computation, Machine Learning, and Cognitive Research, pp 79–84
6. Ren, S., He, K., Girshick, R., Sun, J.: Faster R-CNN: towards real-time object detection with region proposal networks. IEEE Trans. Pattern Anal. Mach. Intell. **6**, 1137–1149 (2017)
7. Redmon, J., Divvala, S., Girshick, R., Farhadi, A.: You Only Look Once: Unified, Real-Time Object Detection, https://arxiv.org/pdf/1506.02640.pdf, last accessed 2018/05/04

8. Redmon, J., Farhadi, A.: YOLO9000: Better, Faster, Stronger, https://arxiv.org/pdf/1612.08242.pdf, last accessed 2018/05/04
9. YOLOv2 in Keras and Applications, https://github.com/experiencor/keras-yolo2, last accessed 2018/05/04
10. Howard A.G., Zhu M., Chen B., Kalenichenko D., Wang W., Weyand, T., Andreetto M., Adam H.: MobileNets: Efficient Convolutional Neural Networks for Mobile Vision Applications. https://arxiv.org/pdf/1704.04861.pdf, last accessed 2018/05/04
11. Everingham, M., Van Gool, L., Williams, C.K., Winn, J., Zisserman, A.: The pascal visual object classes (VOC) challenge. Int. J. Comput. Vis. (2)88, 303–338 (2010)

On Comparative Evaluation of Effectiveness of Neural Network and Fuzzy Logic Based Adjusters of Speed Controller for Rolling Mill Drive

Anton I. Glushchenko(✉) and Vladislav A. Petrov

A.A. Ugarov Stary Oskol Technological Institute (Branch) NUST "MISIS", Stary Oskol, Russia
a.glushchenko@sf-misis.ru

Abstract. The article deals with a problem of a speed control of a DC electric drive of a reverse rolling mill under the conditions of its mechanics parameters drift and influence of disturbances. The analysis of existing methods to solve it is made. As a result, two intelligent methods are chosen: the neural network (proposed by the authors) and the fuzzy logic based tuners of linear controllers, the efficiency of which are to be compared. The neural tuner consists of two neural networks calculating the controller parameters of the electric drive, and a rule base that determines at what moments and speed to train these networks. A general description of the fuzzy tuner is also provided. Experimental studies are made using a model of the electric drive of the rolling mill under the above mentioned conditions. The obtained results show that the neural tuner, contrary to the fuzzy one, keeps the speed overshoot within the required limits and also reduces the time of disturbance rejection by 30%.

Keywords: Rolling mill · DC drive · PI controller · Fuzzy adjuster
Neural tuner

1 Introduction

At present, DC motors are still widely used in energy-intensive industries, particularly in the metallurgy [1]. Their main advantages are the ease of a control system structure, significant starting torque and the possibility of a smooth speed control within a wide range. Considering processing units of the metallurgical industry, equipped with the DC drives, it is necessary to single out rolling mills. Their power can reach 10 MW, and they are controlled in accordance with a widespread principle of a cascade control [2] using linear proportional (P) and proportional-integral (PI) control algorithms with constant parameters. The problem with this approach is that the electric drive of the rolling mill is nonlinear, because the parameters of its mechanical (change of a moment of inertia due to a replacement of forming rolls, wear of the mechanics) and electrical (heating and wear of an anchor winding) parts change. The neglect of such non-linearity leads to a deterioration in the control quality and an increase in an energy consumption and a percentage of defective products.

B. Kryzhanovsky et al. (Eds.): NEUROINFORMATICS 2018, SCI 799, pp. 144–150, 2019.
https://doi.org/10.1007/978-3-030-01328-8_15

The least expensive solution of this problem is an adaptive control system development, which does not require a complete replacement of the above mentioned control algorithms, but provides a real-time adjustment of their parameters [3]. The analysis of existing methods of this problem solving is given in the mentioned research, as well as in [4, 5]. In general, it should be noted that for an application of these approaches it is necessary to have a model of the plant or a reference model. Obtaining these models under the conditions of a real production is rather a difficult task. At the same time, this problem is successfully solved without the use of such models by a process automation engineer using their knowledge, experience and abilities to be taught and generalize. These can be taken into account to some extent through the use of intelligent methods.

In most cases, model-free tuning of the parameters of the linear controllers is based on the intelligent techniques such as neural networks (NN) [6, 7] and fuzzy logic [8, 9]. In particular, a neural tuner of the linear controllers is proposed by one of the authors in [7] for metallurgical electromechanical plants. It combines the advantages of the neural networks and the expert systems and allows to maintain the required quality of transients both for setpoint value changes and disturbance occurrence. However, the efficiency of its functioning is estimated only in comparison with conventional PI and P controllers. In this research a comparison of its functioning is made with a fuzzy tuner of a speed controller using a model of the rolling mill DC drive.

2 Problem Statement

In this research the P and PI-speed controllers of the rolling mill DC drive are to be tuned during: (1) following the schedule of the setpoint under the conditions of change of the inertia moment of the drive in order to maintain the required quality of transients (given by a process instruction), (2) rejection of a step disturbance (a roll bite) to increase the speed of the rejection and reduce the maximal deviation of the speed from the setpoint. For this purpose, it is proposed to adjust the speed controller parameters online with the help of the neural network and the fuzzy logic based tuners. The rolling mill is functioning at the moment, and the current values of the parameters of its controllers are known. They are not optimal, but allow it to operate stably. The speed setpoint is changed linearly using a power-up sensor.

3 General Neural Tuner Description

The neural tuner is an augmentation to the P or PI-speed controller and consists of two neural networks with the same structure and a rule base (Fig. 1). One of the networks is used during transients, while the other one calculates the parameters of the speed controller at the moments of the disturbances occurrence.

The currently used neural network of the tuner receives information from the sensors and calculates the values of the parameters of the adjustable controllers. The method of its structure selection for different types of the controllers and the setpoint schedules is given in [7]. For a non-stepwise setpoint NN has the structure: (1) for the

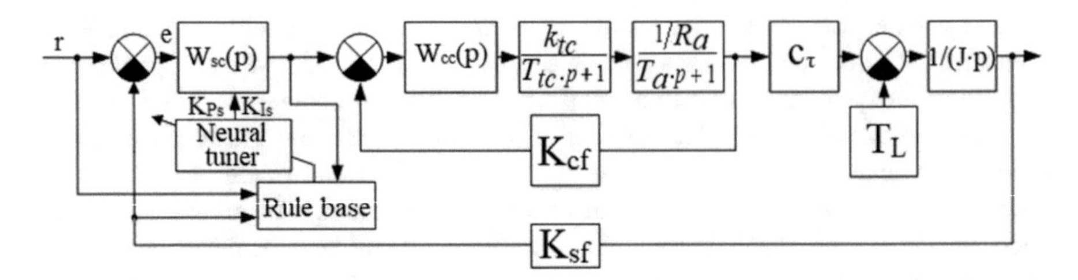

Fig. 1. Block diagram of the DC drive with the neural tuner (K_{sf} and K_{cf} are the speed and the armature current feedback parameters, R_a and T_a are the armature resistance and time constant, J is the mechanics inertia moment, W_{cc} and W_{sc} are the armature current and the speed controllers transfer functions, e is the control error, r is the setpoint, T_L is the load torque, K_{Ps} and K_{Is} are the speed controller parameters, k_{tc} and T_{tc} are the thyristor converter gain and time constant, c_τ is the parameter of conversion of armature current to torque).

PI-controller: 5-14-2, (2) for the P-controller: 2-7-1 (for both cases a sigmoidal activation function is used in the hidden layer, and a linear one – in the output one).

At the time of the tuner start up NN are trained by the method of extreme learning [10] so as to form in their outputs the values of the controller parameters, which have been used before the tuner application. Then they are trained online by the back-propagation method, and the learning rate for each neuron of the output layer is calculated by the base of the rules given in [11]. The tuner is called at discrete moments each Δt seconds [7]. The stability of the control system is estimated according to [12].

4 General Fuzzy Tuner Description

The fuzzy tuner is depicted as block FLT in Fig. 2. It has two input variables: the control error e and its first derivative de/dt. Five criteria to estimate e value (terms) are chosen: highly negative (NH), negative (NL), zero (optimal) (ZO), positive (PL), highly positive (PH). The same set of terms are used for the input variable de/dt.

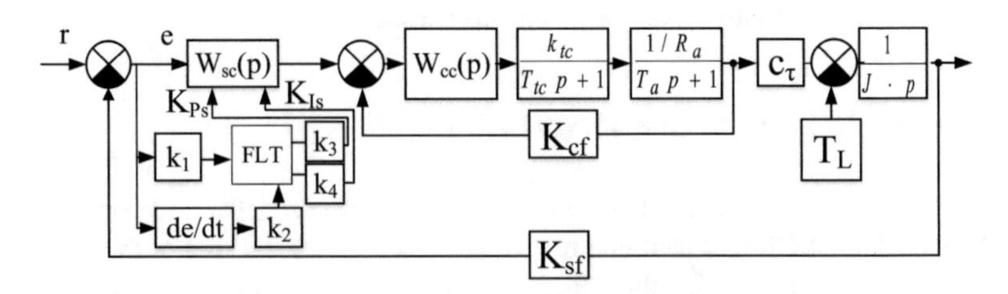

Fig. 2. Block diagram of the DC electric drive with the fuzzy tuner in the speed loop (FLT is the fuzzy logic based tuner, k_1, k_2, k_3 and k_4 are the normalization coefficients).

Four terms are used to estimate output variables (K_{Ps}, K_{Is}): close to zero (ZO), low (L), medium (M), and high (H). For all input variables, the first and last terms have the form of trapezoids, the rest are triangles. All terms for the output variables are in the shape of a triangle. The fuzzy logic system of Mamdani-Zadeh is used. The parameters of the membership functions and rules are taken from [8]. All input variables are normalized into the range [−1; 1], all output variables – into the range [0; 1] with the help of k_1, k_2, k_3, k_4 (Fig. 2). Their values are found empirically.

5 Results of Numerical Experiments

A simplified model (Fig. 3) of the main electric drive of a two-high rolling mill 1000 has been developed in the Matlab Simulink software on the basis of a nameplate data of a DC motor 1JW5539-5DK07-Z-001 (its power is 3.5 MW).

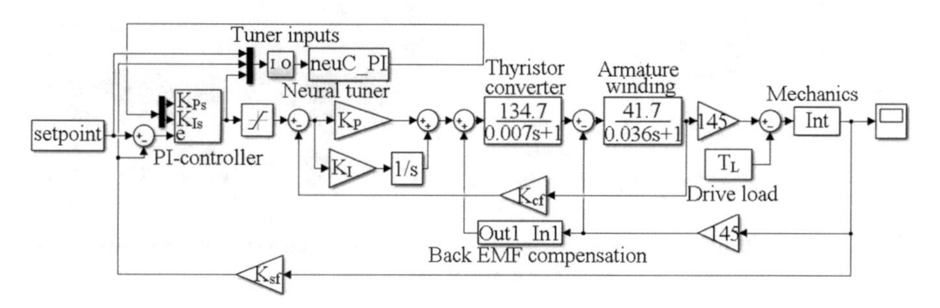

Fig. 3. System with the neural tuner (K_P, K_I are the armature current controller parameters).

$K_P = 0.489$ and $K_I = 13.649$ were calculated in accordance with the technical optimum. Using the same optimum, K_{Ps} was calculated as 1.745 for the speed P-controller ($K_{Is} = 0$). It was used to follow the setpoint value. At the time when the transient had been finished, the tuner switched to the PI-speed controller ($K_{Ps} = 1.745$, $K_{Is} = 31.157$), and used it to reject disturbances. The mechanics of the electric drive was represented by one mass described as an integrator link $(1/J \cdot p)$ ($J = 4.798 \cdot 10^3$ kg $\cdot$ m^2) implemented as S-function to change J value during the modeling. $K_{sf} = 0.637$; $K_{cf} = 9.407 \cdot 10^{-4}$. The neural tuner inputs included: the speed setpoint at the moment and Δt seconds ago, the speed value (n) at the moment and Δt seconds ago, the output value of the speed controller. The outputs of the tuner were K_{Ps}, K_{Is}.

Speed setpoint was implemented using S-function *setpoint*. Firstly, the drive was accelerated from zero rpm to 60 rpm, then it was braked to zero rpm and reversed to minus 60 rpm, finally it was braked again. In accordance with the acceleration of the electric drive under consideration, the setpoint signal was changed linearly with the speed $\Delta r = 0.008$ V/ms (8 V/s). Δt value was calculated as 5 ms.

The system with the fuzzy tuner was implemented according to the scheme in Fig. 2 based on the diagram in Fig. 3. The following values of the normalization coefficients were chosen experimentally to comply with the requirement on the speed overshoot σ 0.35%–0.55% of 60 rpm – $k_1 = 3$, $k_2 = 2$, $k_3 = 3.45$, $k_4 = 62.5$.

5.1 Setpoint Schedule Following

The experiment was conducted with the neural and the fuzzy tuners, in which a smooth change (drift) of J value within 50% ÷ 150% of its nominal value with a speed $\Delta J = 24\ \text{kg} \cdot \text{m}^2/\text{s}$ was simulated. During entire experiment the load torque was not applied ($T_L = 0$), so in fact, the P-speed controller was adjusted. The task was to keep σ within the required limits of 0.35%–0.55%. The results are shown in Fig. 4 (for the neural tuner – Fig. 4A–D, for the fuzzy tuner – Fig. 4E–H).

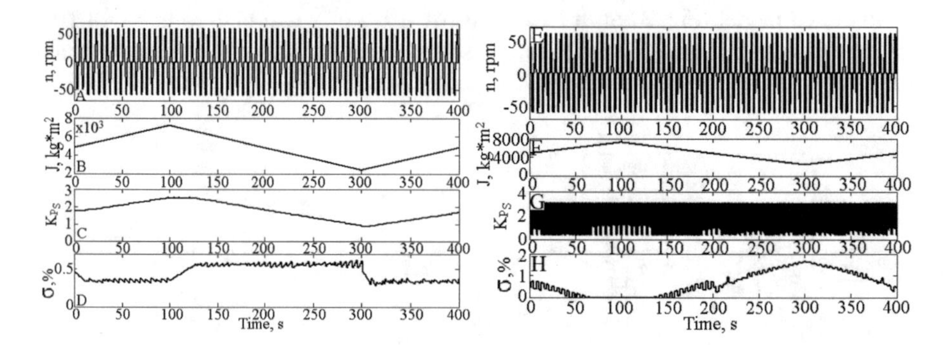

Fig. 4. The results of the experiments with the drift of the parameters of the drive mechanics: (1) the neural tuner (A–D), (2) the fuzzy tuner (E–H).

As for the neural tuner, σ value was in the required limits during the whole experiment (Fig. 4D). The fuzzy tuner did not allow to achieve such results – minimal and maximal values of σ were 0% and 1.4% respectively (Fig. 4H).

5.2 Disturbance Rejection

The main disturbance for the rolling mills is the roll biting (the stepwise disturbance acting the drive torque). Compensating it, the speed PI-controller was adjusted. The model in Fig. 3 was used (the drive parameters were nominal). The value of the load torque (T_L) was chosen according to the weight and geometric dimensions of the rolled billets (300 kN · m). The results of the experiments are shown in Fig. 5.

During the experiments a cycle of the rolling mill operation was simulated (Fig. 5A): acceleration of the rolls to the nominal speed (60 rpm), the roll biting, rolling the billet, "release" of the billet, stop, reverse.

In case of the disturbance occurrence the neural tuner changed K_{Ps} (Fig. 5B) and K_{Is} (Fig. 5C). This allowed to reduce the maximal deviation of the drive speed from the setpoint by 18% and the time of the disturbance rejection by almost 30% (Fig. 5E) comparing to the fuzzy tuner (Fig. 5J).

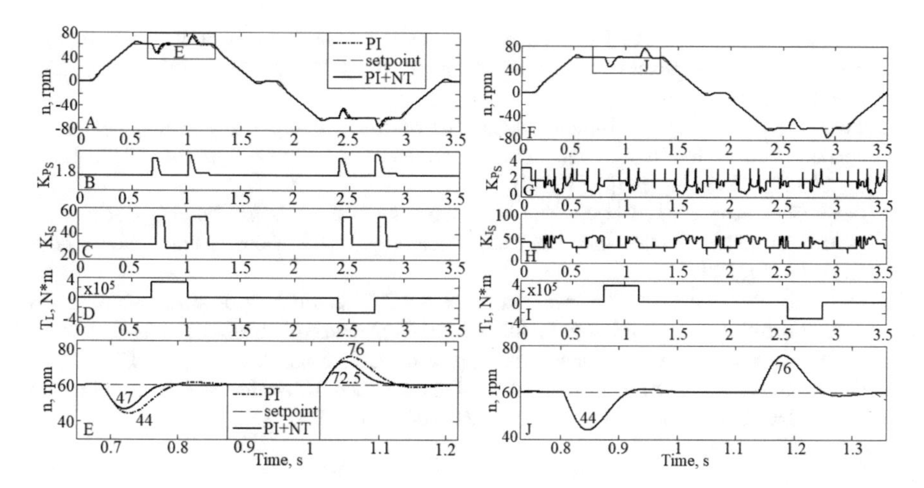

Fig. 5. Simulation results for the DC drive of the rolling mill with the neural tuner (A–E) and the fuzzy tuner (F–J) (NT is the neural tuner).

6 Conclusion

The results of the research show that the neural tuner allowed to achieve more efficient compensation of the nonlinearity of the DC drive both during the setpoint following and the disturbances rejection in comparison with the fuzzy tuner.

Acknowledgments. This work was supported by the Ministry of education and science of the Russian Federation. Grant No. 14.575.21.0133 (RFMEFI57517X0133).

References

1. Muruganandam, M., Madheswaran, M.: Stability analysis and implementation of chopper fed DC series motor with hybrid PID-ANN controller. Int. J. Control Autom. Syst. **11**, 966–975 (2013)
2. Leonhard, W.: Control of Electrical Drives. Springer, Berlin (2001)
3. Astrom, K.J., Wittenmark, B.: Adaptive Control, 2nd edn. Dover, New York (2008)
4. Berner, J., Soltesz, K., Hagglund, T., Astrom, K.J.: An experimental comparison of PID autotuners. Control Eng. Pract. **73**, 124–133 (2018)
5. Ribeiro, J.M.S., Santos, M.F., Carmo, M.J., Silva, M.F.: Comparison of PID controller tuning methods: analytical/classical techniques versus optimization algorithms. In: 2017 18th International Carpathian Control Conference, pp. 533–538. IEEE, Sinaia (2017)
6. Riverol, C., Napolitano, V.: Use of neural networks as a tuning method for an adaptive PID. Chem. Eng. Res. Des. **8**(78), 1115–1119 (2000)
7. Glushchenko, A.I.: Neural tuner development method to adjust PI-controller parameters on-line. In: 2017 IEEE Conference of Russian Young Researchers in Electrical and Electronic Engineering, pp. 1–6. IEEE, Saint-Petersburg (2017)

8. Ahmed, H., Rajoriya, A.: Performance assessment of tuning methods for PID controller parameter used for position control of DC motor. Int. J. u-and e-Serv. Sci. Technol. **7**(5), 139–150 (2014)
9. Yesil, E., Guzay, C.: A self-tuning fuzzy PID controller design using gamma aggregation operator. In: 2014 IEEE International Conference on Fuzzy Systems (FUZZ-IEEE), pp. 2032–2038. IEEE, Beijing (2014)
10. Feng, G., Huang, G.B., Lin, Q., Gay, R.: Error minimized extreme learning machine with growth of hidden nodes and incremental learning. IEEE Trans. Neural Netw. **20**(8), 1352–1357 (2009)
11. Eremenko, Y.I., Glushchenko, A.I., Petrov, V.A.: DC electric drive adaptive control system development using neural tuner. In: 2017 IEEE Conference of Russian Young Researchers in Electrical and Electronic Engineering, pp. 1–6. IEEE, Saint-Petersburg (2017)
12. Glushchenko, A.I.: Method of calculation of upper bound of learning rate for neural tuner to control DC drive. Stud. Comput. Intell. **736**, 104–109 (2018)

Disease Detection on the Plant Leaves by Deep Learning

P. Goncharov[1(✉)], G. Ososkov[2], A. Nechaevskiy[2], A. Uzhinskiy[2], and I. Nestsiarenia[1]

[1] Sukhoi State Technical University of Gomel, Gomel, Belarus
kaliostrogoblin3@gmail.com
[2] Joint Institute for Nuclear Research, Moscow Reg, Dubna, Russia
auzhinskiy@jinr.ru

Abstract. Plant disease detection by using different machine learning techniques is very popular field of study. Many promising results were already obtained but it is still only few real life applications that can make farmer's life easier. The aim of our research is solving the problem of detection and preventing diseases of agricultural crops. We considered several models to identify the most appropriate deep learning architecture. As a source of the training data, we use the PlantVillage open database. During research, the problem with PlantVillage images collection was detected. The synthetic nature of the collection can seriously affect the accuracy of the neural model while processing real-life images. We collected a special database of the grape leaves consisting of four set of images. Deep siamese convolutional network was developed to solve the problem of the small image databases. Accuracy over 90% was reached in the detection of the Esca, Black rot and Chlorosis diseases on the grape leaves. Comparative results of various models and plants using are presented.

Keywords: Machine learning · Statistical models · Siamese networks
Plant disease detection · Transfer learning

1 Introduction

Increasing number of smartphones and advances in deep learning field opens new opportunities in the crop diseases detection. Probably, the most famous mobile application allowing users to send photos of sick plants and get the cause of the illness is Plantix (plantix.net). The application is developed by PEAT, a German-based AgTech startup. Currently Plantix can detect more than 300 diseases. Unfortunately, Plantix image database is closed and we cannot find any information about technologies used for disease detection. The quality of plantix detection is hard to measure, but we make a special study processing different types of images from our self-collected database. It allows us to conclude that Plantix identification of the plants type is rather good: 60 of 70 images (87%) were recognized as grapes. At the same time the

The reported study was funded by RFBR according to the research project № 18-07-00829.

B. Kryzhanovsky et al. (Eds.): NEUROINFORMATICS 2018, SCI 799, pp. 151–159, 2019.
https://doi.org/10.1007/978-3-030-01328-8_16

disease detection ability is rather limited. The most of the healthy leaves were identified as healthy or healthy was on the top of suggestions. 10 of 20 leaves with Chlorosis were also detected as healthy. Leaves with Esca and Black Rot were recognized as sick but the correct disease names were not on the top of suggestions. Only few images were detected correctly. Perhaps, our dataset does not match some requirements of the Plantix application. We used original images from the Internet and preprocessed those in which problems were obvious, but the result was quite similar.

The aim of our research is to facilitate the detection and preventing diseases of agricultural plants by both deep learning and programming services. The idea is to develop multifunctional platform that will use modern organization and deep learning technologies to provide new level of service to farmer's community. However, in this paper we focus on the deep learning issues only. We would like to reach same functionality as Plantix but with better accuracy in the diseases detection. We will also going to provide an open access to our image database and share experience of our deep learning architectures.

The key issue for the implementation of the plant disease detection platform (PDDP) is an appropriate deep learning architecture. We considered different models used in related works to understand what the best option is. In [1] authors reached high accuracy in detection to 99.7% on a held-out test set. They have used PlantVillage well known public database of 86,147 images of diseased and healthy plant leaves. We have reproduced their experience using the same approach but different software. As a result, we obtained identical high accuracy when using PlantVillage data for training and testing, but the results obtained by applying our trained network on real-life images were quite unsatisfactory (for about 40% only). The problem lies in the type of the used images. PlantVillage photos were collected and processed under special controlled conditions, so they are rather synthetic and differ from real-life images, as it is shown in Fig. 1. Influence of the different image types on the accuracy of the diseases detection with field images is well shown at [2]. Some experiments were done with background modification and other optimizations [3], but it could improve the accuracy slightly.

This proves that if we want a good result, we need a real-life database. Although many related successful studies are known using their self-collected databases [4–7], but there are no references to used databases, unfortunately. We collect our own database of the grape leaves from open source images and then reduce their size and extract only meaningful parts. Eventually, we have a set of 256×256 pixel images consisting of 130 healthy leaves, and 30–70 images with Esca, Chlorosis and Black Rot diseases. The number of images is very small but we are going to refill our database with users' images of correctly detected diseases when the public part of PDDP will be developed. The current database is available at http://pdd.jinr.ru/db. We used this database to test some models and to try some new approaches.

Fig. 1. Top three photos are from the PlantVillage database. Bottom photos are real photos of sick leaves from the Internet.

2 Unsuccessful Results of Our Study

2.1 Transfer Learning

Our first attempt was similar to [1], we applied transfer learning approach to train deep classifier on the PlantVillage images and then evaluated classifier on a test subset of images, collected from the Internet.

To find the most appropriate pretrained network (further base network) for the transfer learning we compared four models the weights of which were formed to solve the ILSVRC 2015 (ImageNet Large Scale Visual Recognition Challenge) [8], they are: VGG19 [9], InceptionV3 [10], ResNet50 [8] and Xception [11].

The comparative scheme of each classifier is to compose all layers of trained networks except final classification layer and to add the global average pooling operation [8] on the top of each base network to reduce the spatial dimensions of a three-dimensional output tensor. Further, we appended a densely connected layer with 256 rectified neurons with dropout having rate of 0.5. At the end of such network, softmax classification layer was utilized.

We froze all layers in the base networks and trained only last three layers using stochastic gradient descent (SGD) with learning rate equals to 5e−3, momentum 0.9 and weight decay with value of 5e−4 for 50 epochs. The best result of classification accuracy with the value of 99.4% on a test subset of the PlantVillage dataset was obtained using ResNet50 architecture. We applied this model to deduce the classification efficiency on a test subset collected from the Internet. Obtained results were very poor – 48% accuracy on a set of 30 images.

2.2 Advanced Transfer Learning and Data Augmentation

Supposing that pretrained on the ImageNet dataset network does not extract meaning features from leaves images, we decided to unfreeze more layers. There is no reason to train the whole network from scratch, as we do not have a dataset with the suitable size to train such a deep network. Authors in [1] got very low accuracy on the Internet images. Thus, we unfroze all layers except first 140 in the base network and trained the remaining 39 defrozen layers with the help of Adam optimizer [15] with learning rate equals to 5e−5 and weight decay with value of 1e−6 for 30 epochs. We used only three plant classes: Esca, Black rot and healthy. We cannot use Chlorosis, because there are no images of that class in the PlantVillage dataset.

Next, we proposed to apply a strong data augmentation by adding random transformations such as shifts, rotations, zooming etc., because the classification network overfits when we train more than 30 epochs. Also, we supposed that only a central part of leaf is required to recognize disease. Therefore, we tried to expand our dataset by using only parts of initial images. We started from 128×128 central square of each initial image and generated then at least 5 new parts around it (as shown in Fig. 2).

We supposed that this approach of data augmentation by splitting original images into parts allows also to decrease the influence of a background, so we cut bound parts of leaves using only their middle part for classification. However, accuracy on the test dataset of images collected from the internet was 49% only. The network overfits after 10 epochs and even our strong augmentation does not help. Surprisingly, it works quite well for binary classification like recognizing of healthy and disease leaves – we obtained 85% accuracy on the test dataset.

It turns out that, when we crop images, we lose the information about location of spots and it produces very noisy data. Besides some parts of diseased leaves look healthy and automatic split to the parts produces incorrect examples. We tried to use parts of images for multi-class classification, but experiments showed that such approach is not effective. Using the images with their original sizes is more reliable and more suitable for real application.

3 Siamese Networks for Learning Data Embeddings

It becomes obvious that features extracted during training network models that we attempted to try, are not adequate for further classifying. These features cannot represent input images in a new multidimensional space, where diseased and healthy images should be separated into distinctive clusters. Probably, it depends on the synthetic nature of the available training set, so a question arises: how could we learn good features from very small amount of data collected from the Internet? We address this problem to so-called one-shot approach [12] offering a solution by siamese neural networks [12–14].

Siamese network consists of twin networks joined by the similarity layer with the energy function at the top. Weights of twins are tied, thus the result is invariant and in addition guarantees that very similar images cannot be in very different locations in feature space, because each network computes the same function. The similarity layer determines some distance metric between the so-called embeddings, i.e. high-level feature representations of input pair of images (Fig. 3).

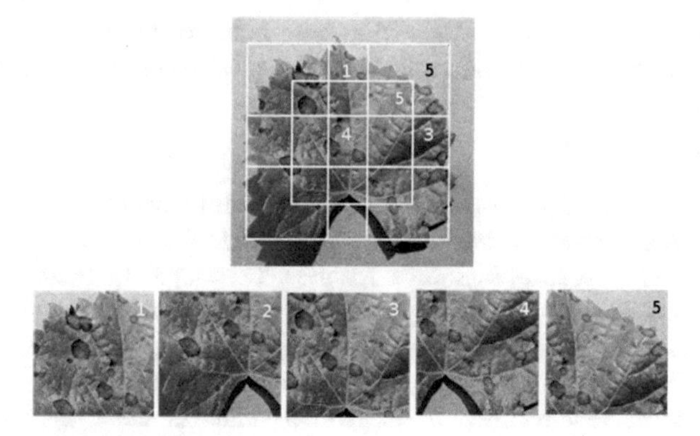

Fig. 2. Splitting original image into training parts

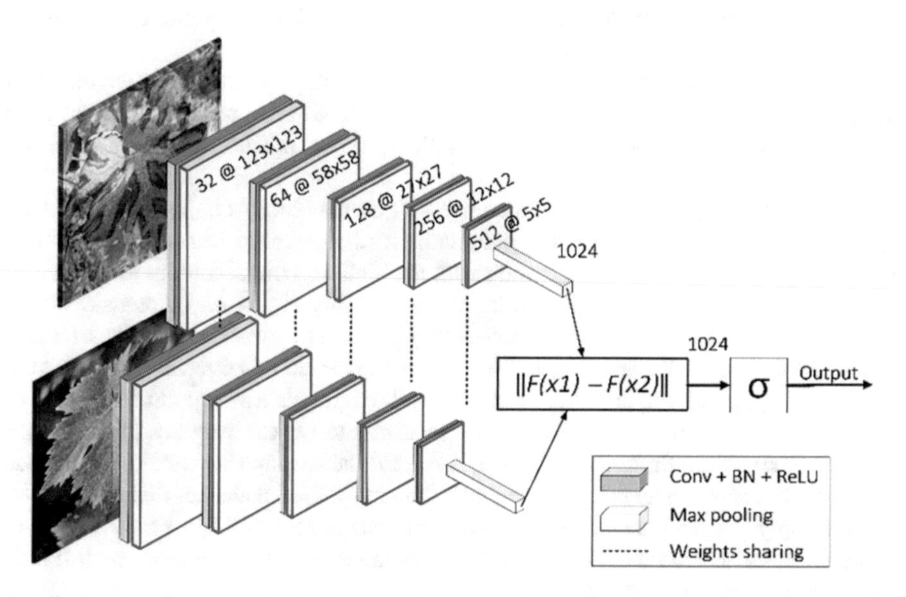

Fig. 3. Our best siamese convolutional architecture. «Conv» means the convolutional operation, «BN» is a batch normalization, «32 @ 123 × 123» – 32 feature maps with the particular size

Training on pairs means that there are quadratically many possible pairs of images to train the model on, thus making it hard to overfit. We can easily compute the number of possible pairs using the combinatorics formula of k-combinations. Thereby, for the smallest class with 31 images (Black rot) we have 1860 pairs, which is a great result.

Our Siamese network is built in the same way as in [12] – it unites the twins within L1 distance layer followed by sigmoid activation in order to train the net with cross-entropy objective. The opposite approach includes application of special contrastive loss, which was first introduced in [13]. Unfortunately, the using of contrastive loss requires a special way of image pairing, otherwise one could face with very low testing accuracy. Our experiments with weighted contrastive loss in tandem with random guessing proved this statement – the maximum value of testing accuracy was 52% only. Other research [14] also confirms the correctness of our conclusions.

3.1 Architecture

Our classification model is the siamese network with convolutional twins, which tie weights between themselves. Each of twins processes one image from input pair of samples to extract a vector of high-level features. Then a pair of embeddings is passed through the lambda layer, actually computing the simple elementwise L1-distance (Fig. 3). We present connection a single sigmoid neuron to the distance layer, thus it becomes possible to train the model with binary cross-entropy loss.

We use exclusively rectified linear (ReLU) units in the twins except the last densely-connected layer with sigmoid activations. The initial number of convolutional filters is 32 and doubling each layer. After the last pooling layer, we added a flatten operation to squeeze convolutional features into vector followed by densely-connected layer with 1024 sigmoid neurons.

We also experimented with adding L2 regularization to each of convolutional layers but have not received visible effect. In addition, we have been trying to vary the size of the embedding layer from 256 to 4096. Out best architecture is illustrated on Fig. 3.

In [16] authors evaluated the trained network on new images in a pairwise manner against the test image. The created pairs consist of test image in each pair and one sample for every class. Then the pairing with the highest score according to the trained siamese network is then awarded the highest probability for the one-shot task. We, in turn, offer to use K-nearest neighbors (KNN) algorithm to solve the classification task in test phase. We set K parameter to 1 nearest neighbor, so it is equivalent to one-shot learning task, except one moment – we utilize all training data as support set instead of random picking from dataset. For the distance metric we can use any of manhattan distance or euclidean distance, but as we have used the absolute distance in the lambda layer of the siamese network, we preferred to manage the manhattan distance.

We apply data augmentation by adding rotations in range of 75°, random shifts in all dimensions, including shift within channel in range of 0.2. In addition, we use little zooming, vertical and horizontal flips.

We have trained the siamese network with the proposed augmentation for 100 batches size 32 per one training epoch for 35 epochs. As the method of optimization we used Adam [15] with learning rate value of 0.0001. For the loss function we utilized a binary cross-entropy loss.

3.2 Classifying by Nearest Neighbors

After training the one-shot model, we left only the encoder network represented as «shoulder» of the one-shot model or so-called twin. We further used this part of network as a feature extractor. After that, we took the training subset of the database, collected by us from the Internet. This database includes four classes of images: healthy images and 3 types of diseases – Esca, Black rot and Chlorosis with 133, 73, 31 and 42 images per class respectively. We split data on train and test with the ratio near 75:25, where 75 means 75% of images of particular class. Accordingly, we have 208 training images and 71 testing images. Then we passed these sets of images through our feature extractor to obtain data embeddings.

The training subset of images is then utilized as training data for the KNN. For verification we used the remaining test set.

3.3 Results

In the Table 1 one can see the confusion matrix[1] of the KNN on the test subset of embeddings data.

Table 1. Confusion matrix on the test subset of internet images

	Black rot	Chlorosis	Esca	Healthy
Black rot	**7**	0	1	1
Chlorosis	0	**11**	0	1
Esca	0	0	**20**	1
Healthy	0	0	0	**29**

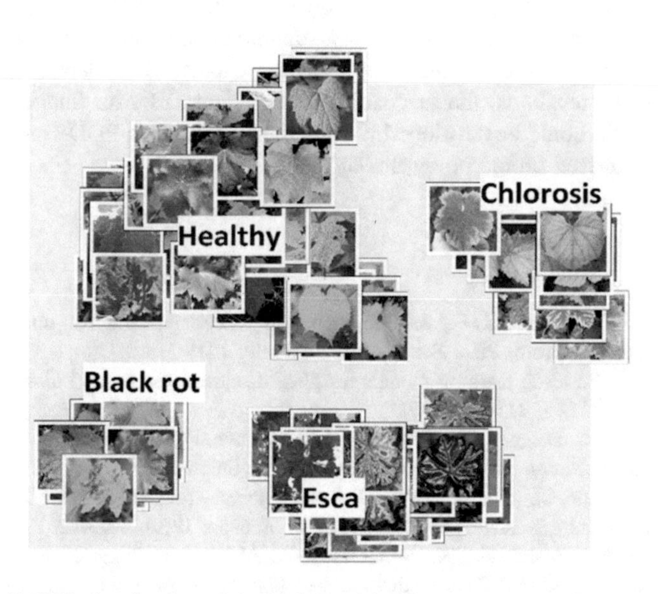

Fig. 4. T-SNE visualization of the high-level features extracted by the siamese twin

[1] This matrix shows how many class i objects were recognized as class j objects [17].

It is simple to deduce that the classification accuracy equals to 94.3% which is acceptable. Besides, we tried to mix the real-life data with the images from the PlantVillage database within train and test subsets. Siamese networks allow to generalize input data to latent high-dimensional embedding space, even for unseen images. The obtained accuracy with the value of 92% (our best result) proves this consideration.

As the low testing accuracy shows, our previous models cannot separate input images into clearly distinguishable clusters of classes. Therefore we use the prepared embeddings to train the T-SNE method [18], which is the common technique to visualize high-dimensional data. We extract two components to plot them in 2D space (Fig. 4). One can see that there are four separate clusters – one per each class. We have signed each cluster with the name of particular class. Although, there are few points, which wrongly got into the different set (see the Table 1), but it is not detractive.

4 Conclusion and Plans

Siamese neural networks are very perspective research field. We are going to use them as a basic deep learning architecture for PDDP and since their power lies in seeking for differences between classes, we are going to add more classes to the train dataset soon. It is clear that unequivocal detection of the diseases is unsolvable task especially at first stages of a plant illness. The list of the most suitable variants should be provided to the users so they could try different treatment schemas. Quality of the database is extremely important for the results so the mechanism of fulfilling DB with images of correctly detected diseases should be developed. We keep on working on PDDP and are going to present web-interface prototype by the end of 2018.

References

1. Mohanty, S.P., Hughes, D.P., Salathé, M.: Using deep learning for image-based plant disease detection. Front. Plant Sci. **7** (2016). Article: 1419
2. Ferentinos, K.P.: Deep learning models for plant disease detection and diagnosis. Comput. Electron. Agric. **145**, 311–318 (2018)
3. Cortes E.: Plant Disease Classification Using Convolutional Networks and Generative Adverserial Networks. Stanford University Reports, Stanford (2017)
4. Fuentes, A., Yoon, S., Kim, S.C., Park, D.S.: A robust deep-learning-based detector for real-time tomato plant diseases and pest recognition. Sensors **17**(9), 2022 (2017)
5. Ramcharan, A., Baranowski, K., McCloskey, P., Ahmed, B., Legg, J., Hughes, D.: Deep learning for image-based Cassava disease detection, frontiers in plant science. Front. Plant Sci. **8**, 1852(2017)
6. Lua, J., Hua, J., Zhaoa, G., Meib, F., Zhanga, C.: An in-field automatic wheat disease diagnosis system. Comput. Electron. Agric. **142PA**, 369–379 (2017)
7. Ronnel, R., Atole, A.: Multiclass deep convolutional neural network classifier for detection of common rice plant anomalies. Int. J. Adv. Comput. Sci. Appl. (IJACSA) **9**(1) (2018)
8. He, K., et al.: Deep residual learning for image recognition. In: Proceedings of the IEEE Conference on Computer Vision and Pattern Recognition, pp. 770–778 (2016)

9. Simonyan, K., Zisserman, A.: Very deep convolutional networks for large-scale image recognition. arXiv preprint arXiv:1409.1556 (2014)
10. Szegedy, C. et al.: Rethinking the inception architecture for computer vision. In: Proceedings of the IEEE Conference on Computer Vision and Pattern Recognition, pp. 2818–2826 (2016)
11. Chollet, F.: Xception: deep learning with depthwise separable convolutions. arXiv preprint (2016)
12. Simonyan, K., Vedaldi, A., Zisserman, A.: Deep inside convolutional networks: visualising image classification models and saliency maps. arXiv preprint arXiv:1312.6034 (2013)
13. Kotikalapudi, R.: contributors (2017). keras-vis (2017)
14. Ososkov, G., Goncharov, P.: Shallow and deep learning for image classification. Opt. Mem. Neural Netw. **26**(4), 221–248 (2017)
15. Ososkov, G., Goncharov, P.: Two-stage approach to image classification by deep neural networks. In: EPJ Web of Conferences, vol. 173, p. 01009. EDP Sciences (2018)
16. Koch, G., Zemel, R., Salakhutdinov, R.: Siamese neural networks for one-shot image recognition. In: ICML Deep Learning Workshop, vol. 2 (2015)
17. Powers, D.: Evaluation: from precision recall and F-measure to ROC, informedness, markedness & correlation. J. Mach. Learn. Technol. **2**(1), 37–63 (2011)
18. Maaten, L., Hinton, G.: Visualizing data using t-SNE. J. Mach. Learn. Res. **9**, 2579–2605 (2008)

Neural Network Based Algorithm for the Measurements of Fire Factors Processing

A. I. Guseva, G. F. Malykhina(✉), and A. S. Nevelskiy

Peter the Great Saint-Petersburg Polytechnic University, Saint Petersburg, Russia
malykhina@icc.spbstu.ru

Abstract. The objective of study is to develop a neural network algorithm for early warning of fire on a ship, using the results of measuring a multitude of sensors. The sensors are designed to measure the main fire factors such as temperature, concentration of carbon dioxide and the presence of smoke. Optimal coordinates of the sensors of the multi-sensor system were solved using suggested evolutionary algorithm. The study is based on the results of modeling fires on a supercomputer. The measurement of many fire factors allowed to develop an algorithm for determining the type of ignition. The algorithm uses a probabilistic neural network with delays at the input. Knowledge of the ignition type makes it possible to increase the speed of reaction to fire and to reduce damage on the ship. The suggested enhanced genetic algorithm fulfills optimization of the sensors position in the cabin. The algorithm allows to reduce the mean time of fire detection by of 15% in average. This effect can be achieved by increasing of the number of sensors in the cabin from 3 up to 11.

Keywords: Neural network · Evolutionary algorithm · Fire-fighting system Modeling

1 Introduction

Mainstream Fire Alarm Systems (FAS) apply detectors based on thresholding principle. The fire system is triggered if at least one of the dangerous fire factors is exceeded. The experience of existing firefighting systems shows that cases of false triggering are not uncommon. For example, from a heater or a cigarette smoke. A large number of false fire alarms reduce the credibility of fire alarms. In addition, some sensors are disposable. Consequently, due to false positives, the purchase of new sensors is required. This leads to a rise in the cost of the system.

A multi-criteria fire detection algorithm consists in the continuous and simultaneous monitoring of several potential fire factors (smoke, heat, and carbon monoxide, and carbon dioxide, infrared or ultraviolet radiation from a flame) and the formation of a reliable fire alarm signal with a minimum delay after the occurrence of fire. A standard, programmable value, upon which a fire alarm is declared, is the “delta value” – the first finite difference of every fire factor [2]. This approach does not allow to use the fullness

B. Kryzhanovsky et al. (Eds.): NEUROINFORMATICS 2018, SCI 799, pp. 160–166, 2019.
https://doi.org/10.1007/978-3-030-01328-8_17

of the measurement information for constructing more advanced criteria for detecting and forecasting the development of a fire.

Further development of the multi-criteria approach is associated with algorithms of intelligent signal processing applied to multisensory system. Algorithm intends to detect ignition at an early stage, notably to determine the type of ignition, the location of the source in the space of the room, and to predict the development of a fire [3]. An important step in the development of a multi-criteria fire detection algorithm is the finding of optimal arrangement of sensors in a cabin [4]. The timeliness of fire detection depends on correctness of this problem solution.

The objective of study is to develop a neural network algorithm for early warning of fire on a ship, using multitude of sensors measurement results. To achieve an objective of study it is necessary to develop an algorithm for optimal sensors arrangement and to propose algorithm of inflammable staff identification.

An evolutionary algorithm for the multiple sensors location optimization uses the ignition model, performing the different typical fires in cabins. The developed algorithms would be checked on the mathematical model of the fire, taking into account all the factors characterizing beginning, development and diminishing flame. The most important task of multisensory fire system is the timely detection of fires on the ship [5].

2 Evolutionary Algorithm for the Optimal Arrangement of Sensors

The genetic algorithm intended for finding the optimal arrangement of sensors. To apply genetic algorithm it is reasonable to define the type of the basic genetic operators: notably, selection, crossover and mutations [6]. It is also necessary to determine the type of fitness function and type of initial population of sensors.

The initial population of sensors is as follows:

$$S = \left(((X_1), \cdots, (X_M))_1, ((X_1), \cdots, (X_M))_2, \cdots, ((X_1), \cdots, (X_M))_N\right), \tag{1}$$

where M is the desired number of sensors, N is the initial size of the generated sets of sensors to be placed, $X_i = (x_1, x_2)^T$ is a sensor coordinates.

Assessment of the fitness function of each sensor is defined as the mean time of the fire detection uses following equation:

$$f(u_i) = \frac{1}{q}\sum_{j=1}^{Q} P(X_j)\frac{\sum_{i=1}^{M} t(u_i)}{M}; \tag{2}$$

where $u_i = [P(X_1), P(X_2), \ldots, P(X_M)]^T$ is the probability vector for the occurrence of a fire from a set in a neighborhood of the point X_i, $P(X_1)$ is a probability of a fire as in experiment 1, Q is a number of experiments.

Fitness function is obtained for the set of experiments and the set of sensors at current location. Enumerated operators of the genetic algorithm were chosen.

Selection. To choose the most effective method of reproduction, we calculated the time spent searching for a solution for multiplication by the Boltzmann method and ranking. The results are shown in Fig. 1.

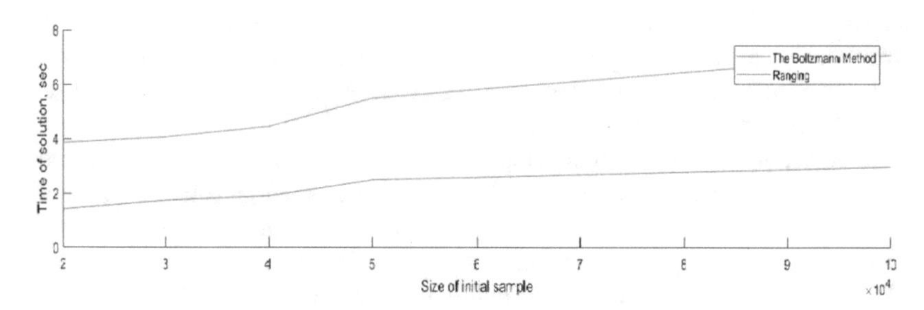

Fig. 1. Comparison of the convergence rate for the algorithm based on the ranking and Boltzmann method

Since the Boltzmann method showed a more rapid convergence rate. The proposed algorithm uses the same approach as when optimizing by annealing. This method for controlling the process introduces the concept of "artificial temperature" T. From a certain moment, we reduce (according to the chosen law) and change the probability of selection of individuals. Thus, the probability of selecting an individual with the fitness function $f(u_i)$: $P_s(a_i) = \frac{1}{N}\left(\frac{e^{f(u_i)/T}}{\overline{e^{f(u_i)/T}}}\right)$, with $\overline{e^{f(u_i)/T}}$ the average value of $e^{f(u_i)/T}$ on the current population [7]. This allows us to narrow down the search in the most promising area of search space at the final stage, while maintaining a sufficient degree of diversity in the population.

Crossover. In the theory of genetic algorithms, there are two basic types of crossing: binary recombination and recombination of real values. In compliance with the specific nature of the input vectors, none of the methods can be directly applied. Therefore, the standard mechanisms of crossing were modified. As a result, the crossing is carried out as follows: we alternately compare the two gene of the parent as in a homogeneous crossing-over, and choose with the best fitness function. In compliance with the intermediate recombination, we choose the coefficient for the coordinates in the space of this sensor. As a result, the set of proximity probes will be slightly different from the parent set.

Mutation. The mutation was performed by a standard method. With probability P_z one of the sensors from the set of sensors was selected and moved to a random value z, selected from the range [−30; 30] sm by length or width of cabin.

3 Identification of Ignition Type

The preliminary simulation of fires in supercomputer center allows to obtain the features for the source of ignition identification. The different factors of fire for the five types of the following staff ignition were simulated:

- gasoline,
- cotton and capron,
- wires in rubber insulation,
- paper,
- ethanol.

The simulation was performed during the first 60 s from the moment of ignition. The model of the cabin characterizes by the cabin dimension, doors, furniture and ventilators positions. With each type of ignition, several tests were carried out having different places of ignition. The sensors are located on a 10 by 10 cm grid. The data is obtained from the sensors once every 1 s

The obtained data of different types of ignition showed that the temperature growth rate, concentration of CO2 growth and visibility (smoke concentration) attenuation in each case of ignition is different. Hereby, if the gasoline, rags and oil are ignited, the temperature and concentration of gases are more sensitive to the fire, but if the diesel fuel and cables were ignited, the smoke concentration factor in the cabin is the first to react. The simulated data of different types of ignition showed that the rate of changing in temperature, concentration of CO_2 in each case of ignition is different. Example of ignition sources location in simulation is shown in Fig. 2.

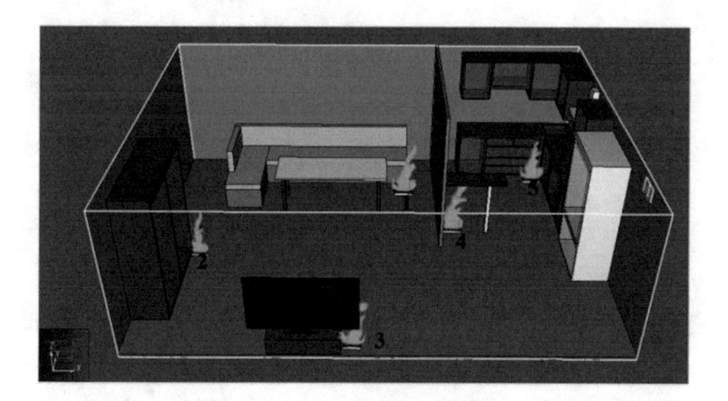

Fig. 2. Location of ignition sources

Figures 3, 4 and 5 show the dynamics of changes in temperature, concentration of CO_2 and visibility for various sources of ignition.

The time delay probabilistic neural network (TDPNN) [8] is used for the ignition identification. Network includes implicit representation of time by short-time memory and multilayer perceptron. Accordingly, 30 samples of the temperature, concentration of CO_2 and visibility were fed the network input. The network outputs present

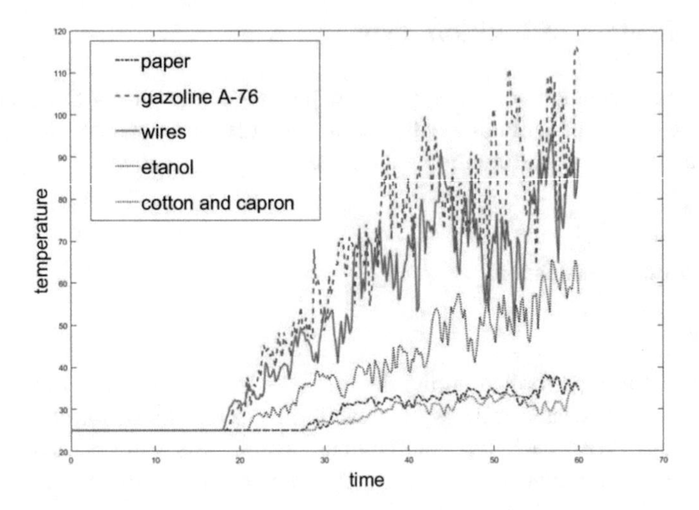

Fig. 3. Dynamics of the temperature growth, depending on the type of ignition

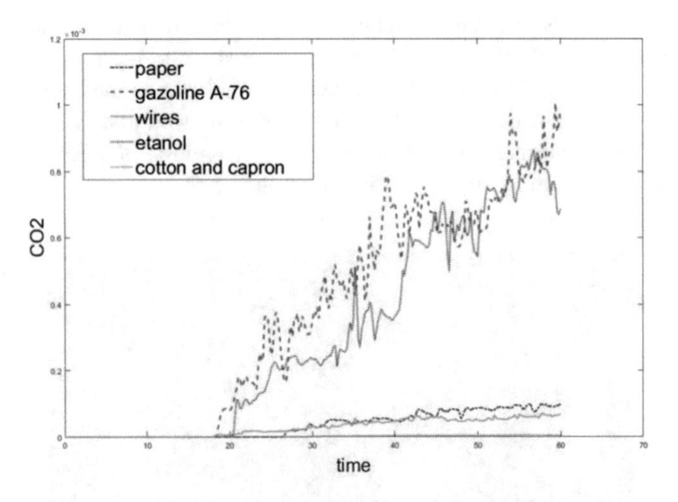

Fig. 4. Dynamics of the CO_2 concentration growth in dependence of ignition type

probabilities of the following situations: absence of the fire, ignition of gasoline or ethanol, ignition of cotton or paper, ignition of high-tension or internal cable (Fig. 6).

The result of ignition identification is shown on Fig. 7.

As can be seen in the Fig. 7, the data of some sensors mislead the network. Since training was conducted on the whole array of sensors (pieces), many sensors should be excluded. The choice of remaining sensors is based on suggested genetic algorithm.

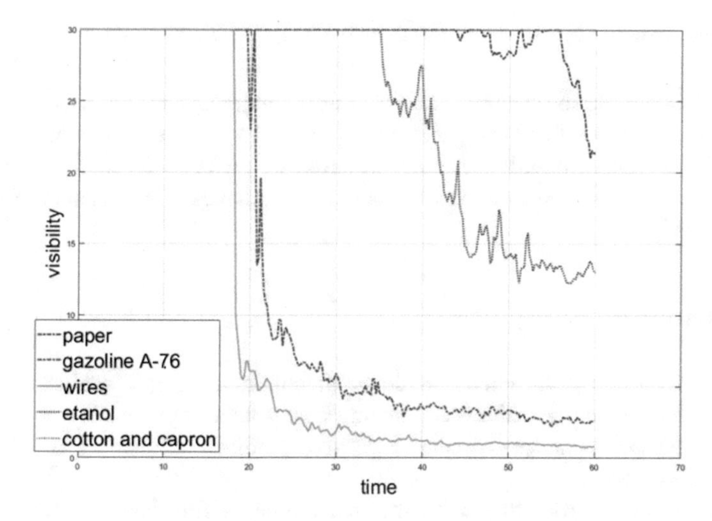

Fig. 5. Dynamics of visibility decreasing in dependence of ignition type

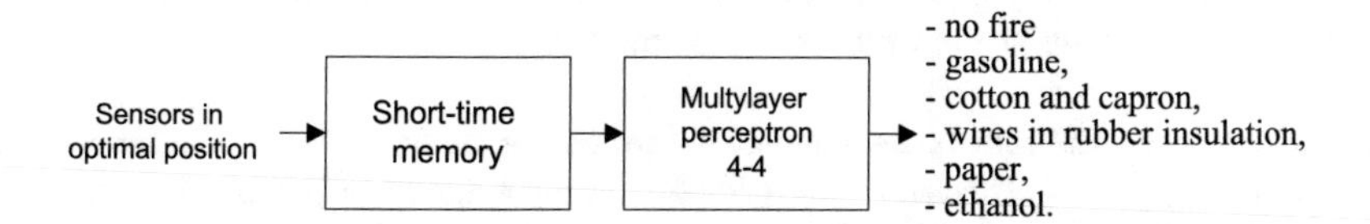

Fig. 6. TDPNN for the ignition identification

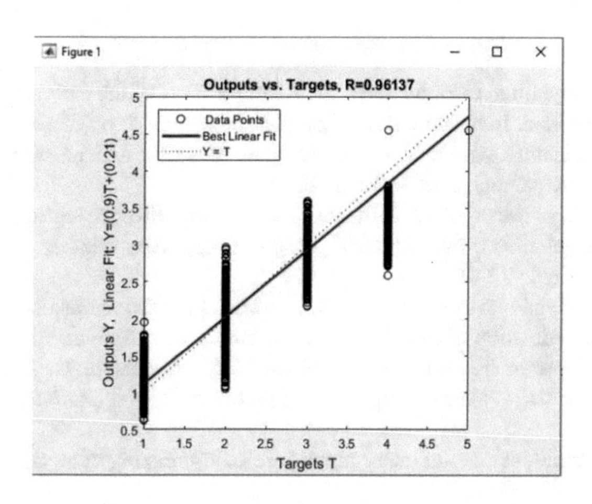

Fig. 7. The result of training the neural network

4 Discussion

We will estimate the influence of the furniture arrangement, which used in the simulated room, on the result of the genetic algorithm and the neural network. The application of these algorithms is assumed on ships where large furniture is not usually rearranged. Furniture, the location of which can be changed, is too small to influence the final result.

5 Conclusion

The study of a fire occurrence and development in the ship's cabins is based on supercomputer modeling without conducting costly field experiments.

The suggested enhanced genetic algorithm fulfills optimization of the sensors position in the cabin.

The algorithm allows to reduce the mean time of fire detection by of 15% in average. This effect can be achieved by increasing of the number of sensors in the cabin from 3 up to 11.

It is reasonability to apply TDPNN for the fire emergency detection and identification of ignition type at the initial stage of the fire.

Acknowledgment. The authors are gratitude the administration of the supercomputer center of Saint Petersburg State Polytechnic University for the opportunity to perform simulations to develop a neural network algorithm for early warning of fire on a ship, using simulation of sensors measurement results.

References

1. Guseva, A.I., Malykhina, G.F., Militsyn, A.V.: Algorithms of the early warning of fire in the premises of the vessel. In the collection: Complex protection of objects of information - 2016 Collection of scientific works of the all-Russian scientific and practical conference with international participation, p. 39–4 (Rus) (2016)
2. Malykhina, G.F., Guseva, A.I., Militsyn, A.V.: Early fire prevention in the plant. In: International Conference on Industrial Engineering, Applications and Manufacturing (ICIEAM), 19 May 2017 (2017)
3. Echegoyen, C., Mendiburu, A., Santana, R., Lozano, J.A.: On the taxonomy of optimization problems under estimation of distribution algorithms. Evol. Comput. **21**(3), 471–495 (2012)
4. Pernin, C.G., Comanor, K., Menthe, M., Moore, L.R., Anderson, T.: Allocation of Forces, Fires, and Effects Using Genetic Algorithms (Technical Report (RAND)) (2008). ISBN-13: 978-0833044792
5. Akbari, Z.: A multilevel evolutionary algorithm for optimizing numerical functions. J. Ind. Eng. Comput. **2**, 419–430 (2011)
6. Goldberg, D.E.: Genetic Algorithms in Search, Optimization, and Machine Learning, 1st edn., 403 pp. The University of Alabama, Addison-Wesley (1989). ISBN-13: 978-0201157673
7. Simon, H.: Neural Networks: A Comprehensive Foundation, 2nd edn. Prentice Hall, Upper Saddle River (1999). 842 p

Artificial Neural Networks for Diagnostics of Water-Ethanol Solutions by Raman Spectra

Igor Isaev[1,2(✉)], Sergey Burikov[1,2], Tatiana Dolenko[1,2], Kirill Laptinskiy[1,2], and Sergey Dolenko[1]

[1] D.V. Skobeltsyn Institute of Nuclear Physics, M.V. Lomonosov Moscow State University, Moscow, Russia
isaev_igor@mail.ru, dolenko@sinp.msu.ru
[2] Physical Department, M.V. Lomonosov Moscow State University, Moscow, Russia

Abstract. The present paper is devoted to an elaboration of a method of diagnosis of alcoholic beverages using artificial neural networks: the inverse problem of spectroscopy – determination of concentrations of ethanol, methanol, fusel oil, ethyl acetate in water-ethanol solutions – was solved using Raman spectra. We obtained the following accuracies of concentration determination: 0.25% vol. for ethanol, 0.19% vol. for fusel oil, 0.35% vol. for methanol, and 0.29% vol. for ethyl acetate. The obtained results demonstrate the prospects of using Raman spectroscopy in combination with modern data processing methods (artificial neural networks) for the elaboration of an express non-contact method of detection of harmful and dangerous impurities in alcoholic beverages, as well as for the detection of counterfeit and low-quality beverages.

Keywords: Neural networks · Inverse problems · Raman spectroscopy Water-ethanol solutions

1 Introduction

At present, when there is a significant amount of danger to health low-quality products on the market, it is necessary to control the quality of alcoholic beverages. Consumption of counterfeit alcoholic drinks is dangerous for life because even a small amount of toxic impurities (methyl alcohol, fusel oils, etc.) can cause intoxication of the human organism – allergenic, immunomodulatory, genotoxic actions. One of the most famous strong alcoholic beverages is vodka produced by fermentation and distillation of grain, potatoes, sugar beet, grape, etc. [1]. Vodka is a water solution of ethanol with a concentration of 37.5–56% vol. (Russia GOST standard 12712-2013); therefore, the development of express and non-contact methods of diagnostics of water-ethanol solutions is an extremely urgent task.

Study performed at the expense of Russian Science Foundation, project no.14-11-00579.

B. Kryzhanovsky et al. (Eds.): NEUROINFORMATICS 2018, SCI 799, pp. 167–175, 2019.
https://doi.org/10.1007/978-3-030-01328-8_18

Control of quality of strong alcoholic beverages assumes the solution of two separate problems: (1) determination of content (concentration) of ethyl alcohol; (2) determination of potentially dangerous impurities and their concentrations. There are many methods of determination of the content of ethyl alcohol in water-ethanol solutions: by measuring the density of the sample [2], refractive index [3], boiling point of the solutions [4], chromatographic method [5], method of nuclear magnetic resonance (NMR) [6, 7] etc. Non-contact means of vibrational spectroscopy – IR absorption spectroscopy [8] and Raman spectroscopy [9] - are also widely used. Most of them are also used to solve the second problem – determination of the concentrations of impurities dangerous to a human in alcoholic beverages. First of all, it applies to chromatography, NMR, methods of vibrational spectroscopy.

It should be noted that the methods based on the measurement of the density of the solution or its refractive index, in the presence of a large number of impurities in the solution are not precise enough. Chromatographic analysis, chemical methods, and NMR method provide high accuracy of determination of substances in solutions, but they are time-consuming and expensive. Besides, these methods are contacting, i.e., they require opening the container and extracting a certain amount of sample from it.

For practical implementation, the method of Raman spectroscopy favorably differs from the other mentioned methods. It is non-contact, express, we can apply it in real time mode, it does not require complicated sample preparation and expensive reagents, so it is widely used for the diagnostics of various drinks. Thus, Raman spectroscopy is used for determination of the content of glucose, sucrose and fructose in food drinks [10], for studying the structure of water-ethanol solutions [11, 12], for determination of the content of ethanol and methanol in alcoholic beverages [13], for identification of various types of whiskey [14].

However, in the presence of a large number of various impurities in drinks, the solution of the inverse problem (IP) of Raman spectroscopy – determination of the type and concentrations of contaminants in multicomponent water-ethanol solutions – is significantly complicated. Therefore, instead of conventional methods of solution of such multi-parameter ill-posed IPs, adaptive methods of data analysis are used. They provide acceptable solutions [15, 16]; e.g., artificial neural networks (ANN) are used to solve such IPs [17]. ANN are actively used in solution of a wide range of problems associated with pattern recognition, forecasting, classification, etc. In particular, ANNs are rather widely used in spectroscopy. So, ANNs were used for determination of the type and concentration of salts dissolved in water by Raman spectra [15, 16, 18], for express determination of components of wine by the absorption spectra [19], for determination of blood glucose by the spectra of IR absorption [20].

In this study, the problem of diagnostics of water-ethanol solutions by Raman spectra included (1) determination of the concentration of ethyl alcohol in the solution and (2) determination of the concentration of harmful impurities in vodka. The method of ANN was used to solve these problems.

2 Problem Statement and Data Preparation

The impurities most commonly occurring in counterfeit and low-quality alcohol – methanol (the main cause of poisoning by counterfeit alcohol), fusel oil, and ethyl acetate, – were used as harmful substances. Since the main components of the fusel oil are isoamyl and isopropyl alcohols, a mixture of 70/30 (by volume) of isoamyl and isopropyl alcohols was used as a model of fusel oil. Methanol, fusel oil, ethyl acetate were dissolved in the mixture of ethanol and water (concentrations 35, 38, 40, 42, 45, 49, 53, 57%). In this way, vodkas of various strengths were simulated. The following impurity concentrations were used: methanol – 0, 0.05, 0.14, 0.4, 1.1, 3.1, 8.6, 24%; fusel oil - 0, 0.025, 0.07, 0.22, 0.66, 2, 6, 18%; ethyl acetate – 0, 0.17, 0.35, 0.7, 1.4, 2.8, 5.6, 11.2%. For the preparation of the solutions, ethyl alcohol category "alpha," isoamyl alcohol category "pure for analysis," isopropyl alcohol category "chemically pure," ethyl acetate category "pure," methyl alcohol category HPLC (suitable for high-performance liquid chromatography) were used.

The Raman spectra were obtained using a laser spectrometer which included an argon laser (488 nm wavelength, 200 mW power) and a registration system consisting of a monochromator (grating 900 grooves/mm, 500 mm focal length) and a cooled CCD camera. Edge-filter was used to suppress the elastic scattering. Spectra were recorded in two ranges with centers at 520 nm (low-frequency region of the spectrum or the region of so-called "fingerprints") and 573 nm (the region of stretching vibrations). As a result, we obtained the full Raman spectrum in the range of 200–3800 cm^{-1}. The useful spectral resolution was 2 cm^{-1}. For each of the ranges, 10 spectra were recorded (two cycles of measurements, 5 spectra each). The acquisition time for one spectrum was 2 s. The obtained 10 spectra were averaged.

The principle possibility of determining the concentration of ethanol and harmful impurities is caused by the fact that ethyl, methyl, isoamyl, isopropyl alcohols and ethyl acetate have their specific lines in the "fingerprint" spectral region (200–1600 cm^{-1}) and in the area of valence vibrations of CH- and OH-groups (2600–3800 cm^{-1}) (Fig. 1). In the case of a single-component solution, one or several Raman bands specific to the component can be selected, and its concentration can be determined by the intensity of this band [13]. In the case of multicomponent solutions, the situation is complicated by the fact that the vibrational bands of different substances overlap.

Among the negative factors that reduce the accuracy of the solution of the problem in real conditions (in the diagnostics of real drinks) are also detector noise, fluorescent pedestal caused by impurities, and fluctuations in the intensity of the Raman signal, produced by the instability of the laser radiation power. The ANN method can successfully overcome these difficulties due to the ability of the ANN to learn, to summarize the provided information and to identify hidden patterns.

The dataset prepared to implement the ANN solution of the problem contained 4046 patterns. We divided it into training, validation and test sets in the ratio of 70:20:10, respectively. Thus, the training set contained 2800 patterns, validation set – 800 patterns, test set – 446 patterns. Training of the ANN was performed on the training set, stop at the minimum error on the validation set was used to prevent overtraining, and independent evaluation of the results was made on the test set.

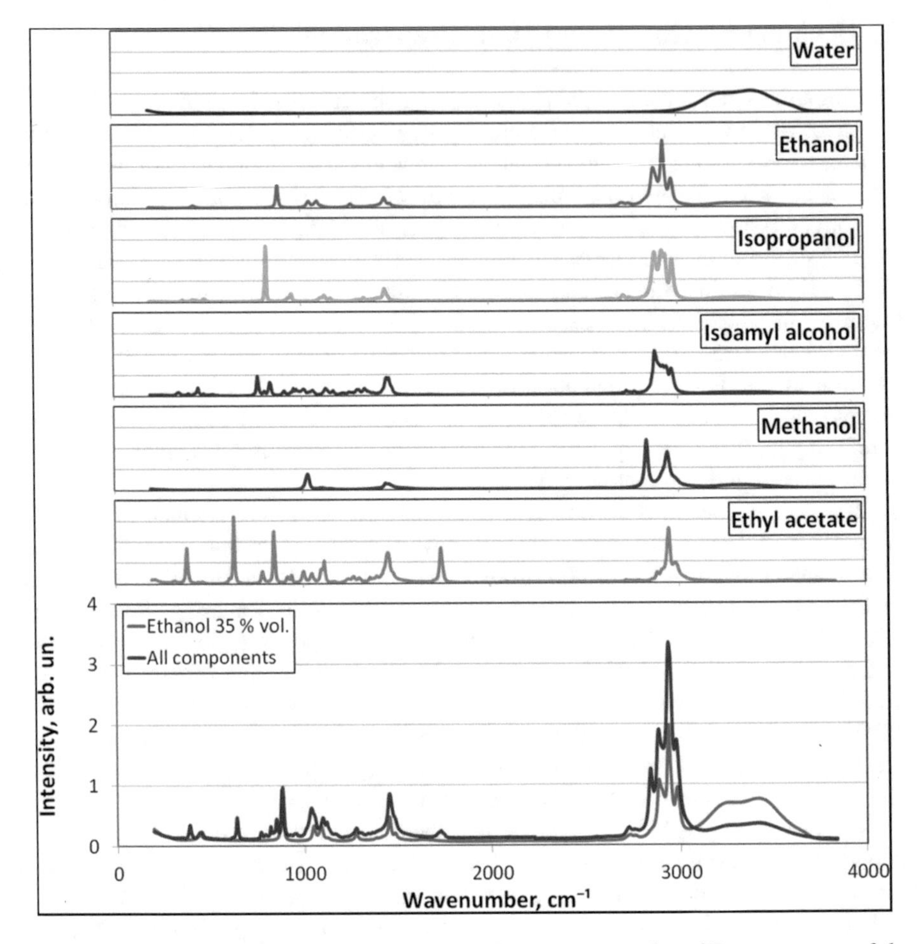

Fig. 1. Top – Raman spectra of pure substances. Bottom – examples of Raman spectra of the studied water-ethanol solutions.

Thus, the regression problem stated in this study was to determine the values of 4 parameters by 2048 input characteristics. The sought-for parameters were the concentrations of fusel oil, ethyl acetate, methanol, and ethanol.

3 Selection of the Optimal MLP Architecture

The ANN architecture used to solve this problem was the multi-layer perceptron (MLP). This study included a search for the optimal numbers of layers and neurons, as well as search for the optimal method of selection of significant input features for the MLP. For each configuration of the parameters, 5 MLPs were trained with various weights initialization; the statistical indicators of their application were averaged.

The multi-parameter IP was solved with two ways of parameter determination. For autonomous determination (ADP), a separate single-output ANN was trained for each parameter. For simultaneous determination (SDP), a single 4-output ANN was built.

At the first stage, the results for MLPs with various numbers of hidden layers (HL) and neurons in them were compared. The results are presented in Fig. 2.

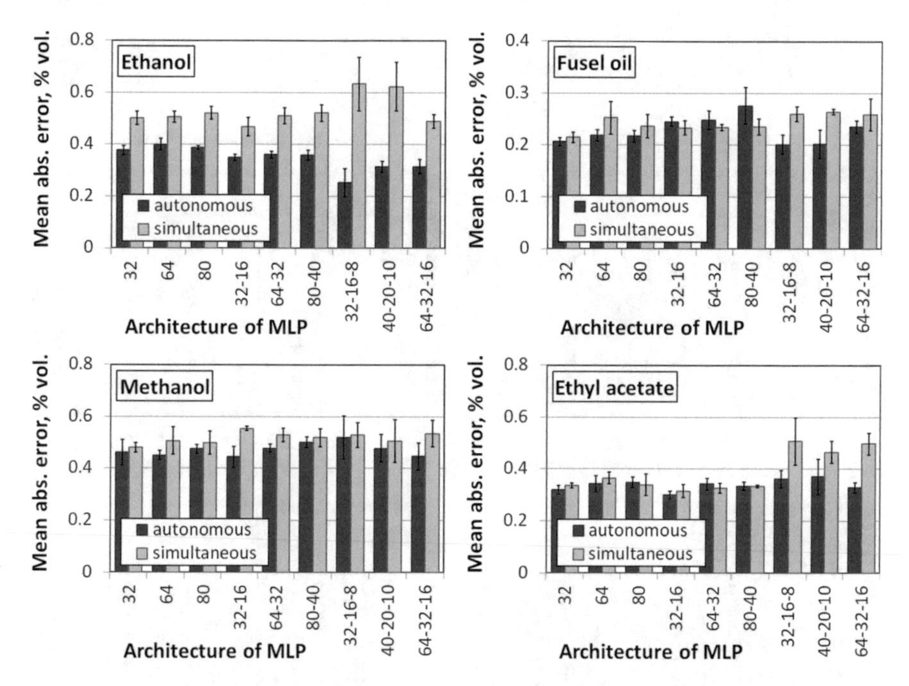

Fig. 2. Mean absolute error, % vol. for each determined component, for various MLP architectures and for different ways of parameter determination.

It can be seen that ADP gives a consistently better result for ethanol and slightly better results for other components of the solutions. All the considered architectures give comparable results except that MLP with three HL perform better for ethanol with ADP, and show some deterioration with SDP.

The obtained results may indirectly indicate that the complexity of the problem is not very high and that the data set has sufficient representativity.

4 Significant Feature Selection

At the second stage, we considered the selection of significant input features by various methods. The selected features were used to train an MLP with 32 neurons in the single HL. The training was also carried out in the modes of ADP and SDP.

Three of the considered methods were based on the calculation of the significance of interrelation between each of the input features and each of the determined

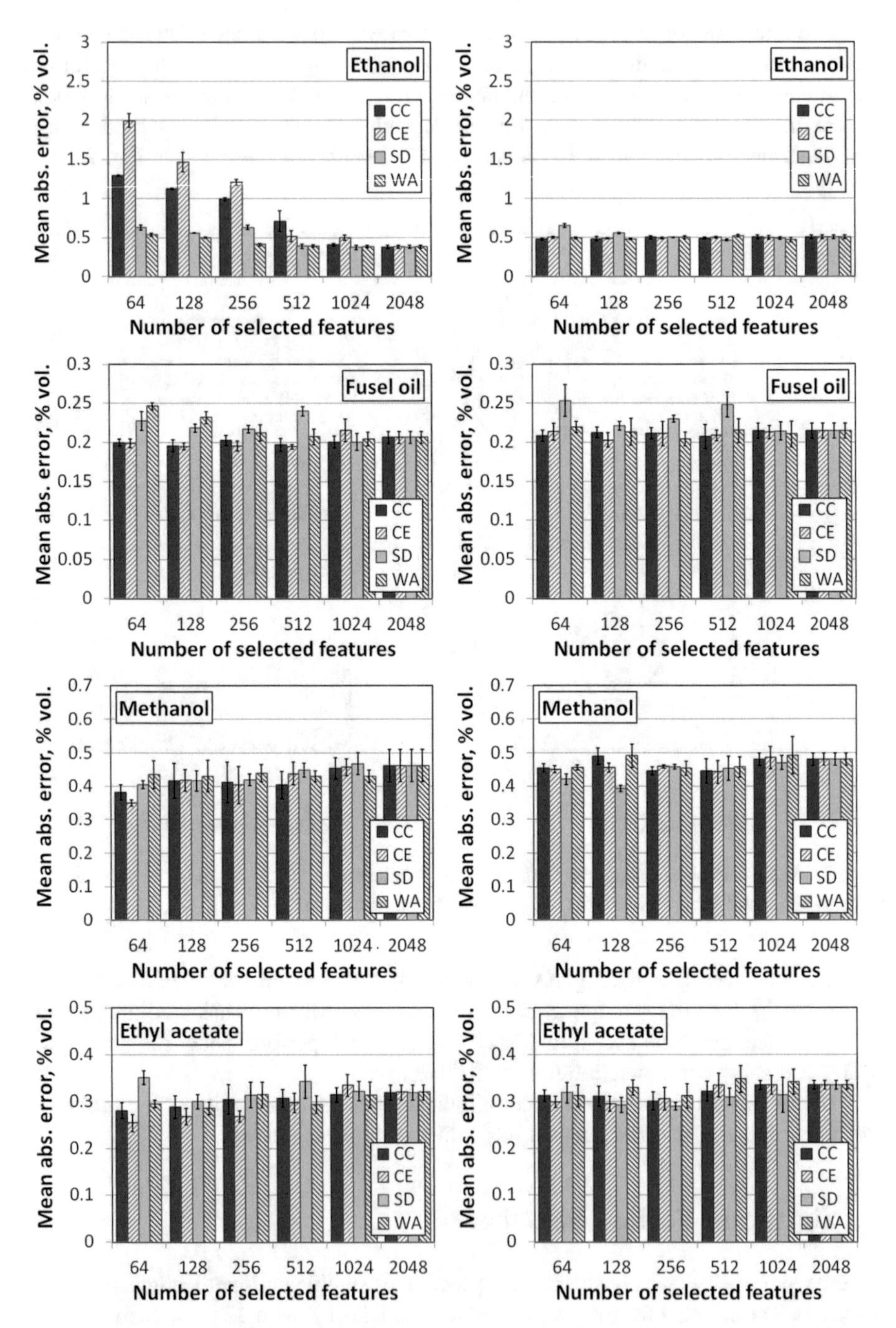

Fig. 3. Mean absolute error, % vol. for each determined component, for various methods of significant feature selection, for various numbers of features. Left – autonomous determination of parameters, right – simultaneous determination of parameters.

parameters. The values of cross-correlation (CC), cross-entropy (CE), and the importance of inputs obtained through the weights analysis of a neural network (WA) were used as the significance value. Those features, the significance value of which exceeded a certain threshold, were selected to be used. The threshold values were set in such a way that a certain pre-defined number of features were selected. For SDP, the network input was fed with all the features selected for at least one of the determined parameters. So, in this case, the number of features was, in fact, higher than for ADP.

The selection of features by the standard deviation (SD) value is based on the assumption that the amount of information carried by the feature is proportional to the entropy, which is, in turn, proportional to SD. Here also the threshold was set in such a way that a certain number of features were selected. As this method uses no information about the outputs, the numbers of features for ADP and SDP were the same.

The results of solving the problem using feature selection are shown in Fig. 3. For fusel oil, methanol and ethyl acetate, feature selection in some cases can slightly reduce the error; the best results are obtained with CE. For ethanol, there is a sharp drop in the quality of the solution when using feature selection through CC and CE in the case of ADP. This fact is since all characteristic lines of the ethanol spectrum overlap with the lines of other components. As a result, without information about the content of other components, it is impossible to accurately determine the concentration of ethanol with a reduced number of features. If such information is received (in the case of SDP), no fall in the quality of the solution is observed.

5 Conclusion

According to the results of the study, the following conclusions can be drawn

- Raman spectroscopy with MLP processing of spectra can be successfully used to detect hazardous impurities in aqueous ethanol solutions.
- Autonomous determination gives a consistently better result than simultaneous for ethanol and slightly better results for fusel oil, methanol, and ethyl acetate.
- All the considered architectures give comparable results except that MLP with three hidden layers perform better for ethanol with autonomous determination, and show some deterioration with simultaneous determination of the parameters.
- Significant feature selection provides a small improvement in the quality of the solution; the best results we obtained with cross-entropy.

The best-obtained accuracy of the determination of component concentrations was: 0.25% vol. for ethanol (enough for practical purposes); 0.19% vol. for fusel oil (less than the maximum permissible concentration (MPC)); 0.35% vol. for methanol and 0.29% vol. for ethyl acetate (several times greater than MPC but more than an order of magnitude less than the lethal concentration for 100 ml of drink).

References

1. Marianski, S., Marianski, A.: Home Production of Vodkas. Infusions & Liqueurs. Bookmagic, LLC (2012)
2. Fuerst, A.: Determination of alcohol by pycnometer. Ind. Eng. Chem. Anal. Ed. **2**(1), 30–31 (1930)
3. Owuama, C., Ododo, J.: Refractometric determination of ethanol concentration. Food Chem. **48**(4), 415–417 (1993)
4. Iland, P., Ewart, A., et al.: Techniques for Chemical Analysis and Quality Monitoring during Winemaking. Patrick Iland Wine Promotions, Campbelltown (2000)
5. Leary, J.: A quantitative gas chromatographic ethanol determination. J. Chem. Educ. **60**(8), 675 (1983)
6. Isaac-Lam, M.: Determination of Alcohol Content in Alcoholic Beverages Using 45 MHz Benchtop NMR Spectrometer. Intern. J. of Spectroscopy 2016, Art.ID 2526946, 8 pp. (2016)
7. Zuriarrain, A., Zuriarrain, J., Villar, M., Berregi, I.: Quantitative determination of ethanol in cider by 1H NMR spectrometry. Food Control **50**, 758–762 (2015)
8. Lachenmeier, D., et al.: Rapid and mobile determination of alcoholic strength in wine, beer and spirits using a flow-through infrared sensor. Chem. Cent. J. **4**, 5 (2010)
9. Burikov, S., Dolenko, T., et al.: Diagnostics of aqueous ethanol solutions using Raman spectroscopy. Atmos. Ocean. Opt. **22**(11), 1082–1088 (2009)
10. Ilaslan, K., Boyaci, I., Topcu, A.: Rapid analysis of glucose, fructose and sucrose contents of commercial soft drinks using Raman spectroscopy. Food Control **48**, 56–61 (2015)
11. Hu, N., Wu, D., et al.: Structurability: a collective measure of the structural differences in vodkas. J. Agric. Food Chem. **58**(12), 7394–7401 (2010)
12. Burikov, S., Dolenko, S., et al.: Decomposition of water Raman stretching band with a combination of optimization methods. Mol. Phys. **108**(6), 739–747 (2010)
13. Boyaci, I., Genis, H., et al.: A novel method for quantification of ethanol and methanol in distilled alcoholic beverages using Raman spectroscopy. J. Raman Spectrosc. **43**(8), 1171–1176 (2012)
14. Kiefer, J., Cromwell, A.: Analysis of single malt Scotch whisky using Raman spectroscopy. Anal. Methods **9**, 511–518 (2017)
15. Burikov, S., Dolenko, S., et al.: Neural network solution of the inverse problem of identification and determination of partial concentrations of inorganic salts in multicomponent aqueous solution. In: Proceedings of XII Russian Scientific-Technical Conference "Neuroinformatics-2010", part 2, pp. 100–110. MEPhI, Moscow (2010)
16. Dolenko, S., Burikov, S., et al.: Adaptive methods for solving inverse problems in laser raman spectroscopy of multi-component solutions. Pattern Recognit. Image Anal. **22**(4), 551–558 (2012)
17. Hassoun, M.: Fundamentals of Artificial Neural Networks, 1st edn. MIT Press, Cambridge (1995)
18. Dolenko, S., Burikov, S., et al.: Neural network approaches to solution of the inverse problem of identification and determination of partial concentrations of salts in multi-component water solutions. In: Wermter, S., Weber, C., et al. (eds.) ICANN 2014, LNCS, vol. 8681, pp. 805–812. Springer, Heidelberg (2014)
19. Martelo-Vidal, M., Vázquez, M.: Application of artificial neural networks coupled to UV–VIS–NIR spectroscopy for the rapid quantification of wine compounds in aqueous mixtures. CyTA J. Food **13**(1), 32–39 (2015)

20. Liu, W., Yang, W., Liu, L., Yu, Q.: Use of artificial neural networks in near-infrared spectroscopy calibrations for predicting glucose concentration in Urine. In: Huang, D.-S., Wunsch, D.C., Levine, D.S., Jo, K.-H. (eds.) ICIC 2008. LNCS, vol. 5226, pp. 1040–1046. Springer, Heidelberg (2008)
21. International Programme on Chemical Safety (IPCS). Environmental Health Criteria 196: Methanol, pp. 1–9 (1997)
22. Hazardous substances in industry. Handbook for chemists, engineers, and physicians, 7th edn. Volume 1. Organic substances. Lazarev, N., Levina, E. (eds.) Khimija Publ., Leningrad (1977). (in Russian)
23. Gosselin, R., Smith, R., Hodge, H.: Clinical Toxicology of Commercial Products, 5th edn. Williams & Wilkins, Baltimore (1984)
24. Golikov, S. (ed.): Emergency treatment of acute poisoning. Meditsina Publishers, Moscow, 312 pp. (1978). (In Russian)

Neural Network Recognition of the Type of Parameterization Scheme for Magnetotelluric Data

Igor Isaev[1,2](✉), Eugeny Obornev[2], Ivan Obornev[1,2], Mikhail Shimelevich[2], and Sergey Dolenko[1]

[1] D.V. Skobeltsyn Institute of Nuclear Physics, M.V. Lomonosov Moscow State University, Moscow, Russia
isaev_igor@mail.ru, dolenko@sinp.msu.ru
[2] S. Ordjonikidze Russian State Geological Prospecting University, Moscow, Russia

Abstract. The inverse problem of magnetotelluric sounding is a highly non-linear ill-posed inverse problem with high dimension both at the input and at the output. One way to reduce the incorrectness is to narrow the scope of the problem. In our case, this can be implemented in the form of a complex algorithm, which first makes the choice of one of the narrower classes of geological sections and then performs the solution of the regression inverse problem within the selected class. In the present study, we investigate the effectiveness of the implementation of the first phase of this algorithm. The neural network solution of the problem of classification of magnetotelluric sounding data was considered. We estimate the maximum accuracy of classification, perform search for optimal parameters, and test the results for resilience to noise in the data.

Keywords: Magnetotelluric sounding · Inverse problems
Artificial neural networks · Multiclass classification

1 Introduction

The inverse problem of magnetotelluric sounding (MTS IP) is to reconstruct the distribution of electrical conductivity in the thickness of the earth by the values of the components of electromagnetic fields measured on its surface [1]. MTS IP is a non-linear, ill-posed or ill-conditioned problem with high dimension both at the input and at the output. In the two-dimensional and three-dimensional cases, it has no analytical solution and is solved numerically. The methods usually used are optimization methods based on multiple solutions of the direct problem with minimization of the residual in the space of the observed values [2], or matrix methods using Tikhonov regularization [3]. These numerical methods have a number of disadvantages. Optimization methods are characterized by high computational cost and the need to set a good first

Study performed at the expense of Russian Science Foundation, project no. 14-11-00579.

B. Kryzhanovsky et al. (Eds.): NEUROINFORMATICS 2018, SCI 799, pp. 176–183, 2019.
https://doi.org/10.1007/978-3-030-01328-8_19

approximation. Besides that, a small residual in the space of the observed values does not guarantee a small residual in the space of determined parameters [4]. For regularization-based methods, the main difficulty is the correct choice of the regularization parameter. Due to the incorrectness of the problem, the accuracy of the solution provided by these methods may be quite low, so when solving it, the role of an expert is very large. The expert may take into account *a priori* knowledge about the object and data obtained by other methods of prospecting. Therefore, as an alternative, in this study we consider neural network (NN) methods, free from many disadvantages inherent in traditional methods of IP solving.

Despite the fact that the use of neural networks was proposed for the interpretation of geophysical data and has been actively developing since the early 1990s [5–8], the application of NN to the solution of MTS IP is complicated. The main problem is the formation of the training set. For this purpose, first of all it is necessary to set the initial distribution of electrical conductivity by a finite number of parameters, i.e. to introduce the so-called parameterization scheme. At the same time, the parameterization scheme should be geologically justified and ensure the correctness of the solution, which is a separate scientific problem. Further, within the framework of the chosen parameterization, it is necessary to generate patterns, and the direct problem should be solved for each pattern. This stage is associated with a considerable computational cost.

In this regard, the studies devoted to the NN solution of MTS IP, consider quite simple cases. Thus, in the study [9], a simple single-dimensional model was considered, which was set by means of 16 parameters – conductivities at a given depth. In [10–12], the so-called "class-generating" model is considered, when parameterization is set for a particular case on the data obtained by other exploration methods. At the same time, parameterization of model data had a simple rigidly defined geometry (constant layer thickness) and a small number (not exceeding 10) of determined macro-parameters (layers width and thickness, their conductivity, etc.). When working with real data, the authors also built parametrization with the initially given geometry of objects, based on previous studies of the area. The disadvantages of this approach coincide with the disadvantages of traditional methods: the problem is solved individually for each case at the cost of a large number of computing resources, and also requires an initial assumption about the structure of the studied area, which implies a high role of the expert and availability of the data obtained by other methods of exploration.

The authors of the article develop an approach based on the use of parameterization schemes with greater versatility and detail in comparison with its predecessors. This approach allows one to get a solution at once, and the computational cost and expertise are shifted from the stage of application of the system to the stage of its design. However, to describe the various geometry of objects, hundreds of parameters will be required, which also leads to an increase in the number of patterns required to train the neural networks, and, accordingly, to an increase in computational costs. In addition, the incorrectness of the problem is also increased.

The incorrectness of the problem in the case of a NN solution leads to the need to use special approaches. One of the ways to reduce ill-posedness is to narrow the scope of the problem. This can be implemented as a complex algorithm, which first makes the choice of one of the narrower classes of geological sections and then performs the solution of the regression IP within the selected class. It is worth noting that in this case the problem of classification is solved not in the parameter space where the classes were defined, but in its complex nonlinear mapping to the space of electromagnetic fields.

This study is a continuation of the study of [13]. In the present study, we investigate the effectiveness of the implementation of the first phase of this algorithm. The NN solution of the problem of classification of MTS data is considered. We estimate the maximum accuracy of classification, perform search for optimal parameters, and test the results for resilience to noise in the data.

2 Problem Statement

In this paper, the two-dimensional MTS IP was considered. The dimension of the input data was determined as follows: the measurement of the module and the phase of two polarizations of the electromagnetic field at 13 frequencies, at 126 points along the profile on the earth's surface (pickets), gave 6,552 features.

The considered parameterization schemes included those containing a different number of alternating conductive (C) and insulating (I) layers (G^C, G^{CI}, G^{CIC}, G^I, G^{IC}, G^{ICI}) in the upper part of the section, and a random distribution of the conductivity in the lower part [13] (Fig. 1, center), as well as a general parameterization scheme (G_0) with a random distribution of conductivity over the entire region (Fig. 1, left). With this method of assignment, classes with fewer layers may include classes with a greater number of layers.

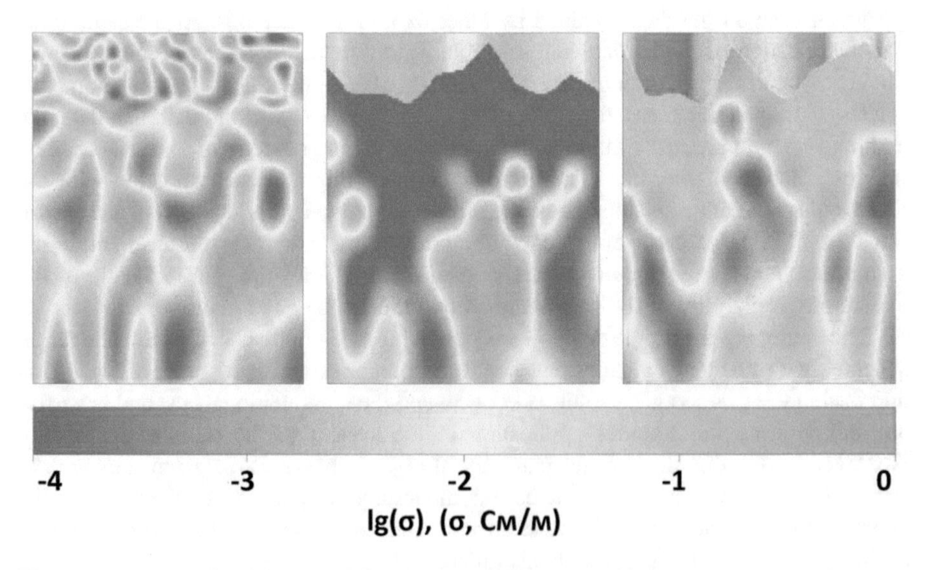

Fig. 1. Examples of geological sections corresponding to the determined classes. Left – class G^0; center – class G^{CI}; right – class G^I, for which recognition refusal is registered: the classifier finds it similar to both classes G^{IC} and G^{ICI}.

However, due to the large number of parameters by which parameterization schemes are set (233 or 336), and to the limited number of patterns, the classes intersect weakly, what will be shown below. The patterns, which cause incorrect operation of the classifier, belong to the boundaries of the concurrent classes, when it is difficult to distinguish which class the pattern belongs to (Fig. 1, right).

The study was carried out with model data obtained by numerical solution of the direct problem. The training set used to train NN consisted of 35,000 patterns (5,000 in each class); the validation set used to stop training by minimum error on it (to prevent overtraining) consisted of 9,800 patterns (1,400 in each class). The test set used to compare the results contained 35,000 patterns (5,000 in each class).

3 Solving the Problem and Results

3.1 Use of Neural Networks

Each NN used the entire set of input features (6,552 features), and had 7 outputs with the desired output values 1 (belongs to this class) and 0 (does not belong to this class). When the NN was applied, belonging to some class was identified if the value at one output exceeded the threshold, while the values at all the other outputs were below the threshold. If the threshold was exceeded for more than one class, or for none of the classes, this was considered a refusal.

The threshold value was chosen in such a way as to provide the greatest number of correct answers, so the value of 0.5 was selected. It should be noted that the quality of the classification depends on the threshold in the range from 0.1 to 0.9 rather weakly (Fig. 2, left). This may indirectly indicate that the classifier has learned well and provides answers with enough contrast for the majority of patterns. It is worth paying attention to the fact that at the value of 0.5 there is also the maximum number of errors, but the minimum number of refusals. The diagram is given for a multi-layer perceptron (MLP) with 8 neurons in the single hidden layer.

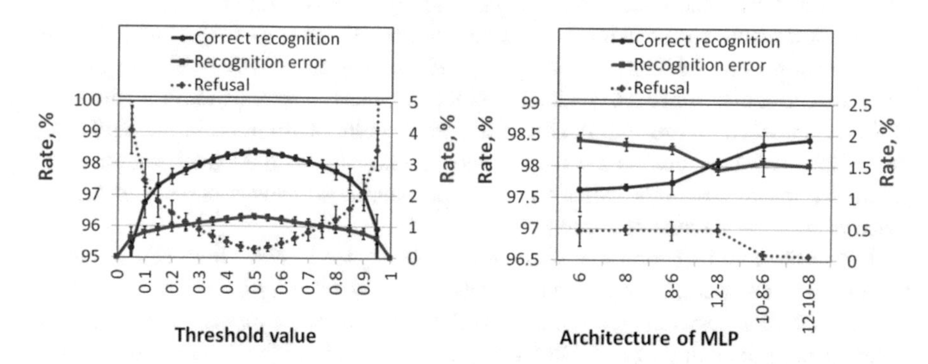

Fig. 2. Left – the results of classification in the field space depending on the threshold value. Right – the results of the classification in the parameter space for various MLP architectures. The left axis refers to the correct recognition rate; the right axis refers to the rates of errors and refusals. The displayed values are average results and standard deviation (error bars) for 5 NN.

For other architectures, the results are similar. The results are also similar for classification in the parameter space (Fig. 2, right).

In all cases, 5 identical NN with various weight initializations were trained, and the rate values were averaged.

3.2 Results of Identification of the Parameterization Scheme by MTS IP Data in the Space of Parameters

At the first stage, the highest possible quality of the classification was estimated. For this purpose, the problem was solved not in the space of observed values (the values of EM fields), but in the space of parameters (the values of electrical conductivity). Since the classes themselves were specified in the parameter space, it was assumed that because of the complex nonlinear mapping of the parameter space into the field space, the quality of identification in the field space should not exceed its quality in the parameter space.

MLPs with one, two and three hidden layers (HL) and different numbers of neurons in them were considered. There was an increase in the correct recognition rate with the complication of the neural network architecture (Fig. 2, right).

The best result was shown by the MLPs having three HL with 12 + 10 + 8 neurons: correct recognition rate −98.43%, refusal rate −0.06%, error rate −1.51%. However, the error rate does not tend to go to zero. This means that within the complex approach, we should also expect some erroneous classifications.

3.3 Results of Solving the Classification Problem on Determination of the Optimal Parameterization Scheme in the Space of Observed Values

At the next stage, the problem of classification was solved in the space of observed values (EM fields). MLPs with one or two HL and different number of neurons in them were considered. Except for the obviously weak architecture (4 neurons in the single HL), all the others show close results which are quite high (Fig. 3, left).

The best result was shown by MLPs having two HL with 16 + 12 neurons: correct determination rate −98.52%, refusal rate −0.29%, error rate −1.19%. This result turns out to be better than for the classification in the parameter space. In addition, the nature of the dependence on the complexity of the architecture is different: in the field space, its complexity does not lead to improved results, unlike the classification in the parameter space. This may indicate that the problem is somewhat simpler in the field space than in the parameter space.

Given the fact that the dimension of the field space is more than 6,500, the following question was raised: how many patterns in the training set is enough to solve the classification problem with a sufficiently small number of errors and refusals? Training sets containing from 50 to 5,000 patterns of each of the 7 classes were tested; MLP with 8 neurons in the single HL was used. The results are shown in Fig. 3, right.

It can be seen that satisfactory results are observed on sets containing 1,000 patterns from each class; further increase in the number of patterns is not necessary due to the saturating nature of the dependence.

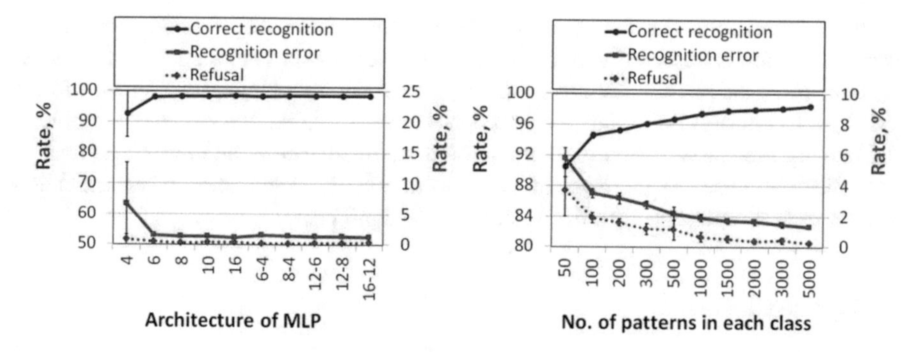

Fig. 3. Left – the results of the classification in the field space for various MLP architectures. Right – the results of the classification in the field space depending on the number of patterns per class in the training set. The left axis refers to the correct recognition rate; the right axis refers to the rates of errors and refusals. The displayed values are average results and standard deviation (error bars) for 5 NN.

3.4 Results of Testing the Solution of Classification Problem for Noise Resilience

Next, we investigated the dependence of the classification results on the level of noise in the test set. The noise of different types and statistics was considered: additive Gaussian (agn), additive uniform (aun), multiplicative Gaussian (mgn), multiplicative uniform (mun) (described in more detail in [14, 15]). Noise levels used were 1, 3, 5, 10, 20%. Each test set contained 1,000 patterns per class in 5 noisy implementations, a total of 35000 examples. The results are shown in Fig. 4.

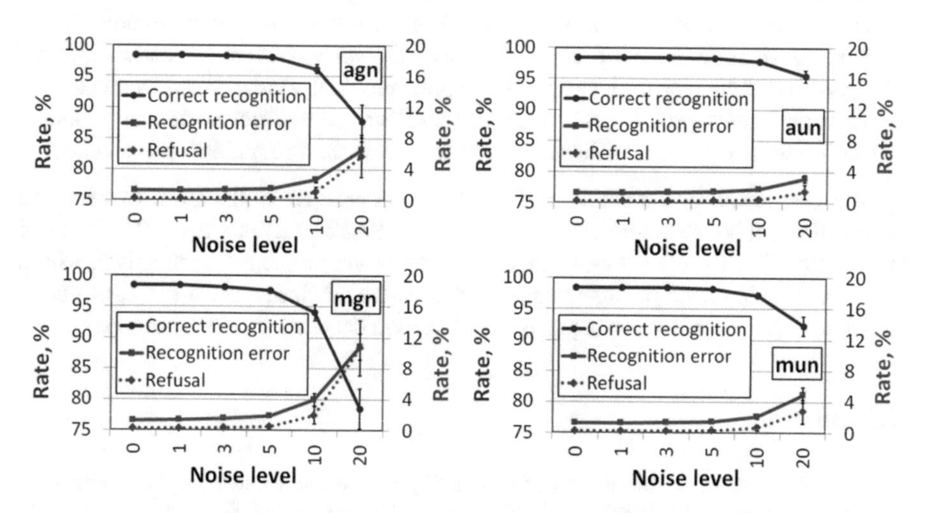

Fig. 4. Results of classification in the field space (7 classes) depending on the level and type of noise in the test set (the types of noise are explained in the text). The left axis refers to the correct recognition rate; the right axis refers to the rates of errors and refusals. The displayed values are average results and standard deviation (error bars) for 5 NN.

It can be seen that the quality of classification decreases with increasing noise level. The worst result is observed with multiplicative Gaussian noise at the level of 20%. At the same time, the NN solution, like that in the regression problem [7, 8], shows greater resilience to uniform noise than to Gaussian. However, unlike the regression problem, NN solving the classification problem show better resilience to additive noise than to multiplicative noise. At the level of noise up to 5%, the degradation of classification quality, regardless of the type of the noise, is insignificant.

4 Conclusion

According to the results of the study, the following conclusions can be drawn:

- When solving the problem of determination of the type of geological section, a certain amount of erroneous classifications are inevitable. This is due to the fact that some patterns can be described with a sufficiently small error within the framework of two adjacent parameterization schemes. However, the proportion of such patterns is low; the best possible quality of the solution that can be obtained from this data set was estimated as 98.52%.
- The problem of classification is somewhat simpler in the field space than in the parameter space: the correct recognition rate is higher, and the complexity of the NN architecture necessary for its solution is lower. At that, the problem in the field space is characterized by a higher input dimension (6552 vs. 233).
- The classification problem is successfully solved by various MLP architectures, including those having a very small number of neurons, and 1000 patterns in each class were enough to train the neural network. In these conditions, the quality of the solution was close to the maximum possible.
- The neural network solution of the classification problem is resilient to noise in the data. At a noise level of up to 5%, the degradation of classification quality, regardless of the type of the noise, is insignificant. At the same time, NN solution of the classification problem shows greater resilience to uniform noise than to Gaussian noise, and greater resilience to additive noise than to multiplicative noise.

The best result for the solution of classification problem in the space of observed values (EM fields) was: correct recognition rate −98.52%, refusal rate −0.29%, error rate −1.19%. It should be stressed that if the regression problem is then solved within the wrongly determined parameterization scheme, the obtained error is not large due to partial equivalence of the adjacent parameterization schemes for such patterns.

References

1. Spichak, V.V. (ed.): Electromagnetic sounding of the earth's interior. In: Methods in Geochemistry and Geophysics, vol. 40. Elsevier, Amsterdam (2006)
2. Zhdanov, M.: Inverse Theory and Applications in Geophysics, 2nd edn. Elsevier, Amsterdam (2015)

3. Zhdanov, M.S.: Geophysical electromagnetic theory and methods. In: Methods in Geochemistry and Geophysics, vol. 43. Elsevier, Amsterdam (2009)
4. Isaev, I., Dolenko, S.: Comparative analysis of residual minimization and artificial neural networks as methods of solving inverse problems: test on model data. In: Samsonovich, A., Klimov, V., Rybina, G. (eds.) Biologically Inspired Cognitive Architectures (BICA) for Young Scientists. Advances in Intelligent Systems and Computing, vol. 449, pp. 289–295. Springer, Cham (2016)
5. Raiche, A.: A pattern recognition approach to geophysical inversion using neural nets. Geophys. J. Int. **105**(3), 629–648 (1991)
6. Van der Baan, M., Jutten, C.: Neural networks in geophysical applications. Geophysics **65**(4), 1032–1047 (2000)
7. Sandham, W., Leggett, M. (eds.): Geophysical applications of artificial neural networks and fuzzy logic. In: Modern Approaches in Geophysics, vol. 21. Springer, Heidelberg (2003)
8. Hajian, A., Styles, P.: Prior applications of neural networks in geophysics. In: Application of Soft Computing and Intelligent Methods in Geophysics, pp. 71–198. Springer, Heidelberg (2018)
9. Hidalgo-Silva, H., Gomez-Trevino, E., Swiniarski, R.: Neural network approximation of an inverse functional. In: Proceedings of IEEE World Congress on Computational Intelligence, vol. 5, pp. 3387–3392. IEEE (1994)
10. Spichak, V., Popova, I.: Artificial neural network inversion of magnetotelluric data in terms of three-dimensional earth macroparameters. Geophys. J. Int. **142**(1), 15–26 (2000)
11. Spichak, V., Fukuoka, K., Kobayashi, T., Mogi, T., Popova, I., Shima, H.: ANN reconstruction of geoelectrical parameters of the Minou fault zone by scalar CSAMT data. J. Appl. Geophys. **49**(1–2), 75–90 (2002)
12. Montahaei, M., Oskooi, B.: Magnetotelluric inversion for azimuthally anisotropic resistivities employing artificial neural networks. Acta Geophys. **62**(1), 12–43 (2014)
13. Shimelevich, M.I., Obornev, E.A.: An approximation method for solving the inverse MTS problem with the use of neural networks. Izv. Phys. Solid Earth **45**(12), 1055 (2009)
14. Isaev, I., Obornev, E., Obornev, I., Shimelevich, M., Dolenko, S.: Increase of the resistance to noise in data for neural network solution of the inverse problem of magnetotellurics with group determination of parameters. In: Villa, A., Masulli, P., Pons Rivero, A. (eds.) ICANN 2016, LNCS, vol. 9886, pp. 502–509. Springer, Cham (2016)
15. Isaev, I.V., Dolenko, S.A.: Adding noise during training as a method to increase resilience of neural network solution of inverse problems: test on the data of magnetotelluric sounding problem. In: Kryzhanovsky, B., Dunin-Barkowski, W., Redko, V. (eds.) Neuroinformatics 2017. Studies in Computational Intelligence, vol. 736, pp. 9–16. Springer, Cham (2018)

A General Purpose Algorithm for Coding/Decoding Continuous Signal to Spike Form

Mikhail Kiselev(✉)

Chuvash State University, Cheboxary, Russia
mkiselev@megaputer.ru

Abstract. An algorithm called Symmetric Integro-Differential Conversion (SIDC) serving to encode continuous real-valued signal to spike form is described. One dynamic numeric signal is converted to 4 spike trains. 2 spiking signal lines serve as a low-pass filter, while 2 other play the role of high-pass filter. The algorithm is computationally efficient and has only one parameter – the desired mean spike frequency in the output spike sequences. This approach allows accurate encoding signals with high variability of spectral properties. A reverse conversion algorithm is proposed which is used to assure that the resulting spike signal preserves information about the original signal to a sufficient extent. Artificial signals being a sum of a sinusoid and a random walk process are utilized to show that the target spike frequency parameter does not require fine tuning - good conversion quality is demonstrated if its value is approximately two orders of magnitude less than the input signal measurement frequency.

Keywords: Spiking neural network · Information encoding
Time series analysis

1 Introduction

Nowadays, many researchers consider spiking neural networks (SNN), as well as convolutional and deep learning networks, as a potential basis for the future breakthrough in IT. One of most promising SNN application is processing dynamic signals such as video streams, sensory data in robotics or signals from technological sensors. However, there are many problems to be solved on the way of their successful practical use. Some of them are rooted in a specific information representation form utilized in SNNs. While continuous dynamical signals to be processed are represented usually in the form of sequences of real numbers, SNN operates with sequences of spikes, short pulses of constant amplitude.

Numerous approaches for conversion of real-valued signals to spike form are considered [1]. They can be subdivided into two groups – rate-based and timing-based [2, 3]. Rate-based methods use mean spike frequency as a value representing signal value. For example, number of spikes emitted by a neuron inside a certain time interval is counted. It is a simple and widely used method but it can be used to code slowly

B. Kryzhanovsky et al. (Eds.): NEUROINFORMATICS 2018, SCI 799, pp. 184–189, 2019.
https://doi.org/10.1007/978-3-030-01328-8_20

changing signal only – the characteristic time of its change should be significantly greater than this interval. This interval can be shorter if spikes of many neurons are counted but in this case big neural ensembles should be dedicated to coding one signal. In timing-based approach, the exact position of a spike on time axis is used to coding. For example, delay between a spike and some global synchronization signal inside fixed time window may represent input signal value. But again, rapidly changing values cannot be encoded by this algorithm – the signal change time scale should be much greater than this time window length. The so called spatio-temporal coding [4], where presence of some stimulus or feature is coded by a sequence of spikes emitted by a certain set of neurons in a strictly fixed order and time delays between spikes, has much better dynamic properties but it can be used to encoding discrete values only.

In this paper, a novel spike coding algorithm is proposed which allows representing accurately signals changing in a very short time scale up to the order of spike duration. This algorithm called Symmetric Integro-Differential Con-version or SIDC using fixed number (4) of neurons per 1 real-valued signal demonstrates the enhanced dynamic characteristics due to the combination of advantages inherent to rate-based and timing-based methods.

This algorithm is described in the next section. The subsequent section is devoted to the inverse algorithm converting 4 spike trains back to 1 real-valued signal. It is used primarily for the measurement of part of information about the original signal remaining in its spike representation. Section 4 illustrates the accuracy of the algorithm and its limitations. At last, Sect. 5 contains discussion and conclusions.

2 Symmetric Integro-Differential Conversion

Let us denote the real-valued signal to be converted to spike form as x. The described algorithm operates in discrete time and it can be assumed that spike duration equals the algorithm time quant (traditionally for SNN research, we take this time quant as 1 ms). Hence, the input signal is represented by a sequence of real values $x(t)$, where t is discrete. We assume that $x(t) \in [0, 1]$. Length of this sequence is L.

The algorithm has only one parameter – the desired mean spike frequency F. The reasonable choice of this parameter is tens of Hertz. The algorithm converts the signal $x(t)$ to 4 spike (Boolean) signals $s_i(t)(1 \leq i \leq 4)$. By default, these spike signal lines are not active (no spikes). Spikes appear in them when certain conditions described below are satisfied.

In order to obtain the signal $s_1(t)$, the internal variable v is used. Its original value is equal to 0 and with every time quant t it is increased by the value of the input signal $x(t)$. If its value becomes greater than the threshold u, it is decreased by this threshold value u, and s_1 becomes active (emits spike). The value of u is determined from specified F by the quite evident formula

$$u = \frac{\sum_t x(t)}{LF}. \qquad (1)$$

The signal s_2 is constructed by the identical rule but using value $1-x(t)$ instead of $x(t)$. The value of u is computed for s_2 using (1) also but with sum of $1-x(t)$ in the numerator. Thus, s_1 and s_2 have an integral nature – these are rate-coded signals conveying information about the mean value of $x(t)$ in the recent past. Two signals are used because the signal s_1 alone contains almost no information about $x(t)$ when its value is close to 0.

To build the signals s_3 and s_4 the two internal variables d^- and d^+ are used. Their initial values are $d^- = \left(\left|x(0)/z\right| - 1\right)z$ and $d^+ = \left(\left|x(0)/z\right| + 1\right)z$, where z is a parameter depending on F. If $d^- < x(t) < d^+$, nothing happens in the time quant t. If $x(t) \leq d^-$, then d^- and d^+ are decreased by z and s_4 generates spike. If $x(t) \geq d^+$, then d^- and d^+ are increased by z and s_3 generates spike. Obviously, the value of z determines the mean frequency of spikes in s_3 and s_4. It is selected on a signal sample to achieve the desired frequency F. We see that the signals s_3 and s_4 describe dynamic, differential properties of the signal – speed of its change at the given moment.

The description of the algorithm clearly illustrates why it is called "integro-differential" and "symmetric" (SIDC). The first pair of the spiking signals is used to accurately reproduce level of slowly changing input signal component while the second pair reflects its fast movements. In order to assess information loss caused by conversion of signals with different spectral characteristics, we consider the reverse conversion in the next section. We will apply the reverse conversion to the spiking signals and compare the original and reconstructed signals. Since the reverse conversion is much more complicated (and slower), has much less direct practical value and due to strict space limitation of this article, we discuss only its general idea.

3 Reverse Conversion – The General Idea

The reconstruction of the signal $x(t)$ from the spike trains $s_i(t)$ is based on the following considerations:

1. We know the value of integral of $x(t)$ between two consecutive spikes in the channels s_1 and s_2. In the absence of spikes in s_3 and s_4, the best we can is to assume that x is constant in this interval. In every time quant, we select for this calculation the shorter inter-spike interval in the channels s_1 and s_2. On the first stage of the algorithm, we use only s_1 and s_2 to reconstruct x. Let us denote this reconstructed signal as $\bar{x}(t)$.
2. On the other side, knowing $x(0)$ we can evaluate x at the moments of spikes in s_3 and s_4 using the iterative formula $x^*(t_i) = x^*(t_{i-1}) + z$ in case of spike in s_3 and $x^*(t_i) = x^*(t_{i-1}) - z$ in case of spike in s_4 (here t_{i-1} is the time of the previous spike in s_3 and s_4 or 0). Of course, we do not know $x(0)$, but we can evaluate it from the requirement that mean of $\bar{x}(t_i) - x^*(t_i)$ should be minimal.
3. The final reconstruction is made using quadratic splines between points $x^*(t_i)$ and (where these points are rare) centrums of the constant value intervals of $\bar{x}(t)$.

In fact, this algorithm is much more complicated - it includes special procedures used in the situations of saturated (very intense) spike signals or high amplitude, high frequency oscillations etc. but we cannot afford to discuss them because of very limited article volume. Instead, we utilize this algorithm to evaluate part of information preserved by SIDC in various cases.

4 Assessment of Accuracy of SIDC and Illustration of Its Operation

There are two factors limiting SIDC accuracy. The first factor is not specific and is rooted in the fact that the algorithm operates in discrete time. The characteristic time of the fastest signal changes cannot be less than time quant value – instead, it should be several times greater (hundreds of Hertz in our terms). The other factor is a balance between spike sparsity and saturation determined by F value. In order to avoid saturation F should be at most tens of Hertz. On the other side, F cannot be much less than the characteristic frequency of the signal.

To illustrate the effect of F change, let us consider 3 different but similar artificial signals being a sum of a sinusoid and a random walk process. One of them is shown in Fig. 1 (the darker line). The signal is converted to 4 spiking signals (dots in the upper part), after that the reverse transformation is applied. The restored signal is shown in Fig. 1 as the lighter line. The signals in the test and their spectrograms are shown in Fig. 2. Signal 1 has high-frequency peak (near 60 Hz), Signal 2 has higher peak but at lower frequency (30 Hz), Signal 3 is smoother and has no powerful high-frequency component.

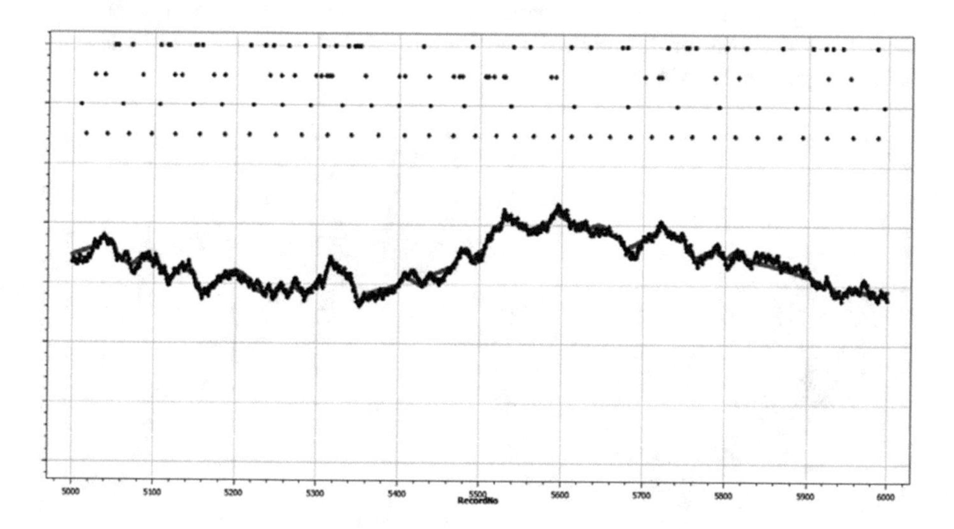

Fig. 1. Example of SIDC and reverse conversion.

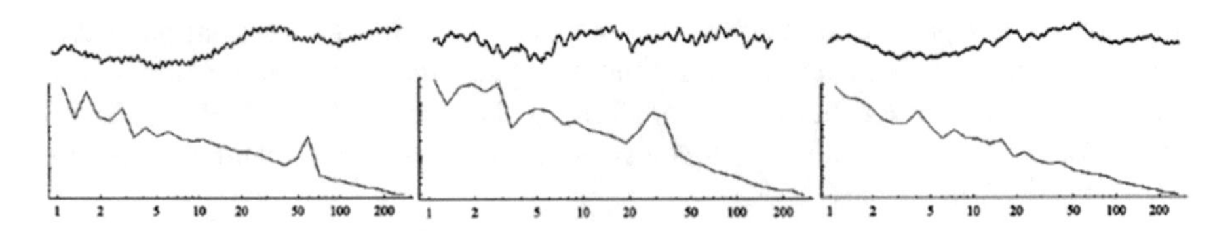

Fig. 2. The signals used for testing and their spectrograms (Hz)

The restored signal helps to evaluate the amount of information about the original signal retained in spike representation as the standard deviation of difference between the restored and original signal divided by the standard deviation of the original signal. The results of this evaluation can be seen in Table 1. They confirm our hypothesis that the optimum value of F for signals with powerful high frequency component is several tens of Hz. In this case only a few percents of information is lost in the process of SIDC.

Table 1. Accuracy of the restored signal for different F

F value, Hz	0.3	1	3	10	30	100	300
Signal 1	0.156	0.114	0.0287	0.0311	0.0164	0.0281	0.177
Signal 2	0.181	0.0983	0.0339	0.016	0.0171	0.0287	0.18
Signal 3	0.164	0.0662	0.031	0.0382	0.0187	0.0287	0.202

It is interesting to see what kind of errors is made by the algorithm in the case when F is too low or too high (Fig. 3). In case of a low F (the darkest line) the algorithm encodes only low frequency signal component, while when F is too high (the light upper line), the form of the original signal is reproduced very well, but its absolute value is significantly shifted.

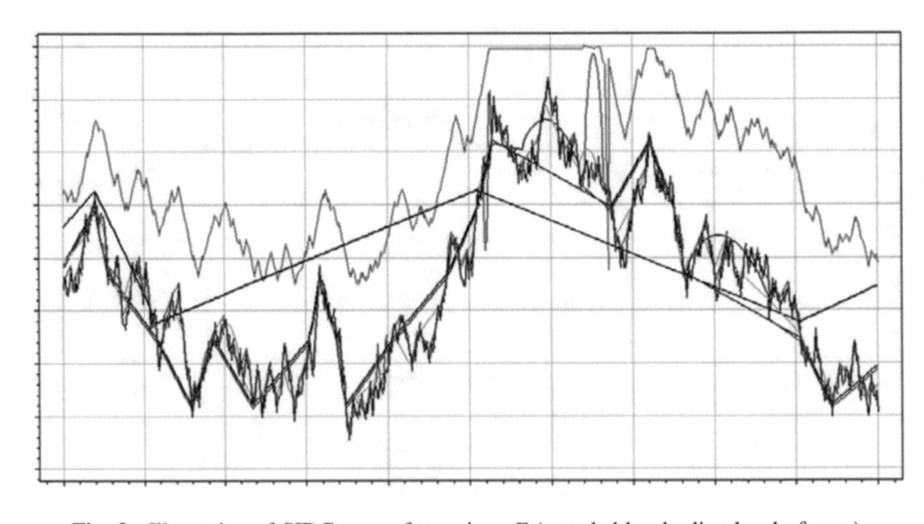

Fig. 3. Illustration of SIDC errors for various F (encoded by the line level of gray).

5 Conclusion

A novel algorithm for conversion of continuous real-valued signal to spike form is proposed in this article. It gives a unified basis for creation of interface of various devices, sources of dynamic signals in technology, science and everyday life, with SNN. Its high efficiency and accuracy are experimentally proven. It is important that the algorithm contains only one parameter, the desired mean spike frequency, and there exist reasonable and reliable recommendations for setting its value.

References

1. Gerstner, W., Kistler, W.: Spiking Neuron Models. Single Neurons, Populations, Plasticity. Cambridge University Press, Cambridge (2002)
2. Brette, R.: Philosophy of the spike: rate-based vs. spike-based theories of the brain. Front. Syst. Neurosci. **9**, 151 (2015)
3. Kiselev, M.: Rate coding vs. temporal coding – is optimum between? In: Proceedings of IJCNN-2016, Vancouver, pp. 1355–1359 (2016)
4. Izhikevich, E.: Polychronization: computation with spikes. Neural Comput. **18**, 245–282 (2006)

Single Trial EEG Classification of Tasks with Dominance of Mental and Sensory Attention with Deep Learning Approach

Irina Knyazeva[1,2,3(✉)], Alexander Efitorov[4], Yulia Boytsova[5], Sergey Danko[5], Vladimir Shiroky[4,6], and Nikolay Makarenko[1,2]

[1] Pulkovo Observatory, Saint-Petersburg, Russia
iknyazeva@gmail.com
[2] Saint-Petersburg State University, Saint-Petersburg, Russia
[3] Institute of Information and Computational Technologies, Almaty, Kazakhstan
[4] Skobeltsyn institute of nuclear physics, Lomonosov Moscow State University, Moscow, Russia
a.efitorov@sinp.msu.ru
[5] Institute of the Human Brain, Russian Academy of Sciences, St. Petersburg, Saint-Petersburg, Russia
[6] National Research Nuclear University MEPhI, Moscow, Russia

Abstract. In this paper, we present classification algorithms based on single-trial ElectroEncephaloGraphy (EEG) during the performance of tasks with the dominance of mental and sensory attention. Statistical data analysis showed numerous significant differences of EEG wavelet spectra density during this task at the group level. We decided to use wavelet power spectral density (PSD) computed in each channel for single trial as the source of feature extraction for the classification task. To obtain a low-dimensional representation of PSD image convolutional autoencoder (CNN) was trained. With this encoded representation binary classification for each subject with multilayer perceptron (MLP) were performed. The classification error varies depending on the subject with the average true classification rate is 83.4%, and the standard deviation is 6.6%. So this approach potentially could be used in the tasks where pattern classification is used, such as a clinical decision or in Brain-Computer Interface (BCI) system.

Keywords: Mental and sensory attention
EEG single trial classification · Deep learning · Neural networks

1 Introduction

Electroencephalograms (EEG) are the multidimensional time series with the recordings of brain activity measured as electrical potential. Analysis of EEG activity is crucial in clinical diagnostics for identification pathologies, for example, epilepsies seizure. Among the many approaches to EEG analysis, one can

B. Kryzhanovsky et al. (Eds.): NEUROINFORMATICS 2018, SCI 799, pp. 190–195, 2019.
https://doi.org/10.1007/978-3-030-01328-8_21

distinguish time-frequency-based analysis. Through the advantages of such methods is that many results from time-frequency- based analyses can be interpreted in terms of neurophysiological mechanisms of neural oscillations. [1]. At the last time, one can observe a growing interest in EEG-based Brain-Computer Interface (BCI). Such interfaces could be considered as an additional way to communicate between people with disabilities. To control BCI, it is necessary to identify different brain activity patterns correctly. This identification based on classification algorithm that could automatically define the type of activity by EEG data. In the 2017 year big review of actual classification algorithms for that time summarized in paper [2]. Recently the same authors released updated actual review of classification algorithm [3]. As the author noted the main difference in classification approaches is that current state-of-the-art approaches based mostly on machine learning, included neural networks and deep learning techniques. Also, authors noted that deep learning methods had not shown convincing improvement over other methods.

In this article, we present a study of a combination of machine learning algorithms for the problem of classifying attention types based on the time-frequency representation of the EEG. The training data contains time-frequency features extracted from single-trial EEG records while subject performed different tasks, in our case different attention tasks.

In psychophysiological literature term attention could be divided on externally vs. internally directed attention [4,5]. We consider sensory attention as externally directed to the incoming sensory information, mental attention we consider as internally directed to operating with the information already in the brain. These two type of attention differ significantly from the neurophysiological point of view, and statistical difference of this process was found at the group level. We decided to check whether it is possible to detect differences for individual subjects based on the time-frequency features of individual trials.

To describe the time-frequency pattern of EEG signals, we used continuous wavelet transform with Morlet kernel. The power spectral density of wavelet transform could be considered as a two-dimensional image for each channel. Usually, for the implementation of EEG pattern classifiers, various statistical features are extracted, which are then used in machine learning algorithms. We used the idea of low-dimensional representation of images using convolutional autoencoders as a features for classification.

2 Experiment Description

28 healthy volunteers (average age 29, 17 women) all right-handed participated in the study. The study described herein was approved by the Ethics Committee of Institute of the Human Brain. All subjects signed written informed consent, in accordance with the ethical standards laid down in the Declaration of Helsinki (1964), prior to their participation in the study. In this paper, we used tasks in which mental attention (MA) or sensory attention (SA) dominates. Tasks were presented in blocks, each block contains 80 trials. Experimental trial overview is presented at Fig. 1.

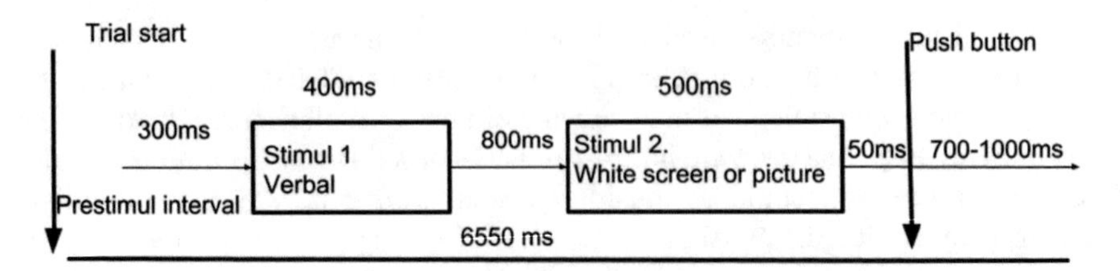

Fig. 1. Experimental trial overview.

In the block of SA task, the trials consisted of pair of stimuli: word (e.g. apple) and corresponding color image, subjects should memorize the image. Mental task consists from two different blocks: visual representation and imagination. In the block of retrieval of visual representations from memory the trials consisted of pair of stimuli: word and white screen, the words used in the previous SA task were presented. Here, the subjects were asked to recall and visualize on a white background an image, corresponding to presented word. In the block of visual imagination the trials consisted of pair of stimuli: 2 words and a white screen. Here, after simultaneous presentation of 2 words (for example: apple, machine), the subjects were asked to invent and visualize a chimera image (for example, an apple-shaped machine, seeds are pour out when the doors are opened). The duration of verbal stimulus presentation is 400 ms, the duration of second stimulus (color image or white screen) - 5 s, the interval between stimuli - 800 ms.

3 Methods

3.1 Preprocessing

19 channels of EEG were recorded using standard 10–20 electrode placement on the scalp by computer electroencephalograph "Mitsar-202" and electroencephalographic caps Electro-Cap. Monopolar montage with average reference electrode from left and right ears was used. Electrode impedances were kept below 5 kOhm. EEG recording and the subsequent removal of eye artifacts were conducted by a software package WinEEG, version 2.83 (copyright V.A. Ponomarev, J.D. Kropotov, RF2001610516, 08.05.2001). Independent Component Analysis (infomax ICA) was used in the package to correct artifacts due to vertical and horizontal eye movements and blinks. Additionally, all EEG records were inspected by operator, samples with strong movement artifacts, noise level and sudden outlets were eliminated from dataset.

Next, for each subject for each type of task, the wavelet power spectrum was calculated, according algorithms adapted for EEG analysis, described in the books [1], Khramov and our previous paper [6]. The code implemented in Python with application examples is available in the https://github.com/iknyazeva/EEGprocessing repository. The wavelet transformation was done for a grid of 20 frequencies from the 4–30 Hz range for each electrode independently, the minimum sample length in the experiment was 3480 counts or 6.96

s. Each power spectrum were normalized to the average power value during the prestimulus interval. The power spectrum for each channel was represented as an grayscale image by the dimension of the number of frequencies per time. Figure 3 shows examples of wavelet transforms for one channel and phase coherence for one pair of channels for one subject (Fig. 2).

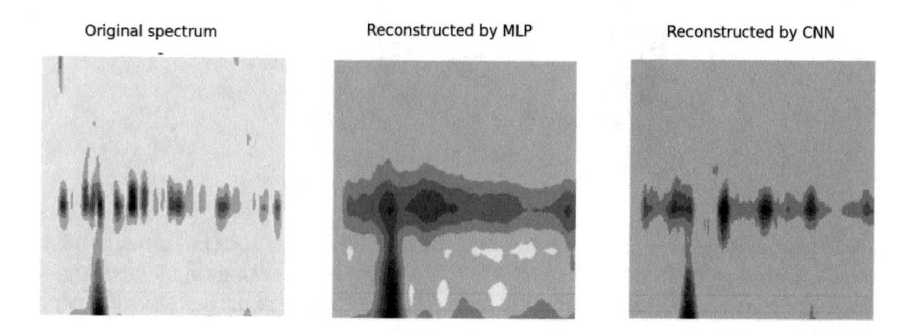

Fig. 2. Wavelet transform and phase coherence example for one subject

3.2 Computational Experiments

Computational experiments of machine learning methods application were done according to the following scheme: compression of wavelet spectra with by auto-encoders and using of low dimensional code-vectors as inputs for some model solving the classification problem. The experiment scheme and the parameters of the auto-encoder are shown in the figure. Obviously, that training neural networks with huge number of parameters (800 000 for CNN autoencoder and 380 000 for dense autoencoder) required large dataset of samples. Since the task of the encoder is the compression and reconstruction spectral patterns, we will assume that the mechanism of their compression and recovery does not depend on the channel number and it is possible to use spectra of all EEG channels independently during training. Thus, the amount of the training set was increased and consist of more than 100,000 samples. The training process was stopped by a validation set (consisted of 10% randomly selected samples with replacement): if the error, calculated on the validation set, did not decrease during 500 epochs. It should be noted, that the solution converges in less than 1000 epochs of training. The next step is solving the classification problem on compressed data. In this case spatial difference between electrodes couldn't be ignored, so the classifiers were built on own small pieces of data, corresponding each electrodes. As classification method shallow a neural network with fully connected layers was used. There were explored two approaches: training universal classifier on data of whole group of subject and training own models for each person. In second case to exclude the the effect of random splitting data to train (80%) and test (20%) sets experiments were returned 5 times: results of solving classification problem, are demonstrated at next chapter, were averaged between these 5 dataset-splits.

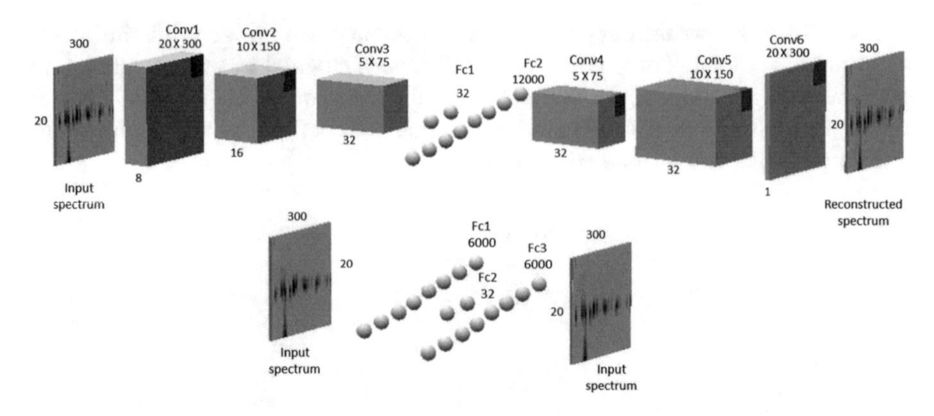

Fig. 3. Schematic representation of auto-encoders. Top: convolutional autoencoder (filter size: 3×3, maxpooling window size: 2×2, activation function: elu), bottom: dense autoencoder (activation function: elu). Loss function: mean square error, optimization algorithm: adamax.

4 Results and Discussion

During the computational experiment, the high efficiency of using convolutional networks for the coding of wavelet spectra was demonstrated. Classical neural networks with dense layers showed a much worse result. The table shows the values of the minimum error for 3 best subject and indicates 3 channels, which demonstrated the minimum error. Classification errors and corresponding channel names presented in cells. It should be noted significant differences in the results of different people: the minimum error obtained value is 1.2%, but maximum value is 31.2%. In this case, the electrodes are also different. The best results are demonstrated on wavelet spectra of the electrodes O1, O2 and T6 (Table 1).

Table 1. My caption

	Best chan1 (error)	Chan 2 (error)	Chan 3 (error)
subj_7	1.2% (O2)	1.8% (T5)	2.4% (P3)
subj_19	7.9% (Pz)	9.3% (P4, O2)	15% (Pz)
subj_16	10% (Pz)	12.5% (T6)	13.8% (F4, C3, C4, P3, P4, O2)
mean_all_subj	16.6% (Fz)	17.9% (T5)	18.6%(T6, O1, O2)

5 Conclusion

In this paper, we presented results of classification of single-trial EEG signals with using convolutional autoencoders and machine learning approach. The experiment consists of two type of tasks with the dominance of mental and sensory attention. We find clear differences not only at the group level but also could reveal them at the single-trial basis. Time-frequency features of EEG patterns, namely power spectral density of wavelet transform were used as a main discriminated factor. We used a low-dimensional representation picture of PSD image for each channel received with a convolutional autoencoder as a feature vector. The classification was carried out by data from individual channels. The quality of classification varies was from 80 to 90% for each subject. The results obtained make it possible to consider this approach as promising for both clinical classification and BCI interface creation. Within the framework of this approach, only the time-frequency features were used. Potentially, other informative features of single-trial EEG patterns, such as common spatial patterns or others [3] can be added for the model improvement.

Acknowledgement. We gratefully acknowledge financial support of Institute of Information and Computational Technologies (Grant AR05134227, Kazakhstan)).

This study has been performed at the expense of the grant of Russian Science Foundation (project no. 18-11-00336).

References

1. Cohen, M.: Analyzing Time Series Data. MIT press, Cambridge, MA (2014)
2. Lotte, F., Congedo, M., Lécuyer, A., Lamarche, F., Arnaldi, B.: A review of classification algorithms for EEG–based brain–computer interfaces. J. Neural Eng. **4**(2), R1 (2007)
3. Lotte, F., Bougrain, L., Cichocki, A., Clerc, M., Congedo, M., Rakotomamonjy, A., Yger, F.: A review of classification algorithms for EEG–based brain–computer interfaces. J. Neural Eng. **15**(3), 031005 (2018)
4. Ray, W.J., Cole, H.W.: EEG alpha activity reflects attentional demands, and beta activity reflects emotional and cognitive processes. Science **228**(4700), 750 (1985). https://doi.org/10.1126/science.3992243
5. Harmony, T., Fernández, T., Silva, J., Bernal, J., Díaz-Comas, L., Reyes, A., Marosi, E., Rodríguez, M., Rodríguez, M.: EEG delta activity: an indicator of attention to internal processing during performance of mental tasks. Int. J. Psychophysiol. **24**(1–2), 161 (1996). https://doi.org/10.1016/S0167-8760(96)00053-0
6. Efitorov, A., Knyazeva, I., Boytsova, Y., Danko, S.: GPU-based high-performance computing of multichannel EEG phase wavelet synchronization. Procedia Comput. Sci. **123**, 128 (2018). https://doi.org/10.1016/J.PROCS.2018.01.021

Neural Network Based Semi-empirical Models of 3D-Motion of Hypersonic Vehicle

Dmitry S. Kozlov(✉) and Yury V. Tiumentsev

Moscow Aviation Institute (National Research University), Moscow, Russia
dmkozlov001@gmail.com, tium@mai.ru

Abstract. We consider the problem of mathematical modeling and computer simulation of nonlinear controlled dynamical systems represented by differential-algebraic equations of index 1. The problem is proposed to be solved in the framework of a neural network based semi-empirical approach combining theoretical knowledge for the object with training tools of artificial neural network field. Special form neural network based semi-empirical models implementing an implicit scheme of numerical integration inside the activation function are proposed. The training of the semi-empirical model allows elaborating the models of aerodynamic coefficients implemented as a part of it. A semi-empirical model using as theoretical knowledge the equations of the full model of the hypersonic vehicle motion in the specific phase of descent in the atmosphere are presented. The results of simulation for the identification task for the aerodynamic pitching moment coefficient implemented as an ANN-module of the semi-empirical model of the hypersonic vehicle motion are presented.

Keywords: Dynamical system · Differential-algebraic equations
Semi-empirical model · Neural network based simulation

1 Introduction

When examining applied problems in such areas as aviation and rocket and space technology, an uneasy problem is the construction of adequate models of controlled dynamic systems. When forming such models, it often turns out that the characteristics of the given object are incomplete or not accurate. One possible way to solve this problem is to create models with the adaptability feature. In [1] we proposed an approach to the realization of such models. Following this approach, we developed a neural network semi-empirical model for controlled nonlinear dynamical systems. A semi-empirical approach allows us to create models containing theoretical knowledge about the modeling object, as well as to allow the refinement of the model from experimental data. In [1], we present the simulation results, confirming the high efficiency of the semi-empirical approach compared with the traditional black-box dynamic neural network models, such as NARX (Nonlinear AutoRegressive network with eXogeneous inputs)

B. Kryzhanovsky et al. (Eds.): NEUROINFORMATICS 2018, SCI 799, pp. 196–201, 2019.
https://doi.org/10.1007/978-3-030-01328-8_22

and NARMAX (Nonlinear AutoRegressive with Moving Average and eXogenous inputs) [2].

The semi-empirical approach assumes the generation of gray-box models using theoretical knowledge about the simulated object in the form of a system of ordinary differential equations (ODE). We transform the initial theoretical model into a semi-empirical one taking into account the ways of integrating the ODE so that neural network methods could modify parts of the model. The difference between the semi-empirical approach and the NARX approach is that in the first case when generating the model, a part of connections between the state variables and control variables of the initial ODE system is embedded into the model without changing. Such approach allows us to reduce the number of the adjusting parameters of the model and improves its generalization properties [1]. However, in some problems, in addition to ODE, the theoretical model includes algebraic equality-type constraints, that is, the basis of the theoretical model is the system of differential-algebraic equations (DAE). An example of such a problem is the controlling of hypersonic vehicle descending in the atmosphere. For DAE systems, the concept of index [3] is introduced. The index value affects the choice of the numerical integration scheme for the DAE system.

In [4] we consider the semi-empirical approach to the generation of models of controlled aircraft based on the explicit conditionally stable methods of the numerical integration. It is impossible to use this approach directly for modeling the systems described by DAE. Therefore, the semi-empirical approach needs to be modified taking into account the specific character of DAE. Further, we will show how this modification can be carried out.

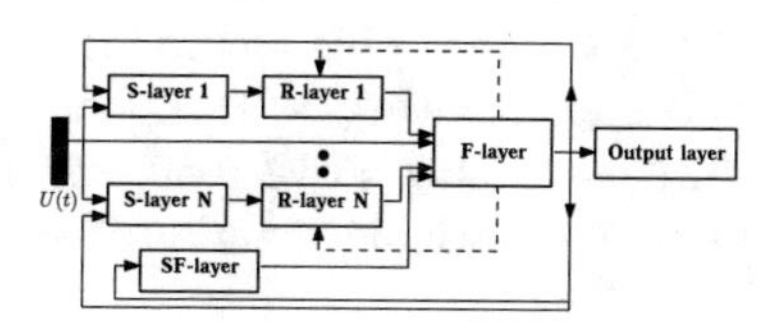

Fig. 1. The structural scheme of the semi-empirical model

2 Semi-empirical Models for DAE Systems

Let us examine the system of the differential-algebraic equations of the semi-explicit form

$$\dot{\mathbf{y}} = \mathbf{f}(t, \mathbf{y}, z, \mathbf{u}), \quad 0 = \mathbf{g}(t, \mathbf{y}, z), \tag{1}$$

where $\mathbf{y} = \mathbf{y}(t)$ is a vector of state variables of the system, $\mathbf{y}$ and $\mathbf{f}$ have the dimension l; $z = z(t)$ is DAE algebraic variable being the same time the state variable of the system (1), $\mathbf{u} = \mathbf{u}(t)$ are the control variables. If the initial values satisfy $0 = g(t_0, \mathbf{y}_0, z_0)$ the initial values are consistent. The system (1) is a DAE system of index 1 if $g_z(t, \mathbf{y}, z)$ is invertible in a neighbourhood of the solution. It is promising to use one-step s-stage methods for the numerical integration of index 1 DAE systems. Implicit Runge-Kutta method (IRK) [3] is the most commonly used method of numerical integration. Using an implicit scheme suggests solving a system of nonlinear equations at each step of integration. We apply Newton's method to solve these equations. We propose to use IRK method based on the quadrature formula Radau IIA [3,5].

The structural scheme of the semi-empirical model is shown in Fig. 1. In the source system of equations, we distinguish variables, which are state variables of the DAE system. Separate layers (R-layers) realize the right-hand sides of the DAE system. The structure of the semi-empirical model takes into account the relationships between the state variables of the initial DAE system. We feed to the R-layers only those values that correspond to the input variables of the original equation. To do this, we include into the model the layers that perform the separation of variables (S-layer and SF-layer). The control signal $U(t)$ is fed to the input of the model. The output layer realizes the observer equation of the dynamic system. In the R-layers, separate neural network modules are built in, corresponding to those parts of the source theoretical model that require tuning. In the process of generation of a semi-empirical model, these modules are subject to structural and parametric adjustment. The network is trained using RTRL (Real-Time Recurrent Learning) algorithm. The training set required to perform the adjustment is generated as a sequence of observed outputs for a given control and initial conditions.

In contrast to neural network semi-empirical modeling in [4], realizing the integration scheme within the neural network structure, we propose an approach where the procedure containing the integration scheme is defined inside the activation function of the F-layer. This technique makes it possible to generate semi-empirical models implementing both explicit and implicit integration schemes. To place procedure containing a difference scheme, it is necessary that the activation function be a multivariable function. We describe the realization of the F-layer in detail in [5–7].

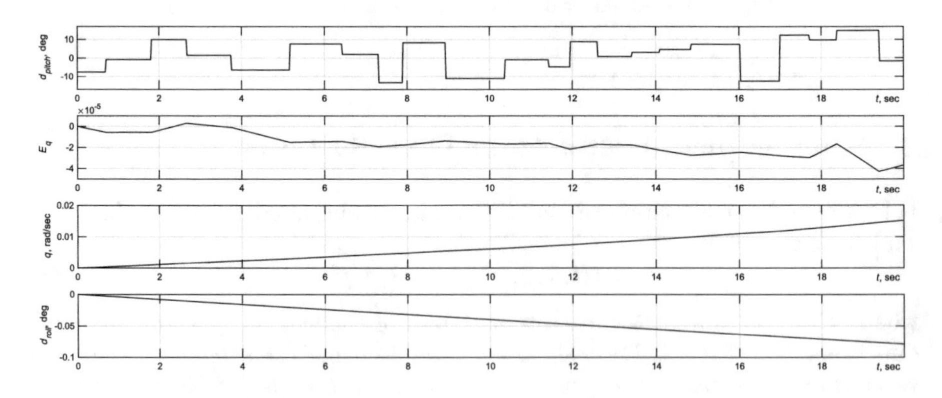

Fig. 2. The semi-empirical model output for values from the test set

3 Simulation Results

We consider here the application of the semi-empirical approach to a model generation for the hypersonic vehicle motion. Then we use this model for trajectory prognosis during aircraft descend in the atmosphere. To estimate the efficiency of the approach we consider the identification problem for the aerodynamic pitching moment coefficient C_m within the model of the hypersonic vehicle motion.

We use a hypersonic vehicle from [8] in the simulation. We consider a full model of the aircraft motion which contains differential equations describing the trajectory and angular motion, as well as the equations of the actuators of the control surfaces. We use the standard model of atmosphere in the absence of perturbations for the calculations. As a result, the full model of the aircraft motion, used for the simulation is typical for the flight dynamics [9]:

$$
\begin{aligned}
&\dot{H} = V\sin\gamma, \quad \dot{\mu} = \frac{V\cos\gamma\sin\psi_W}{r\cos\lambda}, \quad \dot{\lambda} = \frac{V}{r}\cos\gamma\cos\psi_W,\\
&\dot{V} = \frac{-f_{xW}}{m} - g\sin\gamma - \omega_E^2 r\cos\lambda\,(\sin\lambda\cos\psi_W\cos\gamma - \cos\lambda\sin\gamma),\\
&\dot{\psi}_W = \frac{f_{yW}}{mV\cos\gamma} + \frac{V}{r}\cos\gamma\sin\psi_W\tan\lambda - 2\omega_E(\cos\lambda\cos\psi_W\tan\gamma - \sin\lambda)\\
&+\frac{\omega_E^2 r\cos\lambda\sin\lambda\sin\psi_W}{V\cos\gamma}, \quad \mathbf{M} = B\cdot\mathbf{M}_E,\ T_r^2\ddot{d}_r = -2T_r\xi_r\dot{d}_r - d_r + d_{r,act},\\
&f_{xW} = -D, \quad f_{yW} = Y\cos\phi_W + L\sin\phi_W, \quad f_{zW} = Y\sin\phi_W - L\cos\phi_W,\\
&I_x\dot{p} + (I_z - I_y)rq = \bar{L},\ I_y\dot{q} + (I_x - I_z)pr = \bar{M},\ I_z\dot{r} + (I_y - I_x)qp = \bar{N}\\
&\begin{bmatrix}\dot{\psi}\\ \dot{\theta}\\ \dot{\phi}\end{bmatrix} = \begin{bmatrix}0 & \frac{\sin\phi}{\cos\theta} & \frac{\cos\phi}{\cos\theta}\\ 0 & \cos\phi & -\sin\phi\\ 1 & \sin\phi\tan\theta & \cos\phi\tan\theta\end{bmatrix}\cdot\begin{bmatrix}p - M_x\\ q - M_y\\ r - M_z\end{bmatrix}, \begin{bmatrix}M_{xE}\\ M_{yE}\\ M_{zE}\end{bmatrix} = \begin{bmatrix}\cos\lambda & 0\\ 0 & -1\\ -\sin\lambda & 0\end{bmatrix}\cdot\begin{bmatrix}\omega_E + \dot{\mu}\\ \dot{\lambda}\end{bmatrix},\\
&\sin\beta = \cos\gamma[\sin\theta\sin\phi\cos(\psi_W - \psi) + \cos\phi\sin(\psi_W - \psi)] - \sin\gamma\cos\theta\sin\phi,\\
&\sin\alpha = [\cos\gamma[\sin\theta\cos\phi\cos(\psi_W - \psi) - \sin\phi\sin(\psi_W - \psi)] - \sin\gamma\cos\theta\cos\phi]\,/\cos\beta,\\
&\sin\gamma_a = [\sin\gamma[\sin\theta\sin\phi\cos(\psi_W - \psi) + \cos\phi\sin(\psi_W - \psi)] + \cos\gamma\cos\theta\sin\phi]\,/\cos\beta\\
&T_a^2\ddot{d}_a = -2T_a\xi_a\dot{d}_a - d_a + d_{a,act},\ T_e^2\ddot{d}_e = -2T_e\xi_e\dot{d}_e - d_e + d_{e,act},
\end{aligned}
\qquad (2)
$$

where H is the altitude, m; $r = H + a_E$ is the distance from the Earth center to the center of mass of the vehicle, m; $a_E = 20902900$ ft is the Earth radius; μ is the longitude, deg; λ is the geocentric latitude, deg; V is the relative velocity, ft/sec; γ is the relative flight path angle, deg; ψ_W is the relative azimuth, deg; ϕ_W is the bank angle, deg; α is the angle of attack, deg; β is the angle of sideslip, deg; ψ, θ, ϕ are Euler angles: yaw, pitch and roll, deg; $\omega = [p, q, r]^T$ are components of angular velocity vector, rad/sec; B is a matrix transforming vectors from vehicle-carried local Earth reference frame to body-fixed reference frame; D, L, Y are total aerodynamic drag, lift and side forces respectively; $\bar{L}, \bar{M}, \bar{N}$ are aerodynamic rolling, pitching and yawing moments respectively; d_a, d_e, d_r are the deflections of the right and left elevons and the rudder, deg; $d_{a,act}, d_{e,act}, d_{r,act}$ are control signals for right and left elevons and the rudder actuators, deg; d_{pitch}, d_{roll} are pitch and roll motion control signals, deg; $T_a = T_e = T_r = 0.02$ sec are the time constants for right/left elevons and rudder actuators; $\xi_a = \xi_e = \xi_r = 0.707$ are the right/left elevons and rudder actuators damping ratios; $\bar{q} = \rho(H)V^2/2$ is the dynamic pressure; $\omega_E = 0.729211585\text{e}{-}4$ rad/sec is the Earth rotational rate; $g = \gamma_E/r^2$ is the geopotential function, ft/sec^2; $\gamma_E = 0.1407653916\text{e}{+}17$ ft^3/sec^2 is the gravitational constant; $m = 191902.002646$ lb is the mass of the vehicle; $\rho = \rho(H)$ is the air density, lb/ft^3; S is the reference wing area, ft^2; b is the reference wing span, ft; $\bar{c}$ is the mean aerodynamic chord, ft; x_{cg} is the longitudinal distance from the moment reference center to the vehicle center of gravity, ft; I_x, I_y, I_z are the roll, pitch and yaw moments of inertia respectively, slg·ft^2; $M = V/a$ is Mach number; a is the sound velocity, ft/sec.

We examine the equilibrium glide phase of the descending in the atmosphere [5]. It is the zero-thrust phase of the flight. We use right/left elevons and rudder as control surfaces. When moving along this phase of the flight, the flight path angle must remain constant ($\gamma = \gamma^*$). The control signals are preprogrammed to ensure motion along a given trajectory. We enclose the model (2) by an algebraic equality-type relation of the following form:

$$\begin{aligned} 0 = & \frac{-f_{zW}}{mV} + \frac{\cos\gamma}{V}\left(\frac{V^2}{r} - g\right) + 2\omega_E \cos\gamma \sin\psi_w \\ & + \frac{\omega_E^2 r \cos\lambda}{V}(\sin\lambda \cos\psi_W \sin\gamma + \cos\lambda \cos\gamma). \end{aligned} \tag{3}$$

We can treat the system of Eqs. (2), (3) as a DAE system of index 1.

Pitch and roll motion control signals d_{pitch}, d_{roll} are control variables. In the DAE system the state variables are $H, \mu, \lambda, V, \psi_W, \psi, \theta, \phi, p, q, r, d_a, d_e, d_r$, the DAE system algebraic variable is d_{roll}. Following the (α–ϕ_W)-technique for control of aircraft descending in the upper atmosphere [5,6] d_{roll} values are calculated at each step of the numerical integration of the DAE system. The deflection of the elevons influences both the longitudinal motion of the aircraft and the lateral movement, so the transformation is used to control the elevons: $d_{a,act} = d_{pitch} + d_{roll}$, $d_{e,act} = d_{pitch} - d_{roll}$. To reduce sideslip the rudder cross-connection with the lateral control surfaces is used:

$$d_{r,act} = \frac{(C_{n,da}d_a - C_{n,de}d_e) - I_z/I_x \tan\alpha(C_{l,da}d_a - C_{l,de}d_e)}{2(C_{n,dr} - I_z/I_x \tan\alpha C_{l,dr})}. \tag{4}$$

During simulation the neural network module that realizes the pitching moment coefficient C_m is retrained. We use C_m model for the Mach number values bigger than when maneuvering ($M + 5$) as a new dependency. In the training set, we use the d_{pitch} sequences of a particular form as input data [5]. The output data are the values of the pitch rate q. During training procedure, the neural network module weights are changed to reproduce a new relationship.

During the simulation 1000 iterations were performed with the integration step $\Delta t = 0.02$ s. The initial values were $H = 1.2894\text{e}{+}5$ ft, $\mu = 183.8°$, $\lambda = 34.4°$, $V = 7.2364\text{e}{+}3$ ft/s, $\gamma = -3.8895°$, $\psi_W = 35.77°$, $\psi = 50.236°$, $\theta = 9.98°$, $\phi = 46.778°$, $\omega = 0$ rad/s, $d_{roll} = 0°$ $\dot{d} = 0$, $d_a = d_e = 0°$, $d_r = 1°$. The hypersonic vehicle characteristics $I_x, I_y, I_z, x_{cg}, S, \bar{c}, b$ and aerodynamic force and moment coefficient models $(D, L, Y, \bar{L}, \bar{M}, \bar{N},)$ are given from [8].

To implement the model of the hypersonic vehicle motion we used a semi-empirical model realizing order 3 IRK method of numerical integration based on Radau IIA quadrature formulas. We used a perceptron type network with 7 neurons in the hidden layer as an ANN-module for C_m. In Fig. 2 we show the values of the pitch control signal (d_{pitch}) from the test set, the values of the pitch rate q calculated using the semi-empirical model, the values of the algebraic variable (d_{roll}) and the relevant absolute error (E_q) of the q values reproduced by the semi-empirical model. The root mean square deviations for the training, the validation, and the test sets are respectively $6.0305e-6, 8.4164e-6, 1.9358e-5$. We implement the semi-empirical models and computer simulations, using the MATLAB system and the Neural Network Toolbox package.

4 Conclusions

We realize a semi-empirical model using as theoretical knowledge the equations of the full model of the hypersonic vehicle motion in the specific phase of descent in the atmosphere. We present the system of equations in the form of the DAE system of index 1. The identification task for the aerodynamic coefficient implemented as an ANN-module of a semi-empirical model is solved to verify the training properties. The obtained results show that the semi-empirical approach is applicable for ANN-modeling of complex dynamical objects.

This research is supported by the Ministry of Education and Science of the Russian Federation under Contract No. 9.9124.2017/VU.

References

1. Egorchev, M.V., Kozlov, D.S., Tiumentsev, Y.V., Chernyshev, A.V.: J. Comput. Inf. Technol. (9), 3 (2013). (in Russian)
2. Chen, S., Billings, S.: Int. J. Control. **49**(3), 1013 (1989)
3. Hairer, E., Wanner, G.: Solving Ordinary Differential Equations II: Stiff and Differential-Algebraic Problems. Springer Series in Computational Mathematics, 2nd edn. Springer, Heidelberg (2002)
4. Egorchev, M.V., Tiumentsev, Y.V.: Opt. Mem. Neural Netw. (Inf. Opt.) **24**(3), 201 (2015)
5. Kozlov, D.S., Tiumentsev, Y.V.: Proceedings of XVIII International Conference "Neuroinformatics–2016", vol. 3, pp. 61–71. NRNU MEPhI, Moscow (2016). (in Russian)
6. Kozlov, D.S., Tiumentsev, Y.V.: Opt. Mem. Neural Netw. (Inf. Opt.) **24**(4), 279 (2015)
7. Kozlov, D.S., Tiumentsev, Y.V.: Proceedings of 8th Annual International Conference on Biologically inspired cognitive architectures, BICA 2017, vol. 128, pp. 252–257 (2018)
8. Shaughnessy, J.D., et al.: Hypersonic vehicle simulation model: winged-cone configuration. Technical report, NASA (1990)
9. Bochkariov, A.F., Andreevsky, V.V.: Aeromechanics of Airplane: Flight Dynamics. Mashinostroenie, Moscow (1985). (in Russian)

Categorical Data: An Approach to Visualization for Cluster Analysis

Olga Mishulina(✉) and Maria Eidlina

National Research Nuclear University MEPhI
(Moscow Engineering Physics Institute), Moscow, Russia
mishulina@gmail.com, mashaeidlina@mail.ru

Abstract. The problem of studying the cluster structure of a set of objects with qualitative (categorical) features is considered. We propose an approach to visualization of source data and categorical data groups in a form that is convenient for human analysis and decision-making. We generalized Andrews' idea of numeric data visualization for the case of categorical data set. The developed approach can be applied in the case when the frequency distribution of the joint appearance of feature pairs in the data sample is known. For visualization, it is proposed to use not the primary features of the data set, but new paired features that have a strong statistical relationship. In addition, we have corrected the spectral representation of Andrews curves, limiting the maximum frequency of harmonic functions. The proposed visual representation of categorical data makes it possible to estimate the number of clusters in a data set and show their differences. The technique is demonstrated on a model example in which the decision on the number of clusters is taken in conjunction with two other ways of visualizing data clusters: a silhouette and a heat map.

Keywords: Categorical data · Cluster analysis · Visualization
Andrews curves · Statistical features · Number of clusters

1 Introduction

One of the most important stages of the cluster analysis is the interpretation of the data groups. If there is no clear and understandable configuration of data groups in the feature space, the researcher cannot give them a clear interpretation and is forced to repeatedly perform the clustering procedure using different algorithms and their parameters.

The most powerful tool for interpreting the data structure is, undoubtedly, the human eye. The brain has the ability to quickly identify the image objects and extract their quality characteristics and related patterns. To take advantage of this unique opportunity, it is necessary to find a way to represent objects that is most convenient for human perception.

For objects that are characterized by numerical features, there is a wide range of visual representation methods for source data and clustering results [1]. However, the problem of visualization is much more complicated for objects with categorical features [2]. The proposed study is devoted to this problem.

B. Kryzhanovsky et al. (Eds.): NEUROINFORMATICS 2018, SCI 799, pp. 202–209, 2019.
https://doi.org/10.1007/978-3-030-01328-8_23

Andrews [3] proposed a way to visualize objects with numerical characteristics. He considered the vector of numerical object features as a series of Fourier spectral coefficients and applied to it a transformation into a function of a real variable $t \in [-\pi,\ \pi]$:

$$f_x(t) = x_1/\sqrt{2} + x_2 \sin(t) + x_3 \cos(t) + x_4 \sin(2t) + x_5 \cos(2t) + \ldots \tag{1}$$

Thus, each data sample object is represented as a function of the real variable t, and the collection of sample objects is represented by a set of Andrews curves. Those of them that correspond to objects with similar characteristic vectors and belong to the same cluster, appear on the graph as a sufficiently dense group. By the number of such groups and their clear isolation, one can judge the existence of an explicit cluster structure of data and the number of clusters. Andrews' idea aroused the interest of specialists and received its theoretical development and using in applied research [4, 5].

In our work, we use Andrews' idea to visualize categorical data. To this end, we proposed a method for converting the vector of categorical attributes of an object to a number form. The developed approach can be applied in the case when the frequency distribution of the joint appearance of feature pairs in the data sample is known. The conversion from categorical to numerical features allows to construct Andrews curves and visually analyze the structural properties of the object sample.

In carrying out experimental studies, we used known and common methods of clustering categorical data: K-modes, CLOPE, ROCK, LargeItem, LIMBO, TrKMeans [6–11].

Using the Andrews curves, you can perform not only a priori visual analysis of sample data, but also a posteriori qualitative analysis of the clustering results. For this, the Andrews curves are marked with colors that correspond to the clusters to which they belong. A person can easily perceive the similarity or difference of clusters by the arrangement of groups of colored Andrews curves.

The developed approach is illustrated on the model example.

2 Statement of the Problem and the Notation System

Categorical data are characterized by a set of qualitative attributes $r_i,\ i = \overline{1,\ M}$ (for example, color, form, manufacturer, style, etc.). Each of the attributes has a set of possible values $r_{ij},\ i = \overline{1,\ M},\ j = \overline{1,\ M_i}$, where M_i is the number of possible values of the attribute r_i. For example, the attribute $r_i = \ <color>$, $r_{ij} = \ <red>$, $M_i = 4$. Figure 1 shows the structure of the vector of categorical features for the object that has three qualitative attributes $(M = 3)$ with numbers of possible values $M_1 = 3, M_2 = 4, M_3 = 2$, respectively. Values 1 and 0 are used in the table as indicators of the presence or absence of attribute value. Therefore, we consider the feature vector of an object $\mathbf{x}_n = (x_{n1}, x_{n2}, \ldots, x_{nd})$ of dimension $d = \sum_{i=1}^{M} M_i$. The last three lines of the table (Fig. 1) give examples of categorical objects.

r_i	r_1			r_2				r_3	
r_{ij}	r_{11}	r_{12}	r_{13}	r_{21}	r_{22}	r_{23}	r_{24}	r_{31}	r_{32}
$\mathbf{x}$	x_1	x_2	x_3	x_4	x_5	x_6	x_7	x_8	x_9
$\mathbf{x}_n$	x_{n1}	x_{n2}	x_{n3}	x_{n4}	x_{n5}	x_{n6}	x_{n7}	x_{n8}	x_{n9}
$\mathbf{x}_1$	1	0	0	0	1	0	0	1	0
$\mathbf{x}_2$	0	1	0	0	1	0	0	0	1
$\mathbf{x}_3$	0	1	0	1	0	0	0	1	0

Fig. 1. Structure of the vector of categorical features. Examples

Let's assume that for the set of sample data, the observation frequencies of characteristics x_i are calculated:

$$p_i = P\left[X_i = 1\right], i = \overline{1,\ d}, \tag{2}$$

where X_i is a random variable that takes the binary values 0 or 1.

In addition, suppose that the frequency matrix Θ of simultaneous observation of characteristics x_i and x_j is known. We denote p_{ij} elements of the matrix Θ:

$$p_{ij} = P[(X_i = 1) \wedge (X_j = 1)],\ i,\ j = \overline{1,\ d}. \tag{3}$$

Let's put the task of converting the vector of categorical characteristics of objects represented in binary form to numerical features for visualization of objects by the Andrews' method.

3 Development of Andrews Plot for Categorical Data

Categorical object can be represented by a numeric vector $\mathbf{y}$ obtained by element-wise multiplication of vectors $(x_1, x_2, \ldots, x_d)$ and $(p_1, p_2, \ldots p_d)$:

$$(y_1, y_2, \ldots y_d) = (x_1 p_1, x_2 p_2, \ldots, x_d p_d). \tag{4}$$

However, such numerical representation of categorical objects will not lead to the desired result of Andrews curves visualization. The reason for it is that the categorical attributes and, consequently, the numerical features $y_i = x_i p_i,\ \ i = \overline{1,\ d}$, are statistically strongly related. This statistical relationship is natural for samples with a cluster structure and is the reason for the formation of clusters of similar categorical objects. At the same time, it is known that Andrews curves for numerical data give a vivid visual representations only in the case of weakly correlated features [5]. Vector (4) does not meet this requirement.

We propose to create new paired features y_{ij} from initial features $x_i,\ i = \overline{1,\ d}$:

$$y_{ij} = \begin{cases} 1 & if (X_i = 1) \wedge (X_j = 1); \\ 0 & otherwise. \end{cases} \tag{5}$$

Let's estimate the significance of the paired feature y_{ij} by the parameter γ_{ij} of statistical relationship of the initial features which are included in y_{ij}:

$$\gamma_{ij} = {p_{i\,|j}}/{p_i} = {p_{ij}}/{p_i p_j}, \tag{6}$$

where $p_{i|j}$ is a conditional probability: $p_{i|j} = P((X_i = 1)|(X_i = 1))$.

It should be noted that γ_{ij} can have a high value even for features that have a low probability of an observation in the sample. It can be essential for detecting small clusters. We will call the statistical relationship of the pair of features i and j strong if $\gamma_{ij} > \gamma_0$.

Let's select all the paired features $y_{ij}, i,j = \overline{1,d},\ i \neq j$, for which $\gamma_{ij} > \gamma_0$, and sort them in descending order of the values of γ_{ij}. Let's denote the resulting vector as $\mathbf{z} = (z_1, z_2, \ldots, z_q)$, where q is the number of strongly related pairs of features. Thus, each object is now characterized by a binary vector, the ones in which represent all the strongly related features in the object. The vector of parameters of strong statistical relationship for the corresponding pairs of features is denoted by $\boldsymbol{\lambda} = (\lambda_1, \lambda_2, \ldots, \lambda_q)$. Similarly to (4), for each object we construct a vector $(z_1\lambda_1, z_2\lambda_2, \ldots, z_q\lambda_q)$. In addition, we complement this vector from the left with d coordinates corresponding to the values of initial features with the weight coefficients p_i specified in formula (2). Let's sort the considered coordinates in descending order of values $p_i,\ i = \overline{1,\ d}$. The combined vector of dimension $n = (d + q)$ is denoted by $\mathbf{v}$. This vector is used in the construction of Andrews curves.

Another question is related with the choice of the frequencies of harmonic functions in formula (1). With a large dimension n of the vector $\mathbf{v}$, the high frequency of the Andrews curves calculated according to the formula (1) will not allow a person to identify groups of similar curves on the plot. Let's write the spectral transformation in the following form (it is assumed that n is odd):

$$\begin{aligned} f_x(t) &= v_1/\sqrt{2} + v_2 \sin(2\pi\kappa t) + v_3 \cos(2\pi\kappa t) \\ &+ v_4 \sin(2\pi\kappa 2t) + v_5 \cos(2\pi\kappa 2t) + \ldots + v_n \cos(2\pi\kappa \tfrac{n-1}{2} t) \end{aligned}, \tag{7}$$

where $t \in [0,\ 1]$, κ is a parameter that limits the maximum frequency of harmonic functions in the spectral transformation (7). Let the maximum frequency $\omega_{\max} = 2\pi m$. Then the following inequality must hold:

$$2\pi\kappa \frac{n-1}{2} \leq 2\pi m, \quad \kappa \leq \frac{2m}{n-1}. \tag{8}$$

4 An Example of Model Data Analysis

A dataset consisting of 500 randomly generated objects was used. Each object was characterized by $M = 5$ attributes; each attribute could take one of M_i, $i = \overline{1,\ 5}$, different values, where M_i = 2, 5, 3, 2, 3 respectively. Figure 2 shows probability distributions for all features. 41 pair of features are characterized by parameter of statistical relationship value $\gamma_{ij} > 1$ between 1,05 and 3,3.

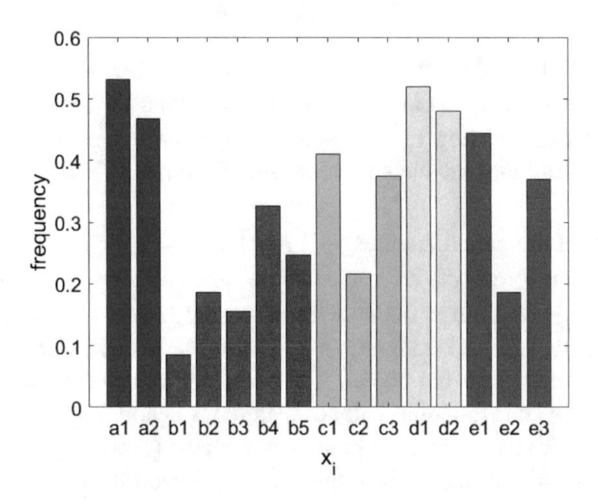

Fig. 2. Frequency distributions for values of five attributes

Cluster structure of the data was identified using only initial feature vector of dimension $d = 15$. Consider result of applying LIMBO algorithm [10] to the dataset (for 4 clusters). The result is characterized by high values of validity indices. In addition to standard validity indices [12], silhouette plot and heat map of distance matrix were used (Fig. 3*a*).

To construct Andrews curves, we used the $n = 21$-dimensional feature vector $\mathbf{v}$ $(d = 15, q = 6)$. The parameter m in inequality (8) was assumed equal to 2.5, $\gamma_0 = 2$ (Fig. 4*a*). Curves belonging to different clusters, have different shape. Curves belonging to the same cluster, are bundled together.

For each cluster C_k, we calculated the central curve (centroid) $c_k(t)$ according to the formula:

$$c_k(t) = {}^1/_{N_k} \sum_{\substack{i, \\ \mathbf{x}_i \in C_k}} f_i(t), \tag{9}$$

where N_k denotes number of objects in cluster k, $f_i(t)$ signifies Andrews curve for dataset vector $\mathbf{x}_i$ (feature vector $\mathbf{v}_i$). Figure 4*b* shows cluster centroids.

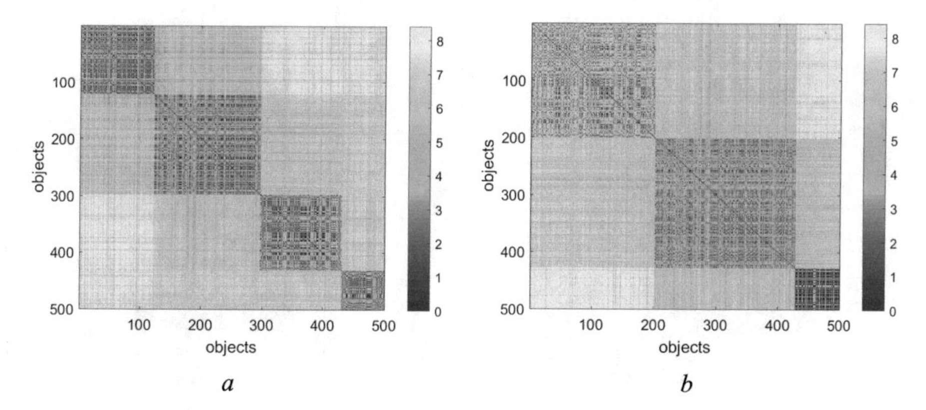

Fig. 3. LIMBO clustering visualization using a heat map for clustering with the number of clusters: *a* – 4, *b* – 3. Distance between objects was evaluated using Euclidean metric

Visual analysis of Andrews curves (Fig. 4*a*) shows that clusters colored in blue and red, contain objects close to each other. These clusters are the first and second on the heat map (Fig. 3*a*). Silhouette plot for "questionable" clusters (Fig. 5*a*) contains negative values, which confirms the need for additional research. Analysis of frequency distributions of features in the first two clusters shows that they have similar statistical properties.

Consider the result of clustering data for three clusters. Silhouette plot (Fig. 5*b*), as in the previous clustering, contains negative values. However, clusters on the heat map are more clearly distinguished (Fig. 3*b*). Andrews curves for this clustering are shown in Fig. 6. Visual analysis shows significant difference in cluster centroids and confirms the preference of the obtained result.

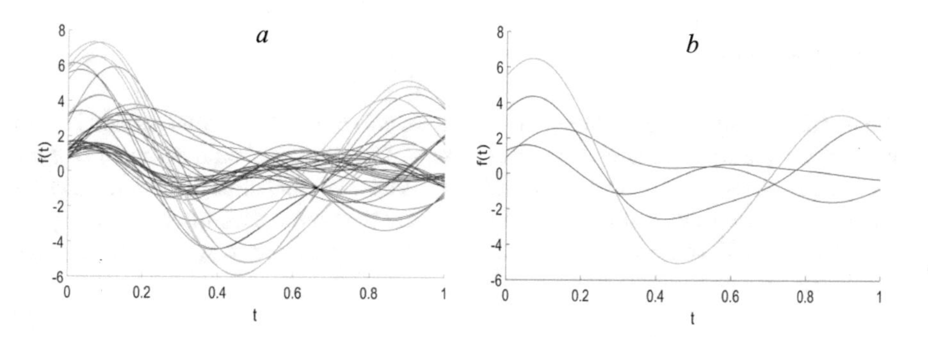

Fig. 4. *a* – Andrews curves painted with colors of four clusters; *b* – cluster centroids $C_k, k = \overline{1,4}$

Nevertheless, the silhouette plot indicates a bad separation of clusters in the space of attributes. The complex structure of the analyzed data is related to their synthetic origin and the lack of an objective pattern inherent in real world data.

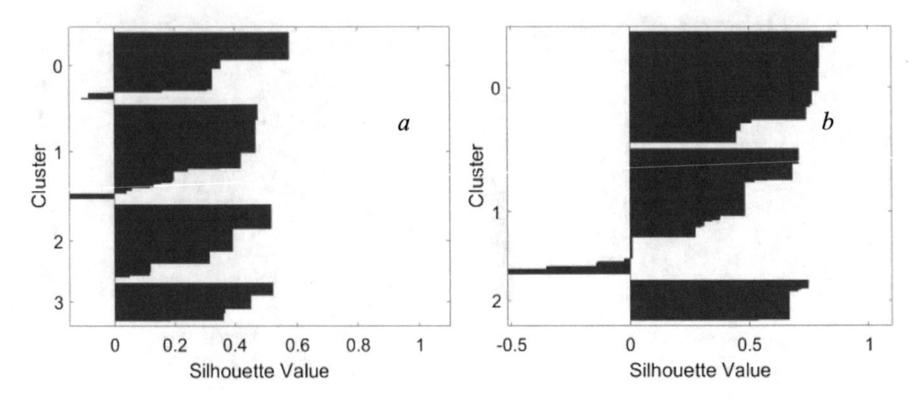

Fig. 5. Silhouette plot for clustering with the number of clusters: a – 4, b – 3

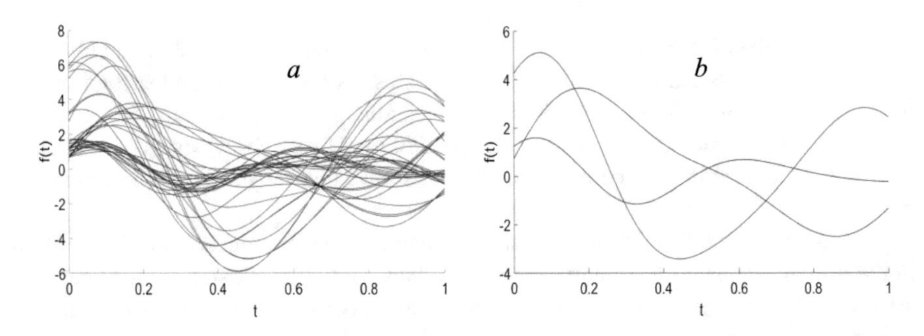

Fig. 6. a – Andrews curves painted with colors of three clusters; b – cluster centroids $C_k, k = \overline{1,3}$

5 Conclusions

The performed studies have shown that Andrews curves can be of considerable help to interpret the cluster structure of categorical data. However, it should be noted that the shape of Andrews curves and clarity of the visual representation of data depend on parameters γ_0, m, statistical feature properties, and the metric used in the feature space of the vector $\mathbf{v}$. Therefore, it is recommended to vary parameters γ_0, m to achieve the desired result.

References

1. Gan, G., Chaoqun M., Jianhong W.: Data Clustering: Theory, Algorithms, and Applications. ASA-SIAM Series on Statistics and Applied Probability, SIAM, Philadelphia, ASA, Alexandria (2007)
2. Blasius, J., Greenacre, M.: Visualization of Categorical Data. Academic Press, New York (1998)

3. Andrews, D.F.: Plots of high dimensional data. Biometrics **28**, 125–136 (1972)
4. Khattree, R., Naik, D.N.: Andrews plots for multivariate data: some new suggestions and applications. J. Stat. Plan. Inference **100**(2), 411–425 (2002)
5. Grinshpun, V.: Application of Andrew's plots to visualization of multidimensional data. Int. J. Environ. Sci. Educ. **11**(17) (2016)
6. Huang, Z.: Extensions to the k-means algorithm for clustering large data sets with categorical values. Data Min. Knowl. Disc. **2**, 283–304 (1998)
7. Yang, Y., Guan, X., You, J.: CLOPE: a fast and effective clustering algorithm for transactional data. In: Proceedings of the Eighth ACM SIGKDD International Conference on Knowledge Discovery and Data Mining, Edmonton, Alberta, Canada, pp. 682–687 (2002)
8. Guha, S., Rastogi, R., Shim, K.: ROCK: a robust clustering algorithm for categorical attributes. In: Proceedings of the 15th International Conference on Data Engineering, pp. 512–521 (1999)
9. Wang, K., Xu C., Liu, B.: Clustering transactions using large items. In: Proceedings of ACM Conference on Information and Knowledge Management (CIKM), pp. 483–490 (1999)
10. Andritsos, P., Tsaparas, P., Miller, R.J., Sevcik, K.C.: LIMBO: a scalable algorithm to cluster categorical data. University of Toronto, Department of Computer Science, Toronto, CSRG Technical report 467 (2003)
11. Giannotti, F., Gozzi C., Manco, G.: Clustering transactional data. In: Principles of Data Mining and Knowledge Discovery: 6th European Conference, Helsinki, Finland, pp. 175–187 (2002)
12. Bai, L., Liang, J.: Cluster validity functions for categorical data: a solution-space perspective. Data Min. Knowl. Disc. **29**(6), 1560–1597 (2015)

The Time Series Forecasting of the Company's Electric Power Consumption

Anastasiya V. Seliverstova(✉), Darya A. Pavlova, Slavik A. Tonoyan, and Yuriy E. Gapanyuk

Bauman Moscow State Technical University, Moscow, Russia
stasic7772@gmail.com

Abstract. The purpose of this paper is to choose the appropriate method for time series forecasting of the company's electric power consumption. The peculiarity of the problem is that it is necessary to make four types of forecasts: ultra-short-term forecasting (less than a day), short-term forecasting (from one day to one week), medium-term forecasting (from one week to one year), long-term forecasting (longer than a year). The subset of CRISP-DM (Cross-Industry Standard Process for Data Mining) methodology was adapted for the company's electric power consumption task solving. The ARIMA, GMDH, LSTM, and seq2seq methods are considered. The MSE, MAE, and MAPE metrics are used for the forecasting quality evaluation. The Python technological stack is used for experiments. The "StatsModels" library is used for ARIMA, "Keras" library is used for LSTM and seq2seq. The GMDH implementation is self-developed. The experiments results show that the best method is GMDH. The ARIMA and the LSTM may be considered as second-place methods.

Keywords: CRISP-DM · ARIMA · GMDH · LSTM
Sequence-to-sequence (seq2seq) · Time series forecasting

1 Introduction

The objective of the forecasting future values of the time series based on its historical values is the basis for financial planning in the economy and trade, planning, management and optimization of production volumes, warehouse control. Currently, companies are accumulating historical values of economic and physical indicators in databases, which significantly increases the amount of input information for the forecasting task. At the same time, the development of hardware and software provides more and more powerful computing platforms, which can implement complex forecasting algorithms. Also, modern approaches to economic and technical management impose increasingly rigid requirements for the accuracy of forecasting. Thus, the problem of time series forecasting becomes more complicated simultaneously with the development of information technologies.

In this paper, we consider the task of the forecasting of the company's electric power consumption. The peculiarity of the problem is that it is necessary to make four types of forecasts: ultra-short-term forecasting (less than a day), short-term forecasting

B. Kryzhanovsky et al. (Eds.): NEUROINFORMATICS 2018, SCI 799, pp. 210–215, 2019.
https://doi.org/10.1007/978-3-030-01328-8_24

(from one day to one week), medium-term forecasting (from one week to one year), long-term forecasting (longer than a year).

It is clear that different methods can show the best results for different types of forecast. Thus, the main purpose of this research is to choose the appropriate method for each type of forecast based on the company's historical values of electric power consumption. Among the methods used, both methods using neural networks and traditional methods of forecasting are considered.

2 The Proposed Approach Description

We use the subset of CRISP-DM (Cross-Industry Standard Process for Data Mining) methodology [1] adapted for the company's electric power consumption task solving. According to [1] the CRISP-DM methodology includes the following stages:

1. Business Understanding;
2. Data Understanding;
3. Data Preparation;
4. Modeling;
5. Evaluation;
6. Deployment.

In this paper, we consider in details only stages 3, 4, 5. The data understanding stage is not necessary, because the data semantics is clear – it is a time-series data. The steps 1 and 6 are out of the scope of our research.

3 The Data Preparation Stage

The data preparation stage includes anomaly detection and time series smoothing. We implemented anomaly detection using Mullineaux & Irwin approach described in [2]. For smoothing the Holt-Winters exponential smoothing method [3] was used.

4 The Modeling Stage

The modeling stage involves the use of one of the time series forecasting methods. We consider the following methods: ARIMA, GMDH, LSTM, and Sequence-to-sequence (seq2seq).

The ARIMA (autoregressive integrated moving average) method [4] is a traditional statistical method for time series forecasting. The other three methods are the neural networks based methods.

The LSTM (long short-term memory) neural networks [5] is a subclass of recurrent neural networks (RNN). LSTM is widely used in problems in which it is necessary to identify patterns of sequence elements, including time series forecasting. The Sequence-to-sequence (seq2seq) model [6] consists of two recurrent neural networks

(RNNs): an encoder that processes the input and a decoder that generates the output. Theoretically, the use of two RNNs can increase the accuracy of the forecasting.

The GMDH (group method of data handling) method was invented in 1968 by Prof. Alexey G. Ivakhnenko in the Institute of Cybernetics in Kiev and has since been improved many times. This is a kind of neural network model that is called polynomial neural networks. One of the LSTM inventors Schmidhuber cites GDMH as one of the earliest deep learning methods, remarking that it was used to train eight-layer neural nets as early as 1971 [7]. And the authors of paper [8] showed that GMDH neural network performed better than the classical forecasting algorithms.

5 The Evaluation Stage

This stage is used to evaluate the quality of the forecasting models. The main purpose of this stage is the selection of quality metrics.

The used quality metrics according to [9] are represented in Table 1 where A_i is the actual value at the time i; P_i is the predicted value at the time i. The column "Metric id" is used for reference in the experiments section.

Table 1. The quality metrics.

Metric id	Metric name	Metric description
MSE	Mean squared error	$MSE = \frac{1}{n}\sum_{i=1}^{n}(A_i - F_i)^2$
MAE	Mean absolute error	$MAE = \frac{1}{n}\sum_{i=1}^{n}\lvert A_i - F_i\rvert$
MAPE	Mean absolute percentage error	$MAPE = \frac{1}{n}\sum_{i=1}^{n}\left\lvert\frac{A_i - F_i}{A_i}\right\rvert$
ACC	Forecast accuracy	$ACC = 1 - MAPE$

It should be noted that the metrics MSE and MAE are unlimited from above real numbers, and the metrics MAPE and ACC are in the range from 0 to 1. The higher values of metrics MSE, MAE, and MAPE indicate a higher error. The ACC metric is the complement of the MAPE metric to one.

6 The Experiments

In this section, we present the results of the application of different forecasting methods to our dataset with different depth of the forecast. The Python technological stack is used for experiments. The "StatsModels" library is used for ARIMA, "Keras" library is used for LSTM and seq2seq. The GMDH implementation is self-developed.

The forecasting methods were discussed in Sect. 4. The used quality metrics were discussed in Sect. 5. The quality metrics values are presented in Tables 2, 3, 4 and 5. We use three independent metrics MSE, MAE, and MAPE (ACC is dependent on

MAPE). Thus, we can conclude which forecasting method is more appropriate for different depths of the forecast. The best values for each table are in highlighted and italic font, while the second-place values are highlighted.

Table 2. The results of ultra-short-term forecasting.

Method	Metric			
	MSE	MAE	MAPE	ACC
ARIMA	**864 763.239**	**929.926**	0.304	0.696
GMDH	***546.068***	***23.368***	***0.007***	***0.993***
LSTM	7 946 136.626	2064.354	**0.179**	**0.820**
seq2seq	10 036 006.400	2440.361	0.195	0.805

Table 3. The results of short-term forecasting.

Method	Metric			
	MSE	MAE	MAPE	ACC
ARIMA	**173 484.315**	**290.194**	0.098	0.902
GMDH	***839.687***	***19.059***	***0.006***	***0.994***
LSTM	8 344 613.558	2 153.637	**0.193**	**0.806**
seq2seq	9 211 173.972	2 328.291	**0.199**	**0.800**

Table 4. The results of medium-term forecasting.

Method	Metric			
	MSE	MAE	MAPE	ACC
ARIMA	**1 088 475.692**	**786.449**	0.243	0.756
GMDH	***43 730.045***	***147.275***	***0.048***	***0.951***
LSTM	6 810 422.474	1 824.823	**0.144**	**0.856**
seq2seq	10 061 823.451	2 492.514	0.208	0.792

Table 5. The results of long-term forecasting.

Method	Metric			
	MSE	MAE	MAPE	ACC
ARIMA	**1 226 256.118**	**874.485**	0.271	0.729
GMDH	***117 209.837***	***241.216***	***0.067***	***0.933***
LSTM	6 647 734.099	1 843.635	**0.143**	**0.856**
seq2seq	11 831 016.627	2 694.297	0.199	0.800

According to the Tables 2, 3, 4 and 5, we can see that the best method is GMDH. The second-place method depends on the metric. According to MSE and MAE metrics, the ARIMA is the second-place method in all cases while according to the MAPE (ACC) metric the LSTM is the second-place method.

Now we will represent the medium-term results of forecast in Figs. 1, 2, 3 and 4 in order to check the metrics visually. The predicted values are shown with red and orange colors. According to the Figs. 1, 2, 3 and 4, we can see that GMDH is the best method indeed.

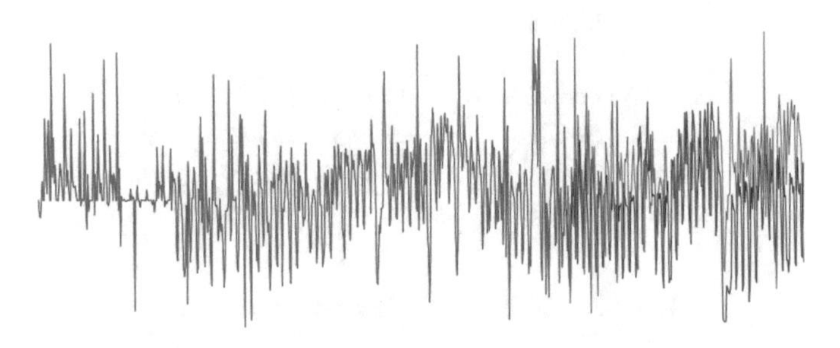

Fig. 1. The results of the medium-term forecasting with ARIMA method

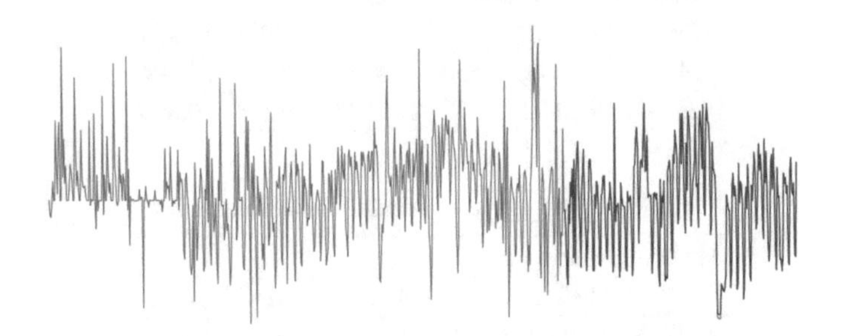

Fig. 2. The results of the medium-term forecasting with GMDH method

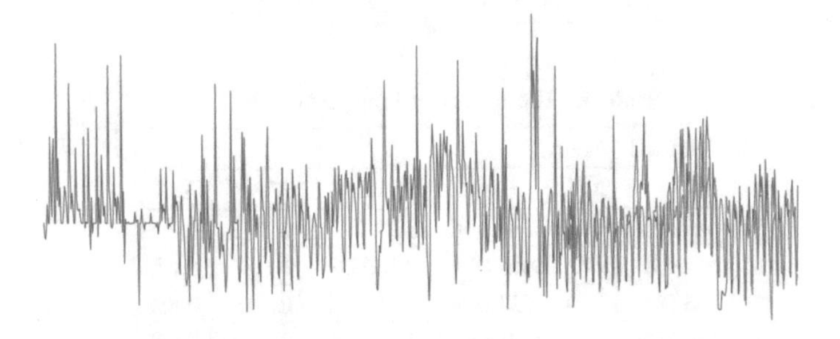

Fig. 3. The results of the medium-term forecasting with LSTM method

The ARIMA and LSTM methods show errors of different types. The ARIMA method tends to increase the amplitude of the time series data, while the LSTM method tends to decrease the amplitude. The results of seq2seq are close to LSTM results but not so accurate.

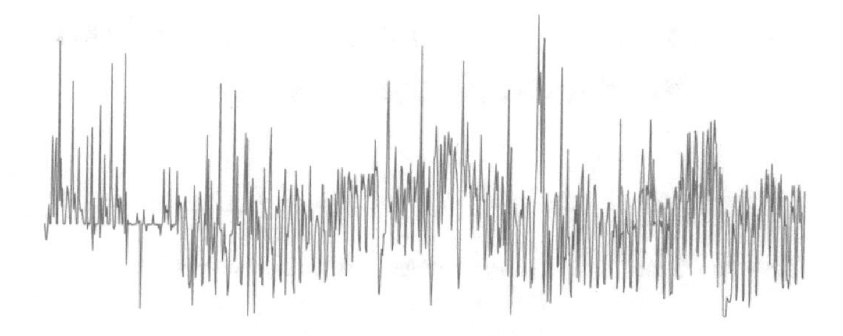

Fig. 4. The results of the medium-term forecasting with seq2seq method

7 Conclusion

The best method for the forecasting of the company's electric power consumption time series data is the GMDH method. It is very interesting that the GMDH method shows the best results for all depths of the forecast: ultra-short-term, short-term, medium-term, and long-term forecasting.

The ARIMA and the LSTM may be considered as second-place methods.

Thus, we can conclude that the forecasting of the company's electric power consumption task may be successfully solved using the GMDH method.

References

1. Shearer, C.: The CRISP-DM model: the new blueprint for data mining. Int. J. Data Warehous. Min. **5**, 13–22 (2000)
2. Mullineaux, D.R., Irwin, G.: Error and anomaly detection for intra-participant time-series data. Int. Biomech. **4**, 28–35 (2017)
3. Kalekar, P.S.: Time series forecasting using Holt-Winters exponential smoothing. In: Kanwal Rekhi School of Information Technology, Mumbai, pp. 1–13 (2004)
4. Asteriou, D., Hall, S.G.: ARIMA models and the box–jenkins methodology. In: Applied Econometrics, 2nd edn., pp. 265–286. Palgrave MacMillan, New-York (2011)
5. Sundermeyer, M., Ney, H., Schlter, R.: From feedforward to recurrent LSTM neural networks for language modeling. IEEE/ACM Trans. Audio, Speech Lang. **23**(3), 517–529 (2015)
6. Sutskever, I., Vinyals, O., Le, Q.V.: Sequence to sequence learning with neural networks. In: Advances in Neural Information Processing Systems (NIPS 2014) Proceedings, Montreal, pp. 3104–3112 (2014)
7. Schmidhuber, J.: Deep learning in neural networks: an overview. Neural Netw. **61**, 85–117 (2015)
8. Li, R., Yi, M., Fong, S., Chong, W.S.: Forecasting the REITs and stock indices: group method of data handling neural network approach. Pac. Rim Prop. Res. J. **23**(2), 1–38 (2017)
9. Ramachandran, K.M., Tsokos, C.P.: Mathematical Statistics with Applications. Elsevier Academic Press, London (2009)

Detection of Initial Moment of Head Motion by Neural Network Modules

Dmitry Shaposhnikov(✉) and Lubov Podladchikova

Research Centre of Neurotechnologies, Southern Federal University, Rostov-on-Don, Russia
dgshaposhnikov@sfedu.ru

Abstract. The results of application of the neural network modules to detect the head motion parameters are presented. Each module has a very simple structure and consists of a pair of excitatory and inhibitory neurons that have common center, different sizes of their receptive fields and time delay. Detection of the initial moment of head motion was evaluated during computer simulation. To test the module performance, synthetic video facial image sequences from SYLAHP database monitoring the head motion were used. It was shown that the U^E amplitude and polarity qualitatively correspond to face motion amplitude. Besides the initial front of U^E quick changes corresponding to quick motion of head was equal to 12 ms in all cases (n = 46). Future steps of research and developments in this direction have been shortly discussed.

Keywords: Head motion · Initial moment · Neural network module

1 Introduction

Real-time and robust estimation of the human head motions, position and orientation is an important task for a wide range of applications such as brain tomography, driver awareness, social interaction, human–computer and human-robot interaction [1, 4, 8–10, 12]. To solve these tasks many approaches, algorithms, and models have been developed, each of them has its own advantages and disadvantages [4, 5, 8, 9, 15]. Moreover, the robustness of the most algorithms for estimation of head motion parameters is not enough for their usage in real conditions. Most prospective approaches to solve the similar problems are concerned with usage depth video cameras [5] and imitation of biological vision mechanisms [2–4, 6, 14].

In the majority of the known approaches and methods, high accuracy is achieved due to a complex structure [1, 3, 4, 7, 9, 15] that result in high computational cost and complicate their practical usage. Besides, they mainly aim to estimate the final head position and ignore the initial moment of motion. Moreover we did not find works in literature devoted to application of neural network models for estimation of the head motion parameters.

In the present paper, the possibility of initial moment evaluation of head motion by neural network module with very simple structure was tested. Earlier [11] it was shown that the similar modules are able to selectively respond to the movements of simple objects (such parameters as the movement velocity, motion direction and the sizes).

B. Kryzhanovsky et al. (Eds.): NEUROINFORMATICS 2018, SCI 799, pp. 216–220, 2019.
https://doi.org/10.1007/978-3-030-01328-8_25

2 Methods

Neural network module which consists of a pair of excitatory and inhibitory neurons (Fig. 1a) was used to detect the initial moment of head motion. They have different sizes of their receptive fields and time delay similar to [11].

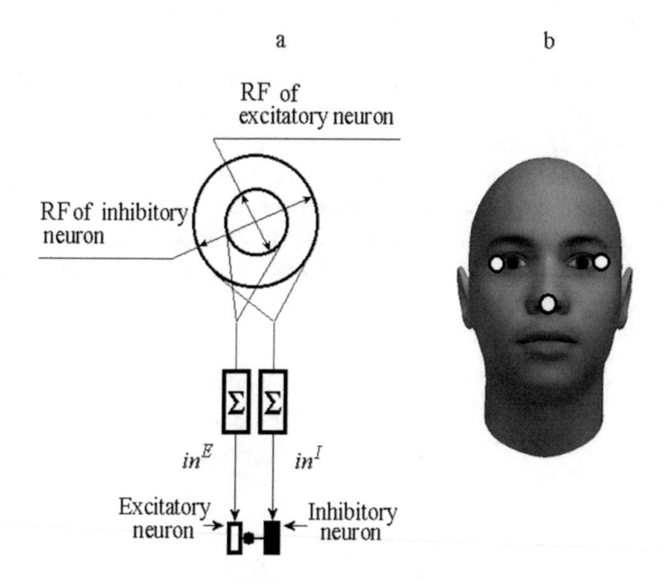

Fig. 1. Structure of neural network module (a) and location of facial landmarks (b).

The behavior of module is described by the system of differential equations:

$$\begin{cases} \tau^E \frac{dU^E}{dt} = -U^E + in^E(t) - w^I \cdot sg(U^I); \\ \tau^I \frac{dU^I}{dt} = -U^I + in^I(t); \end{cases} \tag{1}$$

$$U^E(0) = U^I(0) = 0.$$

where: U is the membrane potential of a neuron; τ^E, τ^I are time constants of excitatory and inhibitory neurons in the module, w^I is a coefficient of the inhibitory connection; t is the time. Here and below, the superscripts E and I designate parameters related correspondingly to the excitatory and inhibitory neuron in the module. The function $sg(\varphi)$ is determined as follows:

$$sg(\varphi) = \begin{cases} 1, & \varphi > 0 \\ 0, & \varphi \leq 0 \end{cases}$$

Functions $in^E(t)$ and $in^I(t)$ are the input signals for the excitatory and inhibitory neurons in the module, respectively, and are determined as follows:

$$in^E = \frac{1}{S^E} \sum_{\substack{0 \le i < R^E \\ 0 \le j < R^E}} I(i,j);\ in^I = \frac{1}{S^I} \sum_{\substack{0 \le i < R^I \\ 0 \le j < R^I}} I(i,j), \tag{2}$$

where: *I(i,j)* determines a gray-scale level of the element *(i,j)* in the sensory layer at the time moment *t*; S^E, S^I, R^E, R^I ($S^I > S^E$) are squares and radii of the receptive fields (RF) for the excitatory and inhibitory neurons, respectively.

According to the previous evaluations using natural faces [7], external eye corners and the middle point of a nose basement are chosen as landmarks which have a set of relatively constant local features, detected from color attributes (lightness, chroma, and hue). Such facial landmarks were used in the present work, too (Fig. 1b). Synthetic SYLAHP database of moving realistic faces created by FaceGen software [13] was used in computer simulation on processing video image sequences. Each moving image in this collection has a resolution of 500x600 pixels. The facial images in each video image sequences vary in wide range of pitch, yaw, and roll (Fig. 2).

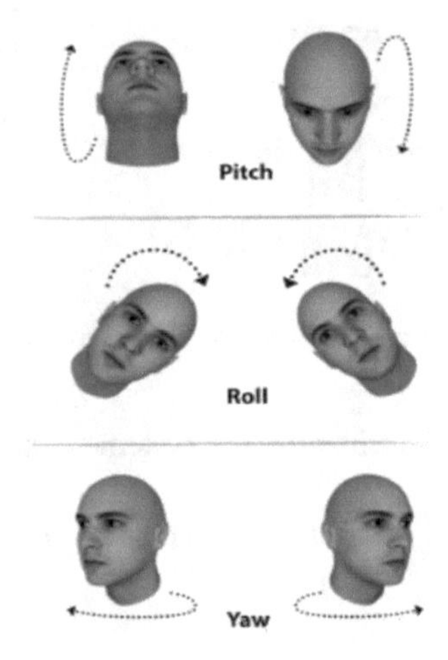

Fig. 2. Orientation of the head in terms of pitch, roll, and yaw movements describing the three degrees of freedom of a human head (modified Fig. 1 from [2]).

3 Results

The goal of computer simulations was to find the module parameters to obtain the output function dynamics similar that in Fig. 3. On the base of testing three video image sequences from SYLAHP database the following module parameters were choose: R^E= 3 pixels, R^I= 7 pixels, w^I = 1.5. The receptive field center of excitatory and inhibitory neurons were manually positioned in the center of a corresponding facial landmark region.

Example of activity dynamics for excitatory neuron of the neural network module while processing video image sequences in real-time is presented in Fig. 3 (lower row).

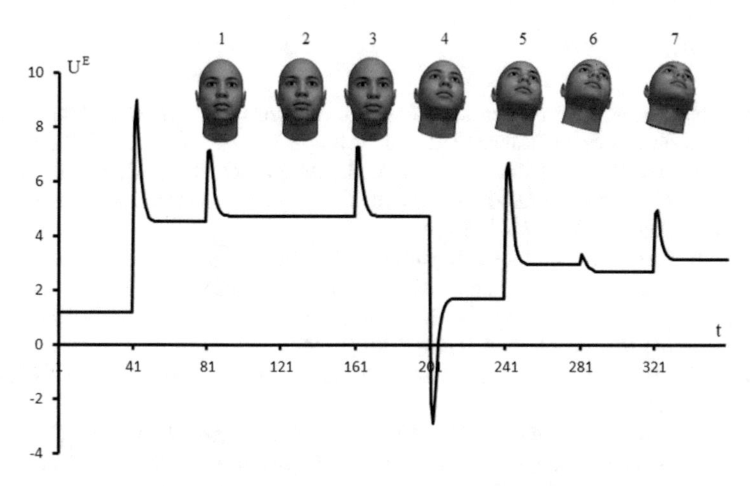

Fig. 3. Example of activity dynamics for excitatory neuron while processing video image sequence. The face pictures during different moments of U^E dynamics are shown in the upper row.

As it can be seen in the figure, the U^E amplitude and polarity qualitatively correspond to the face motion amplitude. In particular, U^E firing is absent if face position do not change (compare face positions 2 and 3). To the contrary, U^E firing has maximal amplitude in response to sharp change of face orientation (compare face positions 3 and 4). It is important that the initial front of U^E quick changes is equal to 12 ms in all cases ($n = 46$).

4 Conclusion

The obtained results indicate that the neural network modules with very simple structure can detect the initial moment of head motion while processing synthetic video facial image sequences from SYLAHP database. It was shown that the U^E amplitude and polarity qualitatively correspond to face motion amplitude.

Evidently sharp changes of output function of module excitatory neuron (U^E) may be used as signal to the modules for automatic determination of facial landmarks and their feature description. This supposition will be tested at next stages of studies by including the module described in the present paper into the full system to detect head motions and calculate head pose developed earlier [7]. Performance of the system as a whole will be evaluated while processing video image sequences of real facial images.

Acknowledgements. The work is supported by the Russian Ministry for Education and Science, projects NN 2.955.2017/4.6 and 6.5961.2017/8.9.

References

1. Alioua, N., Amine, A., Rogozan, A., Bensrhair, A., Rziza, M.: Driver head pose estimation using efficient descriptor fusion. EURASIP J. Image Video Process. **2016**, 2 (2016)
2. Arcoverde, E., Duarte, R.M., Barreto, R.M., Magalhaes, J.P., Bastos, C., Ren, T.I., Cavalcanti, G.: Enhanced real-time head pose estimation system for mobile device. Integr. Comput. Aided Eng. **21**, 281–293 (2014)
3. Caplier, A., Benoit, A.: Biological approach for head motion detection and analysis. In: Proc. 13-th European Signal Processing Conference EUSIPCO-2005, pp. 963–966. Turkey (2005)
4. Chey, J., Grossberg, S., Mingolla, E.: Neural dynamics of motion processing and speed discrimination. Vis. Resolut. **38**, 2769–2786 (1998)
5. Czuprynski, B., Strupczewski, A.: High accuracy head pose tracking survey. Lect. Notes Comput. Sci. **8610**, 407–420 (2014)
6. Derkach, D., Ruiz, A., Sukno, F.M.: Head pose estimation based on 3-D facial landmarks localization and regression. In: 12th International Conference on Automatic Face & Gesture. Recognition, pp. 820–827 (2017)
7. Gao, X.W., Anishenko, S., Shaposhnikov, D., Podladchikova, L., Batty, S., Clark, J.: High-precision detection of facial landmarks to estimate head motions based on vision models. J. Comput. Sci. **3**(7), 528–532 (2007)
8. Jaimes, A., Sebe, N.: Multimodal human computer interaction: a survey. Comp. Vis. Image Underst. **108**(1–2), 116–134 (2007)
9. Murphy-Chutorian, E., Trivedi, M.M.: Head pose estimation in computer vision: a survey. IEEE Trans. Pattern Anal. Mach. Intell. **31**(4), 607–626 (2009)
10. Rahmim, A.: Advanced motion correction methods in PET. Rev. Artic. Iran J. Nucl. Med. **13** (241), 1–17 (2005)
11. Shevtsova, N., Faure, A., Klepatch, A., Podladchikova, L., Golovan, A., Rybak, I.: Model of foveal visual preprocessor. Proc. SPIE **2588**, 588–597 (1995)
12. Steger, T.R., Jackson, E.F.: Real-time motion detection of functional MRI data. J. Appl. Clin. Med. Phys. **5**(2), 64–70 (2004)
13. Werner, Ph.; Saxen, F., Al-Hamadi, A.: landmark based head pose estimation benchmark and method. In: Int. Conference on Image Processing, pp. 3909–3913. China (2017)
14. Wilke, S.D., Thiel, A., Eurich, C.W., Greschner, M., Bongard, M., Ammermuller, J., Schwegler, H.: Extracting motion information using a biologically realistic model retina. In: Proc. on European Symposium on Artificial Neural Networks, ESANN'2001, pp. 323–328. Belgium (2001)
15. Wu, Y., Gou, C., Ji, Q.: Simultaneous facial landmark detection, pose and deformation estimation under facial occlusion. In: Proc. Conference on Computer Vision and Pattern Recognition, pp. 3471–3480. USA (2017)

Semiempirical Model of the Real Membrane Bending

Dmitriy A. Takhov[1], Mariya R. Bortkovskaya[1], Tatyana T. Kaverzneva[1], Daniel R. Kapitsin[1], Irina A. Shishkina[1], Daria A. Semenova[1], Pavel P. Udalov[1], and Ildar U. Zulkarnay[2](✉)

[1] Peter the Great St. Petersburg Polytechnic University, Saint Petersburg, Russia
[2] Bashkir State University, Ufa, Russia
zulkar@inbox.ru

Abstract. The tasks of constructing mathematical models from heterogeneous data, including differential equations, boundary and initial conditions, observational data and other information about the modeled object, are of great practical importance, in particular, in the construction of digital counterparts of complex technical objects. Especially relevant is the search for methods for constructing the above mathematical models in a situation where the physical model and, consequently, the differential equation is known with insufficient precision for modeling purposes. We have developed new methods for constructing mathematical models of the type mentioned above and check them on a model problem with real measurements. In this paper, we consider the solution of the problem of modeling the deflection of a loaded circular membrane, in the center of which the weight of a given mass is located. The accuracy of the models expressing the dependence of the deflection of the membrane from the distance to the center is compared. We constructed the first model on the basis of an analytical solution of the equation of equilibrium conditions. The second model was obtained with the help of the original modification of the refined Euler method. When constructing the second model, it is necessary to select the same number of coefficients as in the construction of the first model. We built the third model in the form of an output of a neural network. The coefficients of the models were selected from the data obtained experimentally. The resulting approximate accuracy models outperform the model based on the exact solution. The neural network model turned out to be the most accurate, but it requires the selection of a larger number of coefficients.

Keywords: Neural networks · Refined Euler method · Round membrane
The dependence of the deflection on the radius

1 Introduction

The wide use of fabric materials for various purposes requires techniques for modeling their behavior under load. The modeling of such phenomena is carried out, as a rule, by differential equations and the accompanying boundary value and initial value, etc. [12].

B. Kryzhanovsky et al. (Eds.): NEUROINFORMATICS 2018, SCI 799, pp. 221–226, 2019.
https://doi.org/10.1007/978-3-030-01328-8_26

In this case, the problem always arises of increasing the accuracy of the description of a physical phenomenon obtained from solution models. There are many approaches to considering computations [2], among which we can use the support vector machine (SVM) [3], the particle swarm optimization [4], the multi fidelity importance sampling [5], Bayesian statistics based methods [6], artificial neural networks (ANN) for solving partial differential equations for both boundary value and initial value problems [7].

In this paper, we mainly follow the last approach, applying our methods [8, 9] to construct the model of the investigated membrane on the basis of the differential equation and experimental data.

A circular membrane of radius R made of Oxford 600 fabric is placed on it, loads of different masses are alternately arranged, the membrane is assumed to be weightless, the load is located in the center of the membrane, its radius is further designated as a, it is assumed that the stretching is isotropic. In this paper, we compare the exact solution of the differential equation and the approximate solution obtained by the two-step Euler method [8, 10], in terms of their correspondence to the experimental data. In addition, a neural network solution is constructed using the methods [11, 12].

2 Methods

Let $u(r)$ – the deflection of the membrane from the equilibrium position. For its description we use the equation:

$$u''_{rr} + \frac{1}{r} u'_r = \begin{bmatrix} -B, & \mathit{if}\ r \in [0,\ a] \\ 0, & \mathit{if}\ r \in (a,\ R] \end{bmatrix} \tag{1}$$

which is the Laplace equation in polar coordinates, where $u(r, \varphi) = u(r)$, that is, the desired function does not depend on the direction, but depends only on the distance r of the point from the center of the membrane [11]. Here $B = A/T$, A - weight of the load, T is the absolute value of the tensile force applied to the edge of the membrane. Since the membrane is assumed to be weightless, its weight in the right-hand side of Eq. (1) is absent.

For a further comparison with the approximate solution, we write out its exact solution of Eq. (1):

$$u(r) = \begin{cases} \frac{1}{2} Ba^2 \ \ln\frac{a}{R} + u_0 - \frac{1}{4} B(r^2 - a^2),\ r \in [0, a] \\ \frac{1}{2} Ba^2 \ \ln\frac{r}{R} + u_0,\ r \in (a, R] \end{cases} \tag{2}$$

Here $u_0 = u(R)$. The solution of (2) is obtained with allowance for continuity of $u(r)$ for $r = a$ and boundedness of the solution for $r = 0$. The choice of the parameter B is made here using the method of least squares, so as to minimize the magnitude $\sum_{i=1}^{10} (u(r_i) - u_i)^2$. Here r_i are the values of r for which the deflection measurements were made, u_i - the results of the corresponding measurements, $u(r_i)$- the values of the function $u(r)$ found by formula (2). Obviously, finding the value of B, we will know the corresponding value of $z_0 = u'(R)$. Taking into account the above formulas, knowing the weight of the cargo from the experiment, and determining the value B, we determine the value of the tensile force T. We reduce Eq. (1) to the normal system of differential equations [10]:

$$\begin{cases} u' = z, \\ z' = -\frac{z}{r} + f(r). \end{cases} \tag{3}$$

Here $f(r)$ is the right-hand side of Eq. (1). After changing the variable $x = R - r$, solving the system (3) by the two-step Euler method, we obtain

$$\begin{aligned} u(x) &= u_0 - xz_0 - \frac{x^2 z_0}{4R}, \\ z(x) &= \left(z_0 + \frac{xz_0}{2R}\right) \cdot \frac{2R}{2R-x}. \end{aligned} \tag{4}$$

The value of u_0, as before, is taken from the experiment. The value of z_0 is not yet defined. Solution (2) is considered for $x \in [0, R-a)$, that is, for $r \in (a, R]$. Now for $r \in [0, a]$ we solve system (3) by the same method, assuming the value of deflection $\tilde{u}_0$ for $r = 0$ unknown, and the value of derivative u_r' for $r = 0$ zero. Then we get

$$\begin{aligned} u(r) &= \tilde{u}_0 - \frac{r^2 B}{4}, \\ z(r) &= -\frac{rB}{2}. \end{aligned} \tag{5}$$

Demanding the continuity of the solution u and its derivative Z at a point $r = a$, we obtain the following conditions:

$$\begin{aligned} \tilde{u}_0 - \frac{a^2 B}{4} &= u_0 - (R-a)z_0 - \frac{(R-a)^2 z_0}{4R}, \\ -\frac{1}{2}aB &= z_0 \frac{2R}{R+a}. \end{aligned} \tag{6}$$

From the continuity conditions (6), we find the expressions for the parameters $\tilde{u}_0$ and B in terms of the value of Z_0, and the last one is determined using the least squares method so as to minimize the value of $\sum_{i=1}^{10} (u(r_i) - u_i)^2$ where we calculate $u(r_i)$ by formula (5) for $r_i \leq a$ and by (4) for $r_i \geq a$, $x_i = R - r_i$.

Now, in the approximate solution u expressed by formulas (4) and (5), all the parameters will be found, and we can compare it with the exact solution in the same way as one did considering the homogeneous differential equation of the membrane.

In addition, methods were used for constructing a neural network model based on the differential equation and experimental data, as described in many of our papers, see, for example [11, 12].

3 Results of Calculations

Let's compare the exact solution (1) and the model obtained with our [1] modification of the two-step Euler method [10]. The value of z_0 for the exact solution is = 0.410, for the approximate value = 0.884, the value of B for the exact solution is 456.028, for the approximate value = 111.173, the value of T for the exact solution is 0.044, for the approximate value = 0.180. Load weight = 2000 gr, load radius = 3 cm, membrane radius = 50 cm. Using the above-mentioned values, approximate and exact solutions were obtained (Figs. 1 and 2):

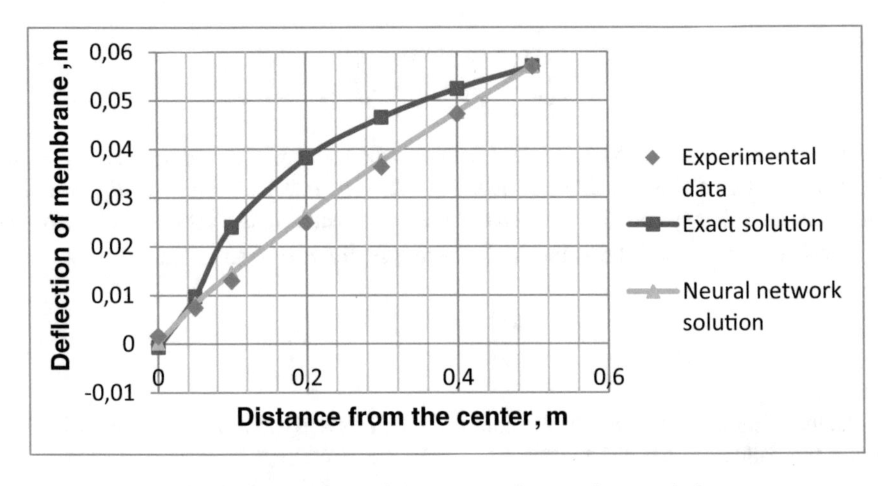

Fig. 1. Comparison of the exact and approximate solution

From the graphs (Figs. 1 and 2) we see that the exact solution deviates more strongly from the experiment than the approximate model. To solve this problem, using the minimization of the error functional by the methods of [11, 12], we constructed a neural network with one hidden layer and a hyperbolic tangent activation function. For seven values of the deflection obtained from the experiment, a neural network was constructed, at all points except for the third and sixth, training was conducted, and the remaining two points were used to test the data obtained.

As a result of computational experiments, we see that for 1 and 2 neurons the error on the test sample is within the error of the training sample, and for 3 or more neurons it exceeds it several times, therefore in this task it is inappropriate to use a neural network with more than 2 neurons (Fig. 3):

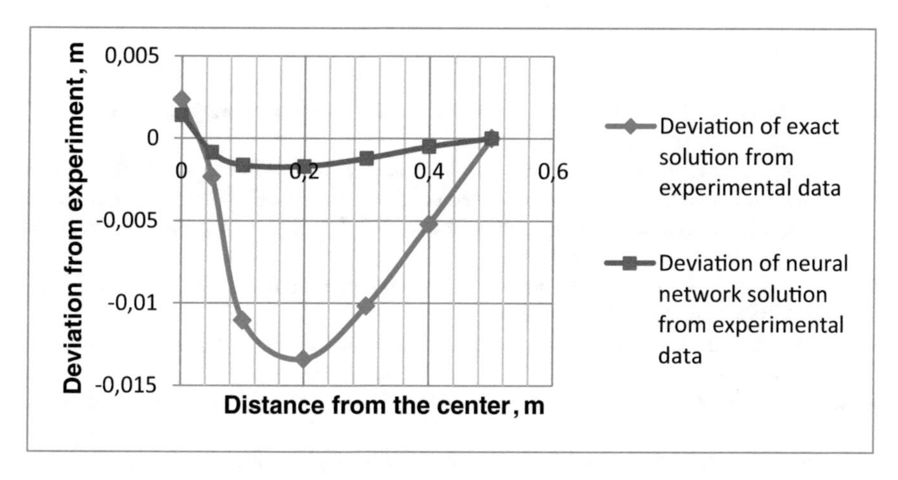

Fig. 2. Graphs of deviation of solutions from experimental values for experience.

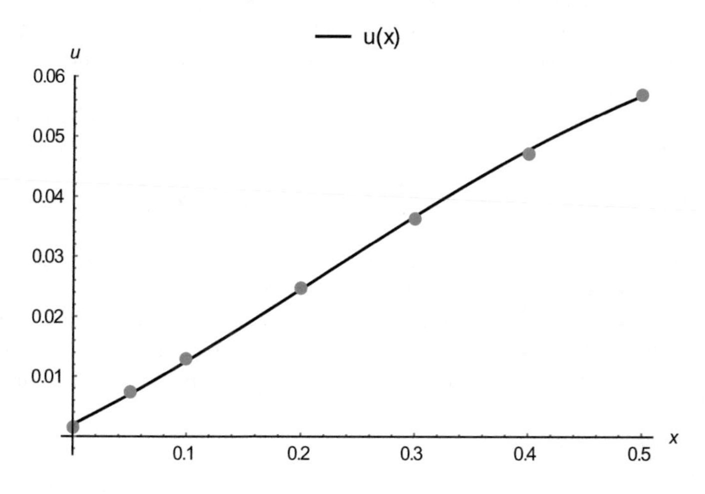

Fig. 3. The graph of the neural network dependence of the deflection (m) on the radius (in m.).

4 Conclusions

The methods of work can be useful in constructing models based on inaccurate information in the form of differential equations and experimental data. In this case, the model constructed by the methods of [8, 9] is less precise, but in its construction we select only the natural parameters of the problem. The neural network model more accurately reflects the experimental data, but has the risk of retraining.

Acknowledgments. The article was prepared on the basis of scientific research carried out with the financial support of the Russian Science Foundation grant (project No. 18-19-00474).

References

1. Kortelainen, J.: Semantic data model for multibody system modelling. Espoo VTT Publications 766 (2011)
2. Glotzer, S., Kim, S., Cummings, P., Deshmukh, A., Head-Gordon, M., Karniadakis, G., Petzold, L., Sagui, C., Shinozuka, M.: International assessment of research and development in simulation-based engineering and science. WTEC panel report, World Technology Evaluation Center, Inc. WTEC (2009)
3. Peherstorfer, B., Willcox, K.: Dynamic data-driven reduced-order models. Comput. Methods Appl. Mech. Eng. **291**, 21–41 (2015)
4. Rosenblatt, F.: The perceptron: a probabilistic model for information storage and organization in the brain. Psychol. Rev. **65**(6), 386 (1958)
5. Maaten, L., Hinton, G.: Visualizing data using t-SNE. J. Mach. Learn. Res. **9**, 2579–2605 (2008)
6. Suzuki, K.: Artificial Neural Networks: Methodological Advances and Biomedical Applications. INTECH Open Access Publisher (2011)
7. Largris, I.E., Likas, A.: Artificial neural networks for solving ordinary and partial differential equations. IEEE Trans. Neural Networks **9**(5), 987–1000 (1998)
8. Lazovskaya, T., Tarkhov, D.: Multilayer neural network models based on grid methods. IOP Conf. Ser.: Mater. Sci. Eng. **158** (2016). http://iopscience.iop.org/article/10.1088/1757-899X/158/1/01206
9. Vasilyev, A.N., Tarkhov, D.A., Tereshin, V.A., Berminova, M.S., Galyautdinova, A.R.: Semi-empirical neural network model of real thread sagging. Studies in Computational Intelligence, vol. 736, pp. 138–146. Springer, New York (2018)
10. Hairer, E., Norsett, S.P., Wanner, G.: Solving Ordinary Differential Equations I. Nonstiff Problem. Springer, Berlin (1987)
11. Lazovskaya, T.V., Tarkhov, D.A., Vasilyev, A.N.: Parametric neural network modeling in engineering. Recent. Pat.S Eng. **11**(1), 10–15 (2017)
12. Lozhkina, O., Lozhkin, V., Nevmerzhitsky, N., Tarkhov, D., Vasilyev, A.: Motor transport related harmful PM2.5 and PM10: from onroad measurements to the modelling of air pollution by neural network approach on street and urban level. J. Phys.: Conf. Ser. **772** (2016)

Abnormal Operation Detection in Heat Power Plant Using Ensemble of Binary Classifiers

Alexander G. Trofimov[1], Ksenia E. Kuznetsova[1(✉)], and Alexandra A. Korshikova[2]

[1] National Research Nuclear University «MEPhI», Moscow, Russian Federation
kuznetsova.ke96@yandex.ru
[2] Interautomatika AG, Moscow, Russian Federation

Abstract. The problem of abnormal operation detection is considered for prediction of malfunctions appearance and their progress in the equipment of power plant. Abnormal operation detection method based on multivariate state estimation technique (MSET) along with machine learning algorithms is proposed. The ensemble of linear regression models is used for feature construction. The ensembles of binary classifiers (logistic regressions) together with the multilayer neural network are used for the abnormal operation index calculation based on the constructed features. The method was applied to abnormal operation detection in turbo feed pump (TFP 1100-350-17-4) at Kashirskaya heat power plant (Moscow region, Kashira). It is shown that the abnormal operation index of the pump starts to increase a few days before accidents appear and stays close to zero during the normal operation periods. The obtained results demonstrate that the developed model can be used to detect and predict operation anomalies in the power plant equipment.

Keywords: Abnormal operation detection · Predictive analytics
Ensemble learning · Logistic regression · Multilayer neural network
Power plant

1 Introduction

During the period of active power plant operation some malfunctions and emergencies inevitably occur. They can have a negative impact on the performance of equipment or even lead to failure. A predictive model of emergencies would allow operators to take actions and repair the equipment on time, thereby increasing its efficiency. Development of such models is the subject of predictive analytics [1, 2].

For almost a century of existence of the Russian power industry, the set of measured parameters that allows successful operating of the power plant equipment without accidents was formed. Preventing malfunctions or accidents in the vast majority of cases does not require any special methods and is based directly on measurements (permissible, warning or emergency alarms are triggered). The accidents that occur are mainly caused either by a violation of rules and regulations or by external factors that cannot be predicted (accidents in the power system, natural phenomena, human factor, etc.). Nevertheless, there is still a certain percentage of malfunctions and accidents that

B. Kryzhanovsky et al. (Eds.): NEUROINFORMATICS 2018, SCI 799, pp. 227–233, 2019.
https://doi.org/10.1007/978-3-030-01328-8_27

do not manifest themselves explicitly in the individual measurements but their progress can be revealed by predictive analysis for some set of measured parameters [3]. In this regard, there is a need to develop models to predict such emergencies in advance.

It should be noted that the use of predictive analysis on the power plant does not require any installation of new sensors (the number of necessary measurements is determined by the equipment suppliers, designers of the power facilities and plant engineers).

The basic idea of predictive analytics is that the occurrence of malfunction and its progress can be predicted with some probability based on data analysis in the period preceding the malfunction (this period can be from several minutes to several years). Modern trends in predictive analytics are directed toward the use of data mining methods and machine learning algorithms, in particular, random forests [4].

Modern software and hardware automatic control systems in power plants (mainly foreign ones) include more complex built-in systems for early failure detection, which are based on statistical models of anomaly detection. They work with according to the principle that if current characteristics of the equipment differ significantly from the nominal values this is a sign of abnormal functioning [5]. The disadvantage of such systems is often late detection of anomalies when there is no time to eliminate them before the accident.

The main players in the market of predictive analytics systems for power plants are Siemens, Clover Group, JSC «ROTEC» . The foreign systems have a significant drawback that they use remote monitoring of the technical state of the equipment: signals from the power plant sensors are collected in a closed mode and sent via Internet to the foreign Expert Center server located abroad.

Predictive analytics system Siemens SPPA-D3000 uses classical outlier detection methods based on the deviation analysis of the equipment parameters from the cloud of points modeled by the multidimensional normal distribution [6].

The JSC «ROTEK» offers the software and technical complex «PRANA» for predictive analytics, which carries out continuous diagnostics of the power plant and predicts the changes in its technical state. The system is based on the multivariate state estimation technique (MSET). The idea of MSET [7, 8] is to construct statistical (in particular, regression) models of equipment responses using the data obtained during the normal operation mode and to analyze the deviations of the observed values from the model outputs (regression residuals). High deviations may indicate the inadequacy of the normal functioning model, i.e. the possible malfunctioning.

In this paper, we propose a method for detecting anomalies in the dynamical system functioning using the idea of MSET in conjunction with machine learning models such as logistic regression and a multilayer neural network.

2 Problem Statement

Let $x(t) = (x_1(t), \ldots, x_m(t))^T$ be the m-dimensional vector of observed indicators of the dynamical system (turbo feed pump) at time t. The measurements of this vector are carried out by a monitoring system with a certain time step Δt (for example, 5 min), resulting in a sequence of vectors $X = (x(1), \ldots, x(T))$ combined into matrix X, where

T is number of measurements. Each moment of time t, $t = \overline{1, T}$, is assigned by the expert to one of two classes as corresponding to the normal functioning or emergency (or pre-emergency) state. Denoting the class label at the time t through $y(t)$ ($y(t) = 0$ for the normal state and $y(t) = 1$ for the emergency) leads to the vector of labels $y = (y(1), \ldots, y(T))$.

One of MSET problems is the choice of input and output variables to build a regression model for the normal operation mode [7]. Predictive properties of regression residuals largely depend on these variables. At the same time, it is reasonable to assume that the output variables depend not only on the observed indicators but also on some derived coefficients that are not explicitly measured. The construction of features $z_1, \ldots, z_M$ from observed indicators $x_1, \ldots, x_m$ requires additional expert knowledge about the accident under consideration.

As soon as the features are constructed it is necessary to determine which ones are inputs and which ones are outputs for the regression model. This question also requires additional knowledge and even experts cannot always give an unambiguous answer because in complex systems often "everything is related to everything".

Another problem of MSET is related to the estimation of abnormal operation degree. In the classical MSET [7] the decision about the anomaly is based on the results of comparing regression model outputs with observed values.

Taking into account the complexity and ambiguity of the answers to these questions we propose to construct the ensemble of regression models [9, 10] using different combinations of input and output variables and ensemble of models to estimate the abnormal operation index (AOI). Each model in the ensemble calculates its particular AOI $p_i(t)$ at time t, $t = \overline{1, T}, i = \overline{1, N}$, where N is the number of models in ensemble, and the final solution $p(t)$ is taken with respect to ensemble decision rule. At each time t the AOI $p(t)$ takes a value from the interval (0; 1), values close to 0 mean normal functioning and values close to 1 mean emergency.

3 Abnormal Operation Index Calculation

The algorithm proposed for AOI calculation consists of the following steps.

Step 1. Construction of features $z_1, \ldots, z_M$ based on observed indicators $x_1, \ldots, x_m$ and calculation of their values $z(t) = (z_1(t), \ldots, z_M(t))^T$ at each time t, $t = \overline{1, T}$. As a result we obtain a sequence $Z = (z(1), \ldots, z(T))$ combined into matrix Z. Thus, the source data D is the $M * T$–dimensional feature matrix Z and the T–dimensional vector of class labels y: $D = (Z; y)$.

Following experts' opinions, it was decided to include observed measures $x_1(t), \ldots, x_m(t)$ of pump's indicators as well as their values normalized to feed water flow into the pump's feature vector $z_1(t), \ldots, z_M(t)$ at time t.

Step 2. Partitioning the sample D into disjoint subsamples: two training D_{tr1}, D_{tr2} samples and one test D_{tst} sample: $D = D_{tr1} \cup D_{tr2} \cup D_{tst}$. We denote the sets of the corresponding time moments as T_{tr1}, T_{tr2} and T_{tst}: $T_{tr1} \cup T_{tr2} \cup T_{tst} = \{1, \ldots, T\}$.

The training samples D_{tr1} and D_{tr2} are used to construct regression models and AOI model respectively and the sample D_{tst} is used for the final testing.

Step 3. Training K linear regression models on sample D_{tr1} and calculation their residuals on sample D_{tr2}.

Let z_{i_k} and ζ_k be a scalar response and a set of regressors respectively for the k-th regression model, $i_k \in \{1, \ldots, M\}$, ζ_k is a subset of the feature set $\{z_1, \ldots, z_M\}$ that does not contain response z_{i_k}, $k = \overline{1, K}$. Then the k-th regression model is:

$$z_{i_k} = \varphi_k(\zeta_k) + e_k, k = \overline{1, K}, \tag{1}$$

where φ_k is a linear regression function, e_k is residual. The response variable z_{i_k} and the set of regressors ζ_k for each regression model are chosen randomly. The random number of regressors is drawn from discrete uniform distribution $U\{1, M–1\}$, the response and regressors are drawn from the set $\{z_1, \ldots, z_M\}$ without repetitions.

Let's denote $e(t) = (e_1(t), \ldots, e_K(t))^T$, where $e_k(t) = z_{i_k}(t) - \varphi_k(\zeta_k(t))$, as a vector of residuals at time t, $t \in T_{tr2}$, and E is the K^*T_{tr2}–residual matrix composed from vectors $e(t)$, $t \in T_{tr2}$.

Step 4. Training the ensemble of N binary classifiers (logistic regressions) on the data from matrix E and the corresponding class labels at time t, $t \in T_{tr2}$. The i-th weak binary classifier is described as follows:

$$p_i = \frac{1}{1 + \exp(-\psi_i(\varepsilon_i))}, i = \overline{1, N}, \tag{2}$$

where ε_i is a subset of residuals $\{e_1, \ldots, e_K\}$, ψ_i is a linear regression function, p_i is the i-th output of weak classifier (particular AOI), $p_i \in (0; 1)$. The set of regressors ε_i for each logistic regression is chosen randomly in the same manner as in step 3.

Let's denote $p(t) = (p_1(t), \ldots, p_N(t))^T$ as output vector of classification ensemble at time t, where $p_i(t)$ is the output of the i-th weak classifier at time t, $t \in T_{tr2}$, and P is the N^*T_{tr2}–dimensional AOI matrix composed from vectors $p(t)$, $t \in T_{tr2}$.

Step 5. Training the decision rule Φ on the data from matrix P and the corresponding labels at times t, $t \in T_{tr2}$: $p = \Phi(p_1, \ldots, p_N)$, where p is the final AOI.

In the simplest case an averaging of weak classifiers outputs can be used as a decision rule. However, given the unequal contribution of individual weak classifiers in the final decision as well as their different accuracies and generalization abilities a feed-forward neural network can be used as ensemble decision rule Φ. The particular AOIs $p_1, \ldots, p_N$ are the network's inputs and the final AOI p is its output.

The scheme of the proposed model is shown in Fig. 1.

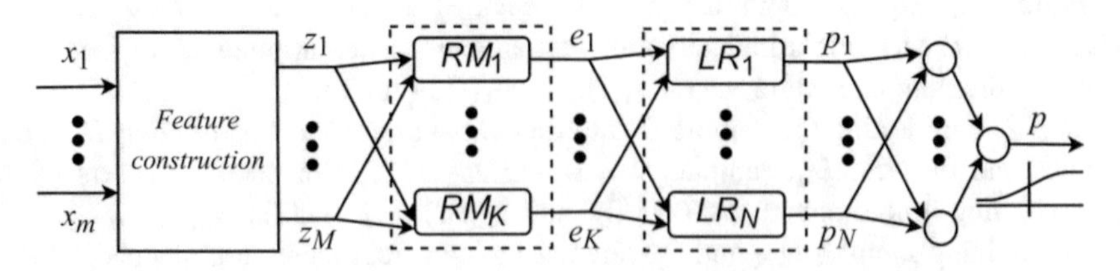

Fig. 1. Scheme of abnormal operation index calculation. RM are regression models, LR are logistic regressions.

As soon as the ensemble was constructed (all regression models, weak classifiers and the neural decision rule are learned), we test it on the test sample D_{tst} and visually evaluate the behavior of the AOI $p(t)$ calculated in the pre-emergency time intervals.

4 Experimental Results

Researches were carried out on the historical measures of the turbo feed pump's indicators (TFP 1100-350-17-4) at Kashirskaya heat power plant (Moscow region, Kashira) during three years. The indicators (total $m = 45$, including oil pressure, bearing vibration speed, steam temperature, etc.) were measured every $\Delta t = 5$ min. After excluding time intervals when the pump was off, the number of time points was $T = 39629$. During the operating period four anomalies were observed (overheating of the drive turbine stop blocks, leakage into the cover on the bearing side, condensate leakage into cap nut and leakage into end cap from the starter side). For each accident, the moment of its occurrence is known (up to one day). The vector of class labels y was formed as follows: the values $y(t)$ were assigned to 1 for all time points t from the noon of the second day before the accident to the noon of the second day after the accident and all other labels were assigned to 0. This labeling strategy is due to possible inaccuracy in the accident registration date.

Figure 2 shows a fragment of the feed water pressure dynamics of the pump preceding one of the accidents.

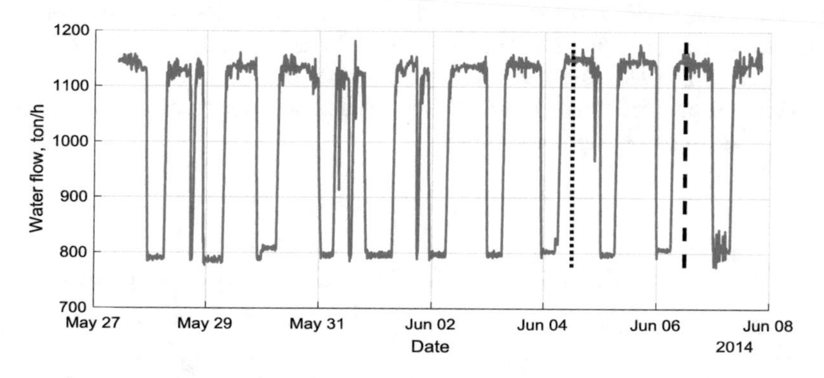

Fig. 2. Dynamics of the feed water pressure of the pump on the time interval preceding the accident (condensate leakage into cap nut). Moment of the accident is marked by a dashed vertical line and the time interval to the right of a dotted vertical line is marked as an emergency interval.

Statistical characteristics of the regression models' determination coefficients R^2 calculated on the training sample D_{tr1} and the sample D_{tr2} (which was the test one for them) are shown in Table 1. The results indicate that the constructed regression models have good generalizing abilities on average.

For each trained regression model the residuals $e_k(t)$, at each time t, $t \in T_{tr2}, k = \overline{1, K}$, were calculated and used to construct $N = 20$ weak classifiers (logistic

Table 1. Statistical characteristics of the regression models' determination coefficients R^2.

Sample	Minimum	Maximum	Mean ± std
D_{tr1}	0.83	1	0.98 ± 0.04
D_{tr2}	0.71	0.99	0.97 ± 0.07

regressions). If the weak classifier's AUC ROC on training sample D_{tr1} was less than 0.6 then such classifier was excluded from the ensemble. The statistical characteristics of weak classifiers' AUC ROC on training sample D_{tr2} and test sample D_{tst} are shown in Table 2. The results indicate that the trained weak classifiers have good generalizing abilities on average.

Table 2. Statistical characteristics of weak classifiers' AUC ROC.

Sample	Minimum	Maximum	Mean ± std
D_{tr1}	0.71	0.95	0.95 ± 0.08
D_{tr2}	0.69	0.93	0.94 ± 0.11

Further, the outputs of weak classifiers on the training sample D_{tr2} were used to train a two-layer perceptron with 5 tanh-neurons in the hidden layer and 1 sigmoid neuron in the output layer. The training was performed with conjugate gradient descent method and the binary cross-entropy loss function. The neural network was trained during 120 epochs, an early stopping was used. The achieved AUC ROC was 0.99 on the training sample D_{tr2} and 0.98 on the test sample D_{tst}.

Figure 3 shows the dynamics of AOI observed at the trained perceptron's output on the time interval preceding the accident.

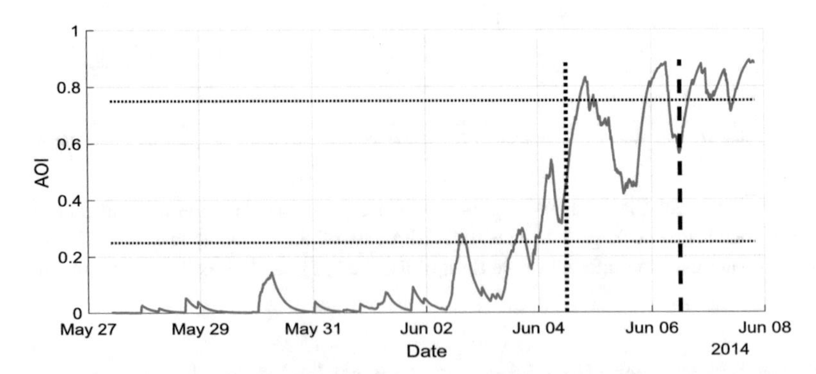

Fig. 3. The modeled AOI dynamics on the time interval preceding the accident. Moment of the accident is marked by a dashed vertical line and the time interval to the right of a dotted vertical line is marked as an emergency interval. Dotted horizontal lines are at levels 0.25 and 0.75.

The figure shows that the AOI starts to increase on the morning of June 2, while the emergency interval in the training sample began only from noon of June 4. There can be observed a trend in the AOI dynamics from 0 to about 0.8 which indicates its predictive ability. We note that the power plant experts haven't found any peculiarities in the dynamics of individual pump's indicators during the pre-emergency time interval up to the moment of the accident.

5 Conclusions

The paper proposes a new method to estimate the abnormal operation degree of a dynamical system based on the idea of MSET combined with machine learning models. An ensemble of regression models was developed to solve the key problem of MSET related to the choice of regression model's inputs and outputs. A logistic regression ensemble with a perceptron-based decision rule was used to calculate the abnormal operation index.

Experimental research of the proposed method carried out on the turbo feed pump TFP 1100-350-17-4 operation data revealed that the constructed model has good generalizing and predictive capabilities (AUC ROC $\approx$ 0.98) which allow it to be used for monitoring the state of the pump at heat power plant and predict its accidents.

References

1. Siegel, E.: Predictive Analytics: The Power to Predict Who Will Click, Buy, Lie, or Die. Wiley, Hoboken (2016)
2. Waller, M.A., Fawcett, S.E.: Data science, predictive analytics, and big data: a revolution that will transform supply chain design and management. J. Bus. Logist. **34**(2), 77–84 (2013)
3. Gromak, E.V., Naumov, S.A., Shishov, V.A.: System for remote monitoring and prognostics of JSC "ROTEC" as an element of energy security. New Russ. Power Gener. **6**, 36–46 (2016)
4. He, L., et al.: Random forest as a predictive analytics alternative to regression in institutional research. Pract. Assess. Res. Eval. **23**(1), 16 (2018)
5. Lipatov, M.: The first complex of predictive analytics for power and industrial equipment in Russia. Exposition. Oil. Gas **3**(49), 82–83 (2016)
6. Aggarwal, C.C., Yu, P.S.: Outlier detection for high dimensional data. In: ACM Sigmod Record, vol. 30, no. 2, pp. 37–46. ACM (2001)
7. Cheng, S., Pecht, M.: Multivariate state estimation technique for remaining useful life prediction of electronic products. In: Proceedings of AAAI Fall Symposium Artificial Intelligence. Prognostics, Arlington, VA, pp. 26–32 (2007)
8. Zavaljevski, N., Gross, K.C.: Sensor fault detection in nuclear power plants using multivariate state estimation technique and support vector machines. Argonne National Laboratory, Argonne, IL (US) (2000). ANL/RA/CP-103000
9. Dietterich, T.G.: Ensemble learning. In: The Handbook of Brain Theory and Neural Networks, 2nd edn., pp. 110–125 (2002)
10. Tresp, V.: Committee machines. Handbook for Neural Network Signal Processing, pp. 1–18 (2001)

Semi-supervised Classifying of Modelled Auditory Nerve Patterns for Vowel Stimuli with Additive Noise

Anton Yakovenko(✉), Eugene Sidorenko, and Galina Malykhina

Peter the Great St.Petersburg Polytechnic University, St.Petersburg, Russia
yakovenko_aa@spbstu.ru

Abstract. The paper proposes an approach to stationary patterns of auditory neural activity analysis from the point of semi-supervised learning in self-organizing maps (SOM). The suggested approach has allowed to classify and identify complex auditory stimuli, such as vowels, given limited prior information about the data. A computational model of the auditory periphery has been used to obtain auditory nerve fiber responses. Label propagation through Delaunay triangulation proximity graph, derived by SOM algorithm, is implemented to classify unlabeled units. In order to avoid the "dead" unit problem in Emergent SOM and to improve method effectiveness, an adaptive conscience mechanism has been realized. The study has considered the influence of AWGN on the robustness of auditory stimuli identification under various SNRs. The representation of acoustic signals in the form of neural activity in the auditory nerve fibers has proven more noise-robust compared to that in the form of the most common acoustic features, such as MFCC and PLP. The approach has produced high accuracy, both in case of similar sounds and with high SNR.

Keywords: Auditory nerve data analysis · Unsupervised learning
Neurogram · Machine hearing · Label propagation
Self-organizing maps

1 Introduction

The central nervous system receive information about the environment through peripheral coding. In the course of evolution, presentation of sensory information has been formed that allows the subject to effectively recognize phenomena in various physical contexts. In particular, the listener can successfully analyze the auditory scene and detect sound events in a wide variety of acoustic conditions.

Automatic speech processing systems are considerably inferior to those of the human in quality of recognition [1], particularly with regard to continuous speech and noisy environments. The performance of automatic speech processing systems can be improved through introducing additional linguistic and contextual information [2], e.g. through prediction of isolated phonemes. However, the

B. Kryzhanovsky et al. (Eds.): NEUROINFORMATICS 2018, SCI 799, pp. 234–240, 2019.
https://doi.org/10.1007/978-3-030-01328-8_28

intrinsic robustness of speech recognition in humans does not inherently depend on language or context [3]. Thus, further development of speech technology can involve integration of knowledge on physiology of auditory perception and neural coding.

The present study proposes a method for analysis of resultant stationary patterns of evoked neural activity, which are generated by a simulation model of the auditory nerve as a response to complex tones represented vowel stimuli, as well as their classification and identification. Generally, such multidimensional data have a complex structure and a large number of observations, which complicates the task and limits the application of many methods. The relevant practical problem is classification in limited prior knowledge, when class labels are only known for some of the observations [4]. In this case, supervised methods cannot ensure the desired accuracy. On the other hand, unsupervised techniques do not enable using prior information and learning data on the basis thereof. To compensate for these shortcomings, the study proposes an approach based on the model of self-organizing maps in the context of semi-supervised classification task. The present study is based on the results of the previous work, in which, processing of the modelled auditory nerve fiber responses revealed a cluster structure in the unlabeled data of the neural activity for voices of different speakers [5]. In order to estimate robustness, the study considers the influence of additive white Gaussian noise (AWGN) on the auditory stimuli. Finally, a conclusion is made about the results obtained versus traditional acoustic features widely used for signal representation in automatic speech recognition, such as mel-frequency cepstral coefficients (MFCC) [6] and perceptual linear prediction coefficients (PLP) [7].

2 Methodology Background

In the auditory periphery, an input sound is converted from acoustic to mechanical oscillations. The latter, in turn, stimulate the electrical activity of the auditory nerve fibers. The signals generated are transported to the corresponding cortex areas via nerve fibers, which results in auditory perception. In this way, humans can effectively identify perceptive qualities of sound (pitch, loudness, timbre, etc.). Various areas of the cortex are characterized by neural maps that ensure spatial representation of sensory information [8]. The tonotopic map, which forms a topographic projection from the cochlea in accordance with signal frequencies, is responsible for processing auditory information.

Self-organizing feature maps (SOM) are a computational model of sensory topographic maps [9]. A unique combination of such properties as approximation of the input space, topological ordering, density matching, and feature selection makes SOM stand out among other artificial neural networks. This method has been widely used in intelligent data analysis in various areas of application.

Two types of SOM are distinguished based on the number of neurons (or nodes, units) in the self-organizing layer. The first type has a small quantity of neurons, each of which after learning becomes a cluster center. This is the most common interpretation of the method. However, in this case, each neuron's area of influence can be regarded as a k-means cluster. Thus, topology preservation in the projection proves to be of little use in this setting. The second type of SOM usage is characterized by a much larger quantity of elements (several thousand), which allows to learn and effectively visualize the data structure. Here, the SOM demonstrates emergent properties that are not confined to the mere sum of individual element actions and are not found in small maps. Consequently, this approach is called Emergent SOM (ESOM) [10]. Approach allows one to construct a boundary between clusters of any complexity. In visualization, the structural properties of data are represented by means of a U-Matrix, P-Matrix or a combination thereof (U*-Matrix). The issues of big multidimensional data processing have become relevant. Thus, given prior uncertainty in data structure, a promising emphasis in the area of knowledge discovery [11] is the development and application of ESOM for automatic cluster identification and unsupervised classification tasks.

3 Algorithm

Let $\mathbf{x} = [\xi_1, ..., \xi_D]^T \in X$ be an input vector. There are a set of SOM neurons $Y_N = \{\mathbf{y}_1 ... \mathbf{y}_i ... \mathbf{y}_n\}, N = \{1, ..., n\}$, described by the model vectors (or synaptic weights), $\mathbf{y}_i = [\xi_{i1}, ..., \xi_{iD}] \in \mathbb{R}^D$. According to the SOM architecture, the input vector $\mathbf{x}$ is applied in parallel to each neuron $\mathbf{y}$ of the output map, which represents a regular grid of interconnected units. SOM can be considered as an undirected graph $G = (V, E)$ consisting of vertices V – indexes of a map units, and edges E – lateral connections between neighboring neurons. In a self-organizing process produced by input samples, the SOM performs a topological mapping $f_{map} : X \subset \mathbb{R}^D \to G \subset \mathbb{R}^M, D > M$. This mechanism is based on the adaptive competitive learning algorithm, introduced by Kohonen [9].

At first, for the input vector $\mathbf{x}$ the winner neuron $\omega(\mathbf{x})$, or best matching unit (BMU), is determined by evaluating the Euclidean distance between $\mathbf{x}$ and each map unit:

$$\omega(\mathbf{x}) = arg\ \min_i \|\mathbf{x} - \mathbf{y}_i\| . \tag{1}$$

The BMU modifies the model vectors of neighboring neurons within the topological neighborhood $h_{i,\omega}$. The adaptive process is accompanied by decreasing $h_{i,\omega}$ with time t, so the neighborhood function is determined by the following exponential dependence:

$$h_{i,\omega} = exp\left(\frac{d_{i,\omega}^2}{2\sigma^2(t)}\right), \tag{2}$$

where $\sigma(t)$ is a decreasing radius of a topological neighborhood, $d_{i,\omega}$ is a lateral distance between the BMU ω and the neighboring unit i. Influence of the BMU is weakened with increasing $d_{i,\omega}$. Updating the model vectors at each iteration occurs according to the following expression:

$$\mathbf{y}_i(t+1) = \mathbf{y}_i(t) + h_{i,\omega}(t)\alpha(t)\left[\mathbf{x}(t) - \mathbf{y}_i(t)\right], \tag{3}$$

where $\alpha(t)$ is a decreasing learning rate. The inputs are fed to SOM in a sequential mode until convergence. Convergence criteria of a training step is determined by the absence of significant changes in the map structure, according to a sufficiently small threshold value δ:

$$O(t)\sum_{i}\left\|\mathbf{y}_i(t+1)-\mathbf{y}_i(t)\right\|,(O(t)-O(t-1))<\delta. \tag{4}$$

With a random initialization of the network weights, and in according to large number of ESOM elements, the probability that some neurons can get into the area of a space with a low data density is increased. So-called "dead" units have a negative impact on the quality of the data interpretation. Therefore, in order to involve all map units, the conscience mechanism [12] with an adaptive activation threshold p_i of each neuron is introduced:

$$p_i(t+1)=\begin{cases} p_i(t)+\frac{1}{n},(i\neq\omega)\\ p_i(t)-p_{min},(i=\omega)\end{cases} \tag{5}$$

where p_{min} is a minimal potential that determines the participation of a given neuron in the competition process. If $p_i < p_{min}$, then the neuron i is temporarily disabling, and the BMU is searched among its nearest neighbors.

The use of the Euclidean measure is due to the fact that then SOM units are represents the set of seed points of a Voronoi tessellation, partitioning the input space. Thus, in accordance with the hexagonal structure of the lateral connections, the SOM is a proximity graph, namely Delaunay triangulation. The proximity graphs are effectively used in semi-supervised learning tasks using the label propagation methods in order to spread class labels to nearby nodes [13]. Consider this approach in the ESOM context. Upon completion of the learning process on the partially-labeled data a set Y_N of neurons consist of a subset of BMUs $Y_L^{(1)}=\{y_1...y_l\}\subset Y_N, L=\{1,...,l\}$, each with a corresponding class label $Z_L=\{z_1,...,z_l\}$, and a subset of the map units $Y_U^{(2)}=\{y_{l+1}...y_{l+u}\}\subset Y_N, U=\{l+1,...,l+u\}, l<u$ for which the exact classification label $Z_U=\{z_{l+1},...,z_{l+u}\}$ is to be determined. Suppose that the number of classes C_k is known. Consider binary classification task $Z_L\in C_k=\{0,1\}$. Formally, in this case, the goal of semi-supervised learning is to construct a classifier function for a finite set of partially-labeled map units $f_C: Y_N\rightarrow\{C_0,C_1\}$. Thus, it is necessary to estimate Z_U from Y_N and Z_L. To determine the values of edges E of a graph G, i.e. the lateral connection w_{ij} between neighboring neurons i and j in the SOM space, u-distance can be used:

$$u(i,j)=\underset{i=j}{min}\sum_{k=1}^{r-1}d(\mathbf{y}_{i_k},\mathbf{y}_{i_{k+1}}), \tag{6}$$

$$w_{ij}(\mathbf{x})=exp\left(-\frac{u^2(i,j)}{\varsigma^2}\right), \tag{7}$$

where $r_i\in\mathbb{N}^2$ is a position of $\mathbf{y}_i$ on the regular grid, ς is a radius that determines how far the class label information will be propagate over the graph from the seed units. The value of the parameter ς is a crucial, the problem of its determining is discussed in [14].

4 Results

Sound oscillations are generally divided into simple and complex. Simple oscillations follow the sinusoidal law and are called pure tones. These sounds, however, are hardly found in nature. A complex sound, by contrast, can be represented as a set of tones different in frequency and amplitude. The same applies to vocal sounds, e.g., used in speech synthesis. Thereby, the complex acoustic signals considered in this paper are represented by a dual-tone multi-frequency model of speech vowels synthesized as a sum of harmonic oscillations of the first two formants. For the information on the mean formant frequencies used to generate the vowel signals, see Table 2 in [15] (age group 20–25 years). The sampling rate and the duration for each signal were 44.1 kHz and 250 ms respectively.

A physiologically-based computational model of the auditory periphery (MAP) [16] has been used to form a response of the auditory nerve to the acoustic signals. Response modeling was performed for the nerve fibers with a high spontaneous rate. The results are presented in a multidimensional data matrix X of observations that defines the auditory neurogram. This is a time-frequency representation that reflects the firing rate of the modeled auditory nerve fiber ensemble responding to the input signal. The matrix columns $\mathbf{x}$ correspond to observations in discrete time, and the rows ξ are the features representing the range of $D = 41$ characteristic frequencies, log-scaled in 250–8000 Hz. Further the matrices obtained for each signal were directly used as input data set for ESOM. The total amount of data was an array of 486200 observations, of which only 25% have a class label.

For data analysis, a hexagonal grid of 4096 neurons with a planar topology was used. ESOM training was carried out on partially labeled input samples for clear auditory stimuli, without AWGN. The quality of the map projection was evaluated using standard metrics, such as final quantization error (FQE) and topographic error (FTE). When convergence is achieved, the BMU nodes are assigned a class label, which is transmitted to the neighboring nodes using the label propagation algorithm. The obtained values of the corresponding learning parameters are presented in Table 1.

Table 1. The resulting learning parameters

Parameters	Group 1	Group 2	Group 3
FQE	0.07	0.06	0.05
FTE	0.03	0.18	0.24
BMU	2881	2839	2780
Sigma (ς)	0.001	0.002	0.02

Table 2. Average classification error (%)

SNR (dB)	Group 1	Group 2	Group 3
Clean	0.7	1.1	20.8
30	1.2	6.9	26.9
20	2.7	14.8	32.4
10	18.5	28.3	39.1

Testing was performed using unlabeled data of clean and noisy auditory stimuli. To evaluate the average quality of the classification, the vowel phonemes were divided into three groups of sounds according to their formant frequencies: Group 1 - different, Group 2 - similar and Group 3 - very similar. The influence of AWGN was verified at 30, 20 and 10 dB SNR. Table 2 shows the average classification error for each group of sound stimuli.

5 Conclusions

According to the obtained results, it can be concluded that the proposed approach has demonstrated high accuracy in solving the task under consideration. The use of emergent self-organizing maps revealed a complex structure of the large multidimensional auditory nerve data and corresponding linearly non-separable clusters representing sound stimuli. Introduction of the adaptive conscience mechanism prevented the appearance of "dead" units, for which it is difficult to determine class affiliation by the label propagation algorithm. Comparison of the sound representation in the form of a stationary response pattern of the auditory neural activity with acoustic features gave the following results: for clean signals, classification accuracy was equally high, however, for signals with AWGN, the quality of MFCC and PLP-based classification was significantly reduced. The weakest results, as expected, were obtained in case of similar vowel phonemes classification (Group 3) with high SNR (10 dB). Under similar conditions, proposed approach was able to provide an accuracy about of 60%, whereas the accuracy of the considered acoustic features did not exceed 30%.

Acknowledgments. The reported study was funded by the Russian Foundation for Basic Research according to the research project 18-31-00304.

References

1. Meyer, B., Wächter, M., Brand, T., Kollmeier, B.: Phoneme confusions in human and automatic speech recognition. In: Proceedings of Interspeech, pp. 1485–1488 (2007)
2. Yousafzai, J., Ager, M., Cvetkovic, Z., Sollich, P.: Discriminative and generative machine learning approaches towards robust phoneme classification. In: Proceedings of IEEE Workshop on Information Theory and Application, pp. 471–475 (2008)

3. Miller, G.A., Nicely, P.E.: An analysis of perceptual confusions among some English consonants. J. Acoust. Soc. Am. **27**(2), 338–352 (1955)
4. Chapelle, O., Schölkopf, B., Zien, A.: Semi-Supervised Learning. MIT Press, Cambridge (2006)
5. Yakovenko, A., Malykhina, G.: Bio-inspired approach for automatic speaker clustering using auditory modeling and self-organizing maps. Procedia Comput. Sci. **123**, 547–552 (2018)
6. Huang, X., Acero, A., Hon, H.: Spoken Language Processing: A Guide to Theory, Algorithm, and System Development. Prentice Hall, Upper Saddle River (2001)
7. Hermansky, H.: Perceptual linear predictive (PLP) analysis of speech. J. Acoust. Soc. Am. **87**(4), 1738–1752 (1990)
8. Imai, T.: Positional information in neural map development: lessons from the olfactory system. Dev. Growth. Differ **54**(3), 358–365 (2012)
9. Kohonen, T.: Self-Organizing Maps, 3rd edn. Springer, Heidelberg (2001)
10. Ultsch, A., Mörchen, F.: ESOM-maps: tools for clustering, visualization, and classification with Emergent SOM. Technical Report, Department of Mathematics and Computer Science, University of Marburg, Germany, p. 46 (2005)
11. Ultsch, A., Lötsch, J.: Machine-learned cluster identification in high-dimensional data. J. Biomed. Inform. **66**, 95–104 (2017)
12. DeSieno, D.: Adding a Conscience to Competitive Learning. In: Proceedings of the Second Annual IEEE International Conference on Neural Networks, pp. 117–124 (1988)
13. Zhu, X.: Semi-supervised learning with graphs. Doctoral dissertation, Carnegie Mellon University. CMU-LTI-05-192 (2005)
14. Herrmann, L., Ultsch, A.: Label propagation for semi-supervised learning in self-organizing maps. In: Proceedings of the 6th International Workshop on Self-Organizing Maps (WSOM). Bielefeld University, Germany (2007)
15. Hawkins, S., Midgley, J.: Formant frequencies of RP monophthongs in four age groups of speakers. J. Int. Phon. Assoc. **35**(2), 183–199 (2005)
16. Meddis, R., et al.: A computer model of the auditory periphery and its application to the study of hearing. In: Proceedings of the 16th International Symposium on Hearing, Cambridge, UK, pp. 23–27 (2012)

Cognitive Sciences and Adaptive Behavior

Modified Exponential Particle Swarm Optimization Algorithm for Medical Images Segmentation

Samer El-Khatib[1], Yuri Skobtsov[2(✉)], and Sergey Rodzin[1]

[1] Southern Federal University, Rostov-on-Don, Russia
samer_elkhatib@mail.ru, srodzin@yandex.ru
[2] St. Petersburg State University of Aerospace Instrumentation, Saint Petersburg, Russia
ya_skobtsov@list.ru

Abstract. Modified Exponential Particle Swarm Optimization algorithm is proposed for medical image segmentation. The main idea of the proposed Exponential Particle Swarm Optimization algorithm is to prevent local solutions and find correct global optimal solutions for medical images segmentation task. The execution time comparison is done with existing segmentation techniques. Found, that proposed method is superior to existing segmentation techniques, including graph-based algorithms. Images from Ossirix image dataset and real patients' images were used for testing. Developed method was tested using the Ossirix benchmark with magnetic-resonance images with various nature and different quality. The results of method's work and a comparison with competing segmentation methods (Fuzzy C-Means, Grow cut, Random Walker, Darwinian Particle Swarm Optimization, K-means Particle Swarm Optimization, Hybrid ant colony optimization-k-means algorithm) are presented in the form of a time table of segmentation methods. In all cases, the algorithm makes a better final segmentation time, comparing to the studied techniques (except Random Walker algorithm, which has lower segmentation quality on 15%).

Keywords: Image segmentation · Particle Swarm Optimization
Artificial intelligence · Medical images · Bio-inspired algorithms

1 Introduction

Image segmentation is one of the most difficult tasks in image processing. Segmentation is the process of the image fragmentation into disjoint parts characterized by spectral or spatial (size, shape, texture etc.) characteristics. This procedure is used to solve a wide range of tasks: finding objects in satellite images (woods, oceans, etc.), face recognition, medical image processing – (magnetic-resonance images (MRI), Computer Tomography) etc. The main difficulty in the process of segmentation is the availability of additional factors inherent to the pictures: the variability of background, the presence of noise in the images, the difference between the parts of images. There

B. Kryzhanovsky et al. (Eds.): NEUROINFORMATICS 2018, SCI 799, pp. 243–249, 2019.
https://doi.org/10.1007/978-3-030-01328-8_29

are two main classes of segmentation methods [1]: automatic methods (participation of the user is not required) and interactive methods (in the process of clarification the additional data are required from the user). Edge detection operators [2] (filter of Roberts, Sobel, Prewitt, Canny), histogram methods and graph-based algorithms are among the most known and widely used automatic segmentation methods. Image segmentation have become one of the actual directions in computer technologies in medicine for image quality improvement and pathology processes recognition. Automated diagnostics of pathology processes using medical images have not been solved yet, as not exist one universal segmentation algorithm.

Recent research results have shown the prospects of application of nature-inspired techniques for image segmentation task such as ant colony optimization, particle swarm optimization and bee colony optimization. Application of bio-inspired techniques, including particle swarm optimization, has not been studied completely and more researches are needed. In the given article proposed modification of particle swarm optimization algorithm – Modified Exponential Particle Swarm Optimization algorithm (EPSO).

2 Particle Swarm Optimization

PSO method uses swarm of particles where each particle introduces own solution of the problem [3, 4]. Particle's behavior each time tuned in the search space using its own and neighbor's experience. Moreover, each particle remembers its own best position with the best fitness-function value and knows the best neighbor's position, where global optima has been achieved at the present moment.

Each particle stores best fitness value and appropriative coordinates. Let denote this fitness value as y_i and name it as cognitive component. Similarly, the best global optima, reached by all particles, let denote as $\hat{y}(t)$ and name it as social component.

Each i-th particle has such characteristics as velocity $v_i(t)$ and position $x_i(t)$ in the moment of time t. Particle's position changes according to

$$x_i(t+1) = x_i(t) + v_i(t+1), \tag{1}$$

where $x_i(0) \sim U(x_{\min}, x_{\max})$.

$$v_{ij}(t+1) = v_{ij}(t) + c_1 r_{1j}(t)[y_{ij}(t) - x_{ij}(t)] + c_2 r_{2j}(t)[\hat{y}_j(t) - x_{ij}(t)] \tag{2}$$

The best position (gbest) at the moment of time (t + 1) can be obtained as

$$y_i(t+1) = \begin{cases} y_i(t) & if\, f(x_i(t+1) \geq f(y_i(t)) \\ x_i(t+1) & if\, f(x_i(t+1)) < f(y_i(t)) \end{cases}, \tag{3}$$

where $f : R^{n_\infty} \to R$ is fitness function, which determines how the current solution is close to optimal. $\hat{y}_j(t)$ (pbest) at the moment of time t can be obtained as:

$$\hat{y}(t) \in \{y_0(t)..y_{n_s}(t)\}|f(\hat{y}(t)) = \min\{f(y_0(t)..f(y_{n_s}(t)))\}, \tag{4}$$

where n_s is total amount of particles in the swarm.

3 Exponential Particle Swarm Optimization Algorithm for Image Segmentation

To obtain better segmentation results hybrid method has been proposed in which all the advantages of k-means and PSO algorithms are used.

Exponential particle swarm optimization algorithm is similar to mixed ant colony optimization-k-means algorithm according to its principles [5].

Each particle x_i represents N clusters such as $x_i = \left(m_{i1}, \ldots, m_{ij}, \ldots, m_{iN}\right)$, where m_{ij} is center of the cluster j for particle i. Fitness function can be calculated as:

$$f(x_i, Z_i) = \omega_1 \bar{d}_{\max}(Z_i, x_i) + \omega_2(z_{\max} - d_{\min}(x_i)), \tag{5}$$

where $z_{\max} = 2^s - 1$ for s-bit image; Z is connectivity matrix between pixel and center of the cluster for particle i.

Matrix indicates whether pixel z_p belongs to cluster C_{ij} for particle i. Constant values ω_1 and ω_2 are determined by user, $\bar{d}_{\max}$ is maximum average Euclidian distance from particles to connected clusters. It can be obtained as:

$$\bar{d}_{\max}(Z_i, x_i) = \max_{j=1..N}\left\{\sum\nolimits_{\forall Z_p \in C_{ij}} d(Z_p m_{ij}/\left|C_{ij}\right|\right\}, \tag{6}$$

$$d_{\min}(x_i) = \min_{\forall j_1, j_2 j_1 \neq j_2} \left\{d(m_{ij_1}, m_{ij_2})\right\}, \tag{7}$$

Formula 7 is minimum euclidian distance between each pair of the cluster centers.

In the following task swarm is used to obtain good clustering using transferred parameters. It has been reached by self-learning. Each particle in PSO algorithm represents pixel. Pixel's intensity has been used as input parameter for PSO algorithm.

Algorithm consists of the following steps:

Algorithm 1. Exponential PSO segmentation algorithm

1. Initialize number of particles in the swarm m, individual and global acceleration rates c_1 и c_2, maximum number of iterations N_{max}, parameters for fitness-function f (5).
2. *For* $i = 1 \ldots m$ (for each particle)
   2.1 Initialize starting position of the particle using vector x_i
   2.2 Starting position of the particle is the best known position $y_i = x_i$.
   2.3 If $f(y_i) < f(\hat{y})$, then update best swarm's value replacing $\hat{y}$ to y_i.
   2.4 Randomly initialize velocities of the particles v_i.
3. Current number of iterations $N = 1$.
4. *For* $i = 1 \ldots m$ (for each particle)
5. *For* $j = 1,\ldots,n$ (fitness function parameters)
   5.1 Update particle's velocity v_{ij} and position according to $x_{ij} = x_{ij} + v_{ij}$
6. If $f(x_i) < f(y_i)$, then replace best local solution for particle $y_i = x_i$, otherwise return to section 4.
7. If $f(x_i) < f(\hat{y})$, update best global swarm's solution $\hat{y} = x_i$, otherwise return to section 4.
8. Number of iterations $N = N + 1$.
9. If $N \le N_{max}$, then return to section 4, otherwise $\hat{y}$ contains best found solution.
10. Initialize K centers of the clusters using best particles positions.
11. Calculate pixel's belonging to cluster (according to distance to the center).
12 Using (5) recalculate clusters centers. If they are not equal to previous, then repeat section 11.
13. Save best individual solution for each particle (pbest (3)).
14. Save best common solution for m particles (gbest (4)).
15. Update clusters centers.
16. If centers have changed, then return to section 12.

Due to particle swarm optimization algorithm, all particles tend to fly directly to gbest position, which was found by best particle. Such approach allows quickly detect possible solutions. Using this mechanism, particles often flock to local minimums instead of global, that leads to find suboptimal solutions. To prevent this effect El-Desouky [6] has proposed to change ω in the linear way, such as:

$$\omega = (\omega - \omega_{\min})\frac{(n_{\max} - n)}{n_{\max}} + \omega_{\min}, \tag{8}$$

where n_{max} is maximum number of iterations, n is the number of the current iteration. Recommended values are: ω_{max} = 0,9; ω_{min} = 0,4. ω can be decreased down to ω_{min} over 1500 iterations. In this article we propose to change ω exponentially. In presented algorithm we propose to change ω in the following way:

$$\omega = (\omega - \omega_{\min})e^{\frac{(n_{\max} - n)}{n_{\max}}} + \omega_{\min} \quad (9)$$

4 Results

To assess the effectiveness of the algorithm, numerical experiments have been provided. Six segmentation methods have been considered. They are: Fuzzy C-Means [7], Grow cut [8], Random Walker [9], Darwinian PSO [10], PSO modification - K-means PSO [10], Hybrid ant colony optimization-k-means algorithm - K-means ACO [5].

Table 1 represents execution time for 3 images from Ossirix image dataset [12] (Fig. 1, 2, 3).

Table 1. Execution time for each tested image using different algorithms.

Image number/ Method name	K-means PSO	Exponential PSO	K-means ACO	Fuzzy C-Means	Grow cut	Random walker	Darwinian PSO
1	7.48	7.34	12.14	9.49	14.78	5.01	11.85
2	0.19	0.18	0.93	0.87	1.35	2.2	16.34
3	17.5	17.5	24.04	12.14	45.30	14.2	15.95

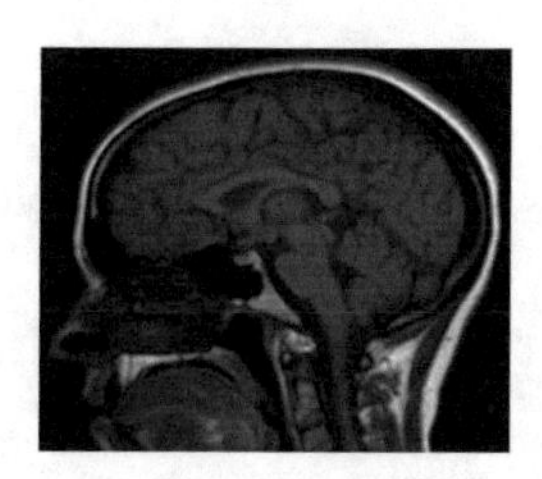

Fig. 1. Brain MRI. Parameters – 420 × 391, 8 bits per pixel

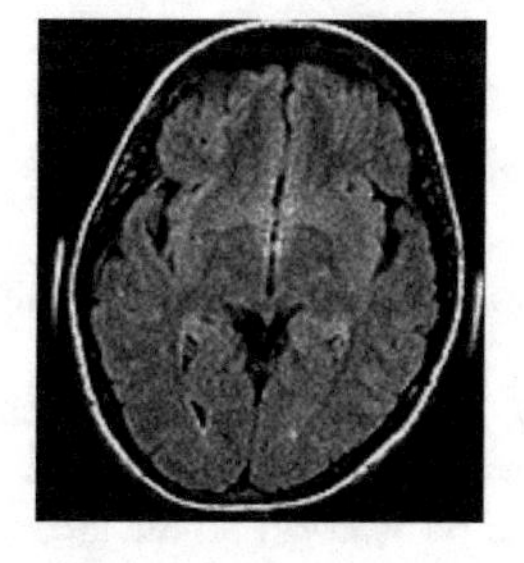

Fig. 2. Noisy brain image with pathology. Parameters: 150 × 166, 8 bits per pixel.

From Table 1 we can observe, that proposed method outperforms all existing ACO and PSO modifications and graph based methods, except only Random Walker (in this case execution time affected to segmentation quality – EPSO segmentation quality outperformed Random Walker on 15%).

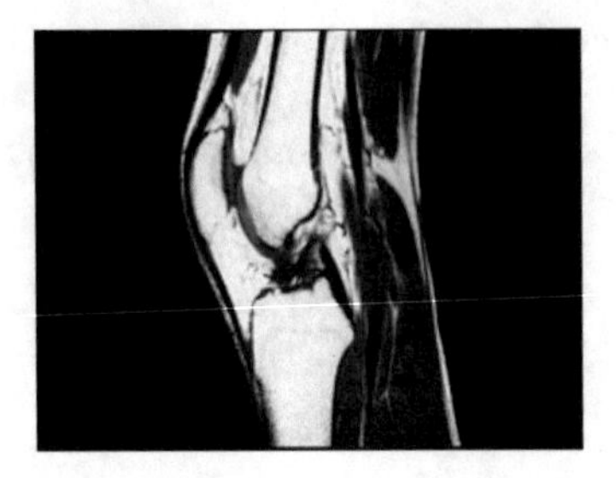

Fig. 3. Contrast sagittal MRI image of the knee. Parameters 800 × 450, 8 bits per pixel.

5 Conclusion

In the presented paper, there has been introduced the modified exponential particle swarm optimization algorithm for MRI images segmentation. A comparison of the algorithms' results with other methods has been presented. In all cases, the algorithm makes a better final segmentation time, comparing to the studied techniques (except Random Walker algorithm, which has lower segmentation quality on 15%). All experimental results have been obtained using Ossirix MRI image dataset and developed software products. Obtained results have showed, that Modified Exponential Particle Swarm Optimization algorithm can be used in digital image processing of medical images.

Acknowledgements. This work was supported by Russian Foundation of Basic Research (RFBR) – project № 16-07-00336 – "Development of the theory and application of meta-heuristic models, methods and algorithms for trans-computational problems of making optimal decisions".

References

1. Gonzalez, R.C., Woods, R.E.: Digital Image Processing, 3rd edn. Prentice-Hall, Englewood (2008)
2. Kennedy, J., Eberhart, R.C.: Particle swarm optimization. In: Proceedings of the IEEE International Joint Conference on Neural Networks, pp. 1942–1948. IEEE Press (1995)
3. El-Khatib, S., Rodzin, S., Skobtcov, Y.: Investigation of optimal heuristical parameters for mixed ACO-k-means segmentation algorithm for MRI Images. In: Proceedings of III International Scientific Conference on Information Technologies in Science, Management, Social Sphere and Medicine (ITSMSSM 2016). Part of series Advances in Computer Science Research, vol. 51, pp. 216–221. Atlantis Press (2016). https://doi.org/10.2991/itsmssm-16.2016.72
4. El-Khatib, S.A., Skobtcov, Y., Rodzin, S.: Hyper-heuristical particle swarm method for MRI images segmentation. In: Silhavy, R. (Ed.) Proceedings of 7th Computer Science Online conference 2018 (CSOC 2018) AISC 764, vol. 2, pp. 256–264. Springer International Publishing AG, part of Springer Nature (2018). https://doi.org/10.1007/978-3-319-91189-2_25

5. El-Khatib, S.: Modified exponential particle swarm optimization algorithm for medical image segmentation. In: Proceedings of XIX International Conference on Soft Computing and Measurements (SCM 2016), St. Petersburg, vol. 1, pp. 513–516, 25–27 May 2016. SPBGETU "LETI" (2016)
6. Saatchi, S., Hung, C.C.: Swarm intelligence and image segmentation. In: INTECH Open Access Publisher (2007)
7. Das, S., Abraham, A., Konar, A.: Automatic kernel clustering with a multi-elitist particle swarm optimization algorithm. Pattern Recogn. Lett. **29**(5), 688–699 (2008)
8. Ossirix image dataset. http://www.osirix-viewer.com/. Accessed 12 July 2018
9. Ghamisi, P., Couceiro, M.S., Ferreira, M.F., Kumar, L.: Use of darwinian particle swarm optimization technique for the segmentation of remote sensing images. In: Proceedings of the 2012 IEEE International Geoscience and Remote Sensing Symposium (IGARSS 2012), pp. 4295–4298. IEEE (2012)
10. Ghamisi, P., Couceiro, M.S., Martins, M.L., Benediktsson, J.A.: Multilevel Image segmentation based on fractional-order darwinian particle swarm optimization. IEEE Trans. Geosci. Remote Sens. **52**(5), 1–13 (2013)

Semantic Space and Homonymous Words

Alena A. Fetisova[1(✉)] and Alexander V. Vartanov[2]

[1] National Research Nuclear University MEPhI
(Moscow Engineering Physics Institute), Moscow, Russia
alenafetisova7@gmail.com
[2] Lomonosov Moscow State University, Moscow, Russia

Abstract. Significant differences were found in the organization of electrobiological responses of the brain to the words-homonyms, presented in different meanings by priming a certain context in the electrophysiological experiment on a sample of 14 people As a result of comparison of series before and after negative reinforcement by the method of Semantic Radical of the word entering the semantic field of one of the two studied meanings of the word-homonym, the fact of indirect long-term influence of the emotional component is revealed. This is manifested not only in the change in the meaning of the homonym word (presented in the relevant context) in the late latency (300–500 MS) in the frontal leads, but also in the reconfiguration of its meaning in a non-relevant context (in the earlier – 200 MS – latencies in the central leads). Thus, the results objectively indicate the possibility of two types of reconfiguration of the semantic map (space) (1) operational reconfiguration to the current context defined in this experiment by the corresponding prime, and (2) long-term reconfiguration determined by the control action.

Keywords: Semantic space · Homonymous words · Evoked potentials

1 Introduction

Semantic space or a map can serve as an important element in the development of virtual agents endowed with social and emotional intelligence [1–3]. There are several ways to build a semantic map of "verbal feelings" - based on formal rules using synonymous-antonymic dictionaries, or empirically, for example, in psychophysical experiments with a human being carried out under the scheme of multidimensional scaling [4]. Vector representation, the possibility of meaningful interpretation of the relevant features and obtaining coordinates, which are characterized by a certain semantics, are an important advantage of the construction of such emotional spaces (maps).

However, there is a significant problem along the way – context (situational) dependent polysemy. Semantics of any natural language contains homonymous words, whose meanings cannot be a priori given uniquely, since it is determined by the context. Therefore, in computer modeling there is a problem of representation of such concepts in the feature space. Indeed, usually the same word has multiple meanings, while it has to be represented by only one point on the map. It was offered [2, 3] to use word combinations, rather than words as elements of a map to solve this problem.

B. Kryzhanovsky et al. (Eds.): NEUROINFORMATICS 2018, SCI 799, pp. 250–256, 2019.
https://doi.org/10.1007/978-3-030-01328-8_30

However, this does not solve the problem completely, but only pushes it away: word combinations can also have different meanings (sense) in different contexts.

Another important problem is the use of a semantic map of "verbal feelings" (or emotional assessments) for the tasks of managing and coordinating local mappings (semantic fields representing, in particular, different meanings of homonyms). The concept of "a Committee of neural networks" (or neurocommittee) is becoming more and more popular in the development of neural networks (the outputs of multiple offline-trained neural networks, independently processing the input data in a certain way to integrate). It is obvious that in the functioning of such parallel working cognizing systems, locally displaying the reality in several different spaces of features, there must be a "switch" or "moderator". Which would coordinate their work based on a universal criterion, which can be an "emotional" map (space) – universal system for assessing the current situation in relation to the needs of the body. This is a «fast» changeover mechanism that explains well the current change in the meaning of a word-homonym that appears in a particular context by switching from one local display to another. However, if such a switch is based on emotions, then it should be possible not only to switch roughly, but also to make a subtle ("slow" and "long-term") transformation of each local semantic representation by changing the emotional attitude to its individual representatives. Is there such a phenomenon in living systems, in particular in the understanding of the meaning of words by human?

Until today, huge number of complex phenomena of human recognition of the word have not yet been explained. In particular, the mechanism of understanding of polysemous words, where the specific meaning of the word varies depending on the speech and non-speech context. However, it is known that the word is more frequent, the more it has meanings [5]. An important role of context in determining the meaning of the word has been shown in a number of electrophysiological studies, where evoked potentials (EP) were written for words that are suitable and inappropriate in meaning to the phrase, as well as neutral. As a result, the P600 component in the centroparietal region was revealed, the amplitude of which increases when the word does not correspond to the context of the sentence, which presumably reflects the complexity of the process of semantic integration [6, 7].

This study is aimed at studying the features of the brain mechanisms of understanding the meanings of homonyms – such signs, in the recognition of which the given context plays a leading role. Signs-homonyms are a unique material, because they are identical in terms of their physical form, but differ in meaning and contextual relevance. This study also considers the possibility of influencing a certain meaning of a homonym through a special emotional impact. Developed By Luria and Vinogradova [8], the Semantic Radical method consists in the analysis of the value by isolating its associative fields using negative reinforcement when the presentation of the target word is accompanied, for example, by an electric shock. Transfer of the conditioned reflex defensive reaction (objectively recorded) from one object to another, semantically connected with it is used as a criterion of semantic proximity of objects. It is known that the generalization of the conditioned reflex reaction is normally carried out by semantic connections (violin—cello), and for the mentally retarded—by phonetic (mouse—house) [8]. The application of this method to the word, which is included in the semantic context of one of the meanings of the word-homonym, opens the opportunity to explore the extent to which it can affect the perception of the meaning of the word-homonym in the appropriate and/or in another context.

Thus, the purpose of this study was to identify the objective indicators of brain activity, as the words-homonyms (polysemous words) are embedded in semantic fields and whether it is possible to influence their value indirectly, through emotional attitude to other words forming a certain semantic context.

2 Method

The experiment consisted of three series: neutral, motivating and testing. In all series, the subject performed the same task, he decided whether the word “alive” or “inanimate” by pressing the right or left button of a computer mouse. The target word was Russian homonym word “лук”, (which means “onion” and “crossbow/arbalest” at the same time) shown on the computer screen. We also used the words necessary to create a context that form two variants of semantic fields, including the meaning of the word-homonym as a weapon (bowstring, quiver, warrior, shooter, arrow, weapon, Indian) and the meaning of the word-homonym as a plant (husks, parsley, carrots, greens, garlic, garden, plant). For each context presented its Prime picture, depicting either a seedbed or two soldiers. This or that Prime had to define the current context and, accordingly, a particular meaning of the word-homonym. The duration of the stimulus is 500 ms, the duration of the Prime-200 ms. Target stimulus were presented 60 times in each meaning of the homonym, the other words-10 times each. The sequence of presentation was random. The neutral series consisted of the presentation of pairs to the test subject (Prime – word), including the homonym word and other relevant Prime words from the above lists.

The motivating series consisted of word presentations representing only one word context as a plant (without presenting primes). At the same time, the presentation of the word “garlic” was accompanied by a frightening unpleasant noise (a loud human cry). It was supposed to form a special (personally significant, emotionally negative) attitude to the given word and other words of the given context, including the word homonym in the meaning of the plant, but not to influence the words of another semantic field, including the other meaning of our target word-homonym. The test series was exactly the same as the neutral one.

Nineteen-channel encephalograph was used for EEG registration. The resulting records were cleared of artifacts based on expert analysis. Then the desired fragments were sorted according to the presentations and were averaged to obtain the EP. Individual and averaged across the whole group subjects evoked potentials for the presentation of homonyms in two different meanings in the control and test series were considered.

The study involved 14 normally developed volunteers aged 20–29 years (8 girls and 6 boys).

All subjects received full information about the study and gave informed and voluntary consent to participate in the study.

The experiment was approved by the Commission on Ethics and Regulations of the Academic Council of the faculty of psychology of Lomonosov Moscow State University.

3 Results

The responses of the subjects in solving the task (alive/not alive) allow us to assert that, depending on the Prime that defines the specific meaning of the homonym, the subjects were well characterized by one value of the word-homonym from another. Thus, the used Prime worked, it created a given context. Figures 1 and 2 show for example the EP obtained in the neutral and test series for both meanings of the word homonym: 1 – plant, 2 - weapon, according to the lead of Cz and Fz.

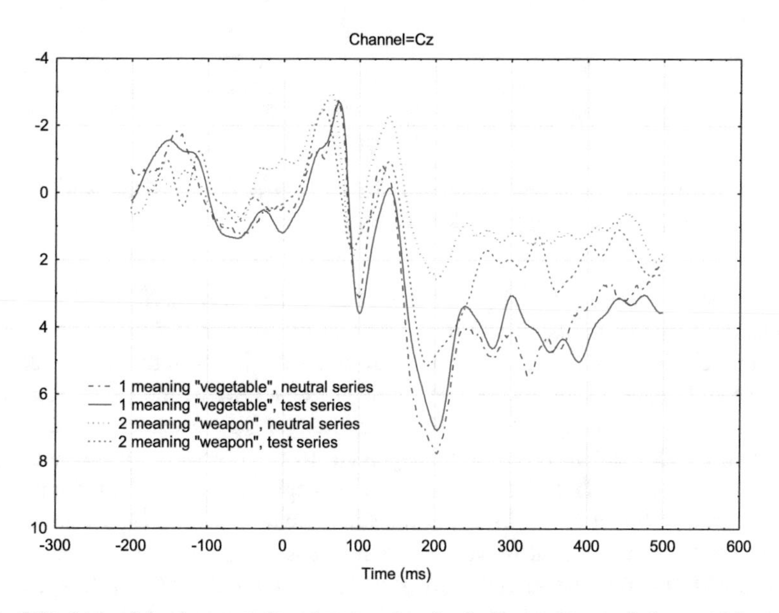

Fig. 1. EP obtained in the neutral and test series for both meanings of the word homonym: 1 – plant, 2 - weapon, according to the lead of Cz.

There are early components N90-100 on these plots, reflecting the perception of the presented objects; there were no significant differences between the stimuli, as they were physically the same. Next, there is the P200 component that is associated with word recognition. Starting from this component and further, there are significant (by T-criterion, $p < 0.05$) differences between homonyms in different meanings, both in central and frontal leads. This later set of waves reflects the final identification of the stimulus, requiring comparison with a sample in memory, and decision-making in relation to the associated action (pressing the mouse button). The late EP period is associated with understanding the meaning of the word and the frequency of its use, this is a component of semantic differences between homonyms [7, 9–11]. Thus, the

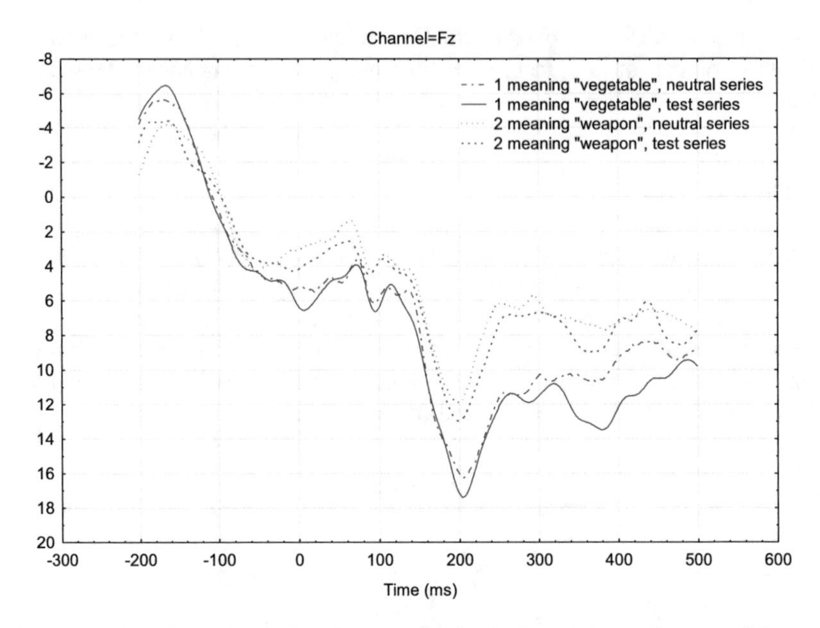

Fig. 2. EP obtained in the neutral and test series for both meanings of the word homonym: 1 – plant, 2 - weapon, according to the lead of Fz.

differences in the amplitude of the EP peaks at this time interval (200–400 ms after the presentation of the stimulus) reflect the change in the meaning of the word-homonym due to its inclusion in different contexts. However, it is in the same time interval that noticeable differences between EP per word are observed in the same context, but presented in different series. At the same time, stronger differences are observed in the early latency (200 ms) in the central assignment to the homonym in the meaning of weapons, despite the fact that the word was negatively supported from another semantic context (plants), i.e. in a non-relevant semantic field (Fig. 1). And the differences due to the emotional impact on the relevant semantic field were found in the late latency (400 ms) in the frontal leads (Fig. 2). Thus, the fact of long-term emotional impact of motivating series on perception of word meaning is objectively shown, as it is localized in the same time interval, in which context-dependent meanings of words-homonyms differ. Since in different series the corresponding words-homonyms were presented in the same context (with the same Prime), i.e. the physically presented stimuli were identical, the observed difference in the amplitude of the EP at this interval can not be explained by anything else, as a change in the meaning of the corresponding word-homonym in the motivating series. At the same time, this influence was indirect – through the modification of the emotional attitude to another word ("garlic"), the semantic field of which is overlapped with only one meaning of the word-homonym. Also, the fact that the motivating effect is manifested initially (in early latency) in the Central leads to a greater extent on the non-relevant meaning of the word (which may be due to its phonetic similarity with the target word) allows us to talk about the global impact on the system of phonetic features, which corresponds to the mechanism of generalization of the conditioned reflex reaction for the mentally retarded people [8]. In addition, to suppress such a reaction normally requires the participation of the frontal lobes of the brain, so the frontal leads revealed another picture-the role of motivating influence is manifested only at a late stage and characterizes only the relevant semantic field. Thus, all of the above proves the global impact of emotions on the definition of the meaning of words and confirms the possibility of functioning of the emotional "map" as a universal "moderator" in the coordination of local representations.

4 Conclusion

Significant differences were found in the electrobiological responses organization of the brain to homonyms, represented in different meanings by priming a certain context. The fact of indirect long-term influence of the emotional component on the meaning of homonyms is revealed, which can be associated with the reconfiguration of the whole context of meanings (the transformation of semantic fields). Differences occur in the late components of the EP, starting with 300 ms. Thus, the obtained results indicate the possibility of two types of semantic map reconfiguration: (1) operational reconfiguration for the current context, determined in this experiment by the corresponding Prime, and (2) long-term reconfiguration, determined by the control action of the emotional system.

Acknowledgments. This work was supported by the Russian Science Foundation, Grant № 18-11-00336.

References

1. Samsonovich, A.V.: The Constructor metacognitive architecture. The AAAI Fall Symposium-Technical report, FS-09-01, pp. 124–134. AAAI Press, Menlo Park (2009)
2. Samsonovich, A.V., Goldin, R.F., Ascoli, G.A.: Toward a semantic general theory of everything. Complexity **15**(4), 12–18 (2010). https://doi.org/10.1002/cplx.20293
3. Eidlin, A.A., Eidlina, M.A., Samsonovich, A.V.: Analyzing weak semantic map of word senses. Procedia Comput. Sci. **123**, 140–148 (2018)
4. Sokolov, Ye.N., Vartanov, A.V.: K issledovaniyu semanticheskogo prostranstva (To the study of semantic space). Psikhologicheskiy zhurnal (Psychol. J.) **2**, 58–65 (1987)
5. Slobin, D.: Psikholingvistika (Psycholinguistics), Moscow (2009)
6. Vartanov, A.V., Pasechnik, I.V.: Mozgovyye mekhanizmy semanticheskogo analiza slov-omonimov (Brain mechanisms of semantic analysis of words-homonyms). Zhurnal vysshey nervnoy deyatel'nosti (J. High. Nerv. Act.). **55**(2), 193–197 (2005)
7. Kaan, E., Harris, A., Gibson, E., Holcomb, P.: The P600 as an index of syntactic integration difficulty. Lang. Con. Process. **15**(2), 159–201 (2000)
8. Luriya, A.R., Vinogradova, O.S.: Ob"yektivnoye issledovaniye dinamiki semanticheskikh sistem. Semanticheskaya struktura slova (Objective study of the dynamics of semantic systems. Semantic structure of the word), Moscow (1974)

9. Sekerina, I.A.: Metod vyzvannykh potentsialov mozga v Amerikanskoy psikholingvistike i yego ispol'zovaniye pri reshenii problemy poryadka slov v russkom yazyke. (The method of evoked brain potentials in American psycholinguistics and its use in solving the word order problem in the Russian language). Trudy mezhdunarodnoy konferentsii Dialog'2002 «Komp'yuternaya lingvistika i intellektual'nyye tekhnologii» (Proceedings of the international conference Dialogue'2002 "Computer Linguistics and Intellectual Technologies"). Nauka, Moscow (2002)
10. Friederici, A.D., von Cramon, D.Y., Kotz, S.A.: Language related brain potentials in patients with cortical and subcortical left hemisphere lesions. Brain **122**(6), 1033–1047 (1999)
11. King, J.W., Kutas, M.: Neural plasticity in the dynamics of brain of human visual word recognition. Neurosci. Lett. **244**(2), 61–64 (1998)

Algorithms of Distribution of General Loads Under the Joint Work of Aggregates

S. I. Malafeev[1(✉)] and A. A. Malafeeva[2]

[1] Joint Power Co. Ltd., Moscow, Russia
sim@jpc.ru

[2] Vladimir State University Named After Alexander and Nikolay Stoletovs, Vladimir, Russia
amalafeeva@rambler.ru

Abstract. The technique and algorithms for searching for the optimal distribution of the total load between parallel units operating under discrete control of their modes are considered. The aggregates are united by a common resource collector and have different technical characteristics (performance and efficiency). The characteristics of the aggregates are given in tabular form for a limited number of operating modes. Such schemes allow to increase the productivity of the process, to save energy and to ensure the reliability of the system. As a function of the goal, the total consumption of energy consumed or the efficiency of the system is used. The description of the method of direct search of variants, the use of the genetic algorithm, the search algorithm by moving along the extremal is given. The search for a solution using a genetic algorithm requires much less computational cost than direct busting. The genetic algorithm, as a rule, gives an approximate solution. The motion search along the extremal allows to obtain a solution with a minimal number of iterations. A variant of the solution of the problem is proposed with variable productivity of the units. In this case, for each capacity value, the minimum number of operating units is determined, which ensures optimal load distribution. The transition from one to another performance value of a group of units is carried out with a minimum number of inclusions or shutdowns. On the basis of the considered algorithms, the software for the automated dispatch control system of the heat and power complex was developed. The use of optimal algorithms for gas boiler houses provides gas savings of up to 5% compared to traditional dispatch control.

Keywords: Energy · Parallel units · Optimization · Genetic algorithm Simulation

1 Introduction

In various industries, for example, in the energy sector, there is often a need for parallel inclusion of technological aggregates, united by a common reservoir for the resource [1–4]. Such schemes allow to increase the productivity of the process, provide energy savings and ensure the reliability of the system [5]. In many cases, the units have different technical characteristics (productivity, efficiency, etc.), so an important task is

B. Kryzhanovsky et al. (Eds.): NEUROINFORMATICS 2018, SCI 799, pp. 257–262, 2019.
https://doi.org/10.1007/978-3-030-01328-8_31

to optimize the distribution of the total load between them. As a function of the target, the total amount of energy consumed or the efficiency of the system is usually taken. If the aggregates allow continuous regulation of the regimes within certain limits, and the load and adjusting characteristics are known, then an analytical solution of the problem is possible [6, 7]. Discrete regulation of the operation modes of aggregates for a given range of deviations from the optimal regime can lead to an ambiguous solution. However, in many cases, the characteristics of aggregates are tabulated for a limited number of modes of operation. For example, in many cases, the number of inclusions and cut-offs of aggregates per unit of time is limited or uniform operation of all units in time is required. At the same time, technological restrictions are imposed on regime switching. Under these conditions, the modeling of the operation of the aggregate group and the computational methods of finding the best solution are used for a rational solution of the problem [8].

The article presents the results of developing and investigation of the algorithms and software for solving the problem of optimal load sharing between parallel units based on simulation.

2 The Problem Description

The technological process is carried out by n parallel connected units, each of them can be in one of the k states. For each element of the states set $\{m_{ij}\}$, $i = 1, \ldots, n$, $j = 1, \ldots, k$ the productivity q_{ji} and energy consumption r_{ij} are assigned. It is required to determine the set $M = \{m_{i^*j^*}\}$, $i^* \in \{i\}$, $j^* \in \{j\}$, of such states of all units, for which the target function $S = \sum_{i=1}^{n} r_{i^*j^*}$ reaches the minimum at given total productivity $Q = \sum_{i=1}^{n} q_{i^*j^*}$ for a given permissible deviation ΔQ.

3 Algorithms for Solving the Problem

3.1 Direct Search

The search for an optimal solution based on direct search is carried out as follows:

- Calculate the required productivity Q.
- Determine the set of L, $L \in M$, containing l possible variants of connecting of the group of aggregates, at which a total productivity $Q \pm \Delta Q$, where ΔQ - permissible derivation, are carried out.
- Calculate values of total energy consumption $R_p, p = 1, \ldots, l$, for all elements of set L.
- Out of set L of variants select one, which assures minimal total energy consumption $R_{p^*} = \min\{R_p\}$.

The algorithm allows to obtain an exact and reliable solution, but it requires large computational costs, which increase with increasing n and k. The decision search time and the result also depend on the value of ΔQ.

3.2 Load Distribution Using Genetic Algorithm

The solution is based on the biological principles of natural selection and evolution [9, 10]. The genetic algorithm repeats a certain number of times the procedure for modifying the population (a set of individual solutions), seeking the formation of new sets of solutions (new populations). At the same time, at each step, "parents" are selected from the population, that is, solutions that are jointly modified (crossed) and lead to the formation of a new individual in the next generation.

The genetic algorithm uses three types of rules, on the basis of which a new generation is formed: the rules for selection, crossing and mutation. To solve the problem, we use the following notation [11]:

- gene is a current state of one unit $\{m_{ij}\}$, $i = 1, \ldots, n$; $j = 1, \ldots, k$;
- chromosome is a set of genes, in this case - a set of aggregate states as one solution;
- generation is a set of N chromosomes (for example, $N = 10$);
- prima chromosome is a chromosome, which has the best value of the quality functional;
- mutation is a random change in the composition of genes in the chromosome;
- migration is a substitution of the chromosomes that have discarded by the casual (external) survival criterion;
- dynasty is a number of iterations, during which the prima chromosome did not change (the principle of elitarism).

The search for a solution using a genetic algorithm requires much less computational cost than direct search. The algorithm, as a rule, gives an approximate solution. On the basis of modeling, it was established that the solution found in this way in the general case is not necessarily the best.

3.3 Search by Moving Along the Extremal

The preliminary processing of the data is as follows. For all elements of the set $\{m_{ij}\}$, the relations

$$d_{ij} = \frac{r_{ij}}{q_{ij}}.$$

For each aggregate the determines the minimum value of $d_{i_0 j_0}$, $i_0 \in \{i\}$, $j_0 \in \{j\}$. The set $\{m_{i_0 j_0}\}$ of states of aggregates corresponds to the best in the criterion of minimum energy consumption for their operating modes and thus forms an extremal. The further distribution of the load is reduced to determining the required number of aggregates, assures productivity $Q \pm \Delta Q$ at minimal $d_{i_0 j_0}$, and correction of the modes of operation of one or several aggregates for accurate keeping of Q. The procedure for solving the problem is as follows.

A new numbering of aggregates is made in accordance with condition $d_{1j_0} \le d_{2j_0} \le \ldots \le d_{nj_0}$.

The number b of included aggregates is determined for which

$$S_1 = \sum_{i=1}^{b} q_{i_0 j_0} \geq Q;\ S_1 - Q \rightarrow \min.$$

For b aggregates, a direct search of the inclusion variants is carried out, ensuring equality $\sum_{i=1}^{b} q_{i^* j^*} = Q \pm \Delta Q$ at the $\sum_{i=1}^{b} r_{i^* j^*} \rightarrow \min.$

To determine the optimal solution after preliminary ranking of aggregate characteristics, the previously considered genetic algorithm can be used.

The search by moving along the extremal allows to obtain a solution with a minimal number of iterations compared to other considered algorithms. For example, for the case n = 6 and k = 6, the search for the optimal solution takes 15 iterations.

3.4 Load Distribution Algorithm for Variable Productivity Q

If the required productivity Q changes during the operation of the units, it is necessary to switch modes. At the same time, some of the units can be switched off or put into operation. The productivity value is selected from the set of $\{Q_f\}$, f = 1, …, F. There are restrictions on switching on and off of the units: for any performance value, the number of units involved must be minimal; when changing the productivity Q, the number of units to be switched on or off must be kept to a minimum. The order of solving the problem is as follows.

The numbering of the elements of the set $\{Q_f\}$ is carried out according to the rule: $Q_1 < Q_2 < \ldots < Q_f$.

For the Q_1 the problem of optimal load distribution solves using the method of moving along the extremal. During this, the set of aggregates $\{i^1\} \in \{i\}$, containing b_1 of elements with minimal values of $d_{i_0 j_0}$.

For Q_2 checking the condition

$$\sum_{i=1}^{b_1} q_{i_0 j_0} \geq Q_2. \quad (1)$$

If (1) is satisfied, then a new solution of the problem of optimal load distribution is made. If (1) does not satisfied, then an element of the remaining $n - b_1$ with the minimum value of $d_{i_0 j_0}$ is added to the set of $\{i^1\}$. For the new set, checked the condition

$$\sum_{i=1}^{b_2} q_{i_0 j_0} \geq Q_2. \quad (2)$$

If condition (2) is satisfied, then the optimal distribution problem for b_2 aggregates are solving. If condition (2) is not satisfied, then the next element of the remaining

$n - b_2$ elements with the minimal $d_{i_0 j_0}$ is added to the set $\{i^2\}$. Then the process repeats.

As a result, for each performance value, the minimum number of operating units is determined, which ensures optimal load distribution. The transition from one to another performance value of a group of units is carried out with a minimum number of inclusions or shutdowns.

4 Conclusion and Future Work

The article presents a solution to the problem of energy saving due to new algorithms for optimal load distribution for parallel aggregates with a discrete set of states. The genetic algorithm and the search for a solution by movement along the extremal have been tested in the dispatch control system of a gas boiler house. Their use provides a gas saving of up to 5% compared to traditional dispatch control.

Further studies are carried out to optimize the operation of energy sources of different types, providing energy to one system, and also research of processes at tests of electric cars in dynamic modes [12].

References

1. Malafeev, S.I., Novgorodov, A.A.: Design and implementation of electric drives and control systems for mining excavators. Russ. Electr. Eng. **87**(10), 560–565 (2016). https://doi.org/10.3103/s1068371216100035
2. Bortoni, E.C., Bastos, G.S., Souza, L.E.: Optimal load distribution between units in a power plant. ISA Trans. **46**, 533–539 (2007)
3. Al-Rababa, K.S.: The operational features of pumping stations equiped with parallel-connected centrifugal pumps for land-reclamation. Am. J. Appl. Sci. **2**(1), 423–425 (2005)
4. Bajaj, S.S., Barhatte, S.H., Shinde, S.U.: Series and parallel demand sharing in Multi-boiler system. Int. J. Curr. Eng. Technol., MIT College of Engineering, Pune, India, AMET 2916, INPRESSCO IJCET Special Issue-4 (March 2016), 4 (2016). http://inpressco.com/category/ijcet
5. Shinskey, F.G.: Energy Conservation Through Control, p. 321. Academic Press, New York, San Francisco, London (1978). ISBN: 0-12-641650-8
6. Nocedal, J., Wright, S.J.: Numerical Optimization, p. 664. Springer, New York (2006). ISBN: 0-387-98793-2
7. Diwekar, U.: Introduction to Applied Optimization, p. 309. Vishwamitra Research Institute, Clarendon Hills (2008). ISSN: 1931-6828. ISBN: 978-0-387-76634-8. https://doi.org/10.1007/978-0-387-76635-5
8. Kothari, D.P., Dhillon, J.S.: Power System Optimization, p. 572. Prentice-Hall of India Private Limited, New Delhi (2007)
9. Kryukov, V.I. An attention model based on the principle of dominanta. In: Holden, A.Y., Kryukov, V.I. (eds.) Neurocomputers and Attention I: Neurobiology, Synchronization and Chaos, Proceedings in Nonlinear Science, pp. 319–351 (1989)

10. Jankowski, T.: Suitable configuration of evolutionary algorithm as basis for efficient process planning tool. In: Katalinic, B. (ed.) DAAAM International Scientific Book 2015, pp. 135–142. DAAAM International, Vienna, Austria (2015). ISBN: 978-3-902734-05-1, ISSN: 1726-9687. https://doi.org/10.2507/daaam.scibook.2015.12
11. Melanie, M.: An Introduction to Genetic Algorithms. A Bradford Book, p. 157. The MIT Press, Cambridge, Massachusetts, London (1999)
12. Malafeev, S.I., Malafeev, S.S.: Dynamic loading of electric machines during testing. Int. J. Eng. Technol. **7**(2.23) Special Issue 23, 184–187 (2018). https://doi.org/10.14419/ijet.v7i2.23.11912

Long- and Short-Term Memories as Distinct States of the Brain Neuronal Network

Evgeny Meilikhov[1,2](✉) and Rimma Farzetdinova[1]

[1] National Research Centre "Kurchatov Institute", 123182 Moscow, Russia
meilikhov@yandex.ru
[2] Moscow Institute of Physics and Technology, 141707 Dolgoprudny, Russia

Abstract. There are two types of memory – short-term and long-term ones. First, the former arises and then the latter one (in the course of the so called consolidation process). Own neuronal networks (engrams) in the brain correspond to each of those memories, and our goal is to understand what is the difference between those networks from viewpoint of their structural properties. It is not about the special biochemical structure of some neurons or synapses arising under the memory consolidation, but about some total topological properties of those brain networks which are associated with the stored pattern. In other words, could the topological reconstruction of the neuronal network promote the memory consolidation and transfer it into the long-term form? The model consideration of that phenomena shows that such a process is quite possible. For that to happen, two conditions have to be met: (i) the neuronal net should be, initially, the scale-free one, and (ii) the memory consolidation should proceed via the building of long-range links that arise at this stage, for instance, by means of new axon-neuron synaptic contacts.

Keywords: Short-term memory · Long-term memory · Consolidation

1 Introduction

Long-term memory in the brain neuronal system could remain unchanged over the course of decades. To keep it up, the activation of gene expression and epigenetic DNA reconstruction in neurons are, likely, essential. Similarly, the memory reconsolidation, occurring over time, could stem from specific long-term molecular processes in cells, which do not exclude some other mechanisms of longstanding conservation of neuronal networks phenotype, modified in the course of learning.

In the present work, another hypothesis is considered – the material structure, fixing the variation and supporting the changed state of the neuronal ensemble, is not the modified gene constitution of neurons and/or synapses, but the changed topology of the relevant neuronal network piece – the engram. It is that new topology enlarges the life-time of the corresponding memory trace and for a

B. Kryzhanovsky et al. (Eds.): NEUROINFORMATICS 2018, SCI 799, pp. 263–273, 2019.
https://doi.org/10.1007/978-3-030-01328-8_32

long time protects it from decaying. To be more specific, we suppose that the difference between the short-long and the long-term memory is due to existence of the long-range inter-neuron connections, characteristic for random scale-free or Small World networks (Experiments show that neuronal networks are, rather, Small Worlds). Such “cementing” links could emerge due to “intergrowing” long axons (while the majority of links is associated with short axons dendrites). The transfer between two memory types results either from originating long-range links (memory consolidation), or their natural or induced destruction (memory obliteration or reconsolidation).

2 Hopfield Model of Neuronal Network – Physical Sense of Parameters

Basic properties of processes, arising in the course of the memory consolidation, could be considered in the framework of two isomorphic models – Hopfield model (Hertz et al. 1991) and Ising one (Baxter 2007). With that, it is necessary to clear what is the role of hypothetical long-range connections. In “magnetic” problem, described by the Ising model, they lift the Curie temperature, i.e. make the system more stable against thermal agitations. In the neuron-network problem, they make the system more stable against various “chemical noises”, leading to the network reconstruction and even – to the violation of the network integrity.

To take advantage of the similarity of two these models, one should find the correlation between involved concepts of pair binding energy of interacting components (neurons – in the Hopfield network and magnetic moments – in the Ising one), as well as the total “energy”, “temperature” and “critical temperature” of both systems.

Considering the Hopfield network, they usually introduce the local dimensional variable u_i ($i = 1, \dots N$), accepting two values $u_i = -1, 1$, which correspond to the active ($u_i = 1$) and the inhibitory ($u_i = -1$) neuron states, and some essential integral parameter – the so called “energy”, to which certain physical sense is not usually assigned. The latter is the quadratic form of variables u_i with weights $w_{ij} \geqslant 0$, characterizing properties of connecting synapses. The weight w_{ij} is proportional to the probability that a spike, produced by the neuron i, leads to generating spike by the neuron j (or vice versa) (Shneidman et al. 2006).

Network dynamics is defined by rules of changing states u_i, w_{ij} of its neurons and weights for the known initial network state, which is prescribed, for instance, in random manner. Specifically, there is postulated the rule, which relates parameters values $u_i(t)$, $w_{ij}(t)$ at the moment t with their values at the next moment $t + 1$. The simplest variant of such a dynamical rule for variables u_i is the McCulloch - Pitts rule (McCulloch and Pitts 1943), and for variables w_{ij} is the Hebb rule (Hebb 1949).

There is shown that changing state of a network tends to proceed towards diminishing the functional

$$E = -\frac{1}{2}\sum_i \sum_j w_{ij} u_i u_j. \tag{1}$$

That suggests using the term "energy" for this functional, since the dynamics of real physical systems passes in the direction of lowering their energy.

What is the meaning of parameters w_{ij} and "energy" E in relation to real neuronal networks? As for the system "energy" E, it is defined by the summary interaction energy of all neurons with each other. That sum depends on neuron states u_i and parameters w_{ij} describing the connectivity between them. Characteristic dimensionless energy for a single pair of interacting neurons is $|w_{ij}u_i u_j| = |w_{ij}|$. From physical point of view, that energy could be, naturally, associated with the true (dimensional) average synapse energy. If the latter is stored in the capacity C_{ij}, formed by pre- and post-synaptic membranes of a synapse being situated in the link between given neurons i, j, then

$$w_{ij} \sim C_{ij} V_{ij}^2,$$

where V_{ij}^2 is the mean-square (time averaged) value of the potential difference between plates of that capacitor.

Estimate for that pair energy could be obtained on the basis of experimental values $C_0 \sim 1\,\mu\text{F/cm}^2$ of specific capacity for synaptic contact and the characteristic voltage amplitude $V_{ij} \sim 30$ mV on this contact (Nicholls et al. 2001). Assuming the contact area is $S \sim 10\,\mu\text{m}^2$, one finds $w_{ij} \sim SC_0V_{ij}^2 \sim 10^{-16}$ J.

This energy is many orders of magnitude more than the thermal energy $k_BT \sim 10^{-20}$ J (k_B is the Bolzman constant), making any network changes due to thermal fluctuations being practically impossible. In fact, the probability of such a synapse transition into another state is defined by the Arrhenius exponent $\exp(-w_{ij}/k_BT) \sim 10^{-4000}$, i.e. is infinitely small.

This means that real neuronal networks are not destroyed by thermal fluctuations and patterns, coded by those networks, are remain unchanged for a long time. The network reconstruction, i.e. modification of its individual elements (u_i – for neurons, and w_{ij} – for synapses), requires energy comparable with the energy $w_{ij} \sim 10^{-16}$ J of the pair interaction and could occur only due to the energy of chemical reactions taking place in synaptic contacts with entering or leaving ions (Na^+, K^+, Ca^{2+}, ets.) and neurotransmitters into/out those contacts. The energy of a single chemical reaction is of $\varepsilon \sim 1\,\text{eV} \sim 10^{-19}$ J. Thus, for the contact reconstructing one needs the total energy which is released in the course of a large number $(w_{ij}/\varepsilon) \sim 10^3$ of elementary reactions. It is the number of ions or neurotransmitter molecules that enter a synapse in a single act of its activation.

3 Isomorphism of Hopfield Neuronal Network and Ising Spin System

The Hopfield network considered is isomorphic to the Ising model for a ferromagnet at zero thermodynamic temperature (we have seen that thermal fluctuations are non-significant) if the interaction energy w_{ij} is identified with the energy J of interaction between magnetic moments. Since in the magnetic problem the critical thermodynamic temperature $k_B T_C \sim J$ coincides, on the order of value, with the energy J of magnetic moment interaction, then the effective critical temperature in the Hopfield problem becomes the averaged (over the engram graph) neuron interaction energy $\langle w_{ij} \rangle$.

As for the thermodynamical temperature T, its analog is the "chemical temperature" T_Φ, being defined by the relation $k_B T_\Phi = \langle \Phi \rangle$, where $\langle \Phi \rangle$ is the chemical analog of thermal fluctuations. In the considered case, those are the "chemical" fluctuations in a synapse (for example, fluctuations of neurotransmitter concentrations in a synaptic contact). This term is purely phenomenal, different processes are grouped together under this same heading which has no implication other than comparison to Johnson-Nyquist noise (Verveen and DeFelice 1974). But, nevertheless, the electric potential of a membrane fluctuates in a random manner (Burns 1968).

Let us generalize the isomorphism idea suggesting that *any* reason for increasing magnetic critical temperature T_C in the Ising problem (*e.g.*, due to changing topology of neuronal graph) results in a more stability of the graph, that becomes less changeable. If, in contrast, T_C lowers, then the graph becomes less stable and its parameters change with time more quickly. It is the essence of our model of the memory consolidation. However, in the framework of such a model one also needs, certainly, to indicate the material source of changing Ising critical temperature T_C and give a quantitative estimate of that changing.

The process of fixing the pattern in the long-term memory proceeds in two stages: first, that image is created in the short-term memory (where it could be stored for a time of $\tau_S \sim 1 - 10$ h $\sim 10^4$ s (Ebbinghaus 1913)), and then (upon the "brain command") it is, supposedly, moved in another topological form, which is more stable and furnishes the storage of such a consolidated image for a much longer time $\tau_L \sim 10 - 100$ years $\sim 10^9$ s. These two states correspond to some local minimums of energy (in the configuration space), separated each of another by the energy barrier Δ, which height just defines the lifetime of the corresponding state – the higher the barrier, the longer the state lifetime. With this, it is naturally to suggest that the higher parameter T_C, defining the stability of the corresponding neuronal graph, the higher the barrier height.

Another analog is the energy gap in the electron spectrum of superconductors, which is simply proportional to the critical temperature. That gap plays the role of the barrier in the process of the thermally induced decay of pair electron states in the superconductivity theory, where $\Delta = 1.76\, k_B T_C$ (Tinkham 2004).

Lifetimes for both considered states of the graph are defined by Arrhenius relations

$$\tau_S = \tau_0 \exp(\Delta_S / \langle \Phi \rangle), \quad \tau_L = \tau_0 \exp(\Delta_L / \langle \Phi \rangle), \tag{2}$$

where the characteristic time τ_0 means the time between successive attempts of the graph to change its state by crossing barriers Δ_S, Δ_L. Absolute values of parameters Δ_S, Δ_L, $\langle\Phi\rangle$ are not known, but ratios $\Delta_S/\langle\Phi\rangle$, $\Delta_L/\langle\Phi\rangle$ (which determine corresponding lifetimes) could be estimated by means of Eq. (2), suggesting that τ_0 is the characteristic "chemical" time of changing the synapse state, which is defined the duration and repetition period of spike potential: $\tau_0 \sim 10$ ms. Then the relation

$$\Delta_S/\langle\Phi\rangle = \ln(\tau_S/\tau_0) \approx 14 \tag{3}$$

corresponds to the adopted lifetime $\tau_S \sim 10^4$ s for the regime of the short-term memory.

How high should be the barrier for the graph lifetime to increase up to the value $\tau_L \sim 10^9$ s? For the new barrier hight one finds

$$\Delta_L/\langle\Phi\rangle = \ln(\tau_L/\tau_0) \approx 23. \tag{4}$$

Thus, relatively small increase (about one and half time) of the barrier hight results in multiple (five order value) increasing the image lifetime. That is the consequence of the exponentially strong dependency of the lifetime on the barrier hight. If one suggest the simple linear relation $\Delta \propto T_C$ between the barrier height in the "neuro-network" problem and the critical temperature in the isomorphic magnetic Ising problem, then for conversing the memory from the short-term form into the long-term one it is sufficient to elevate the critical temperature by half only ($22/14 \approx 1.5$). And this conclusion is absolutely non-critical for adopting an estimate lifetime values τ_S, τ_L, since those values are involved in Eqs. (3), (4) under the logarithm only.

What could be the reason for elevating the graph stability (or the critical temperature in the isomorphic Ising problem) needed for consolidating the short-term memory into the long-range one? We suggest that as such a mechanism (or, at least, one of such mechanisms) could be generating long-range links which are typical for scale-free or Small-World networks (data show that the neuronal networks are, rather, Small-Worlds (Huang and Pipa 2007)). Physically, as such long-rage links could be axons, while the standard "short-range" links, connecting neighbor neurons are dendrites. Conversion of one memory form into another occurs due to growing long-range links – axons (it is the memory consolidation process) or due to their natural or artificial destruction (it is the process of memory forgetting or reconsolidating).

Characteristic time T_{cons} of the memory consolidation is defined by the time of developing inter-neuron axon connections. For the inter-neuron distance $l_a \sim 100\,\mu$m and with the axon growth rate $v_a \sim 1$ mm/day (Bagnard 2007) one finds $T_{\text{cons}} \sim l_a/v_a \sim 2\,$h, that is quite agrees with experiments. One could to validate the considered hypothesis of the memory consolidation by using some special nerve growth factors or axon growth inhibitors. In the latter case, the memory consolidation into the long-term form should be difficult.

In the framework of the considered model, the phenomenon of remembering (which changes the pattern to only a small extent) is accompanied by relatively small changing the topology of a given neuronal network. That occurs due to switching short-range dendrite synaptic links only, driven by "chemical" fluctuations. However, in the process of such a "creep" the network passes from one local minimum of the "energy" landscape to another one (Vedenov 1988). Re-entering in the initial minimum is unlikely, that explains experiments with gradual changing the pattern memory in a series of its rememberings (Bartlett 1932).

4 Enhanced Stability of Neuronal Networks with Long-Range Links

Below we suggest the neuronal network is the so called "Small-World network" (SW) with some specific characteristics (Watts and Strogatz 1998; Albert and Barabási 2002; Huang and Pipa 2007). Usually considered regular networks are graphs whose sites are connected with their near neighbors only. Unlike, in SW there are also random links between remote (in geometrical sense) sites (neurons). Namely those long-range links (shunts, by-passes) are responsible for specific properties of SW.

The Ising problem for SW is formulated exactly as in the traditional case – neurons (magnetic moments) are placed in network sites and interact with their neighbors. However, in SW some geometrically remote neurons become "near" ones, that, naturally, contributes towards enhancing the network stability (increasing the critical temperature in the equivalent Ising problem). Hence, the graph stability depends now on the fraction p of long-range links in the system.

The physical reason for elevating T_{C} with increasing "concentration" p of long-rang links in SW could be understand by considering possible partition of the neuronal graph into domains – regions, within which all neurons are in the same states (for every n neurons of a certain domain, either $u_i > 0$, or $u_i < 0$, where $i = 1, 2 \ldots, n$). In contrast, neurons of adjacent domains are in distinct states (if in one of them all $u_i > 0$, then all $u_i < 0$ in another). Though partitioning the system into domains is energetically unfavorable and increases its energy E (cf. (1)), it does not mean that in the thermodynamic equilibrium the system is uniform (single-domain) and consists entirely of neurons in the same states.

That notice is based on the known statement of the statistical physics, holding that in the thermodynamical equilibrium the system energy is not minimum, but its free energy $F = E - TS$ is minimum, where S is the system entropy, defined by the number G of ways to implement a given energy value E. According to Bolzman, $S = k_{\mathrm{B}} \ln G$. As for the "temperature" T, in the considered case this is the effective parameter corresponding to the average "chemical fluctuation" in synapses. It is not improbable that the magnitude of those fluctuations could be controlled (with drugs, for instance). At $T = 0$, energy coincides with the free energy ($E = F$), so in equilibrium just the energy is minimum, and the system is the single-domain one. With elevating T (that is, with increasing $\langle \Phi \rangle$), the contribution of the entropy term becomes larger, and the system is disintegrated into domains with higher and higher probability. At $T > T_C$, signs of order are disappeared (numbers of neurons in states $u_i > 0$ and $u_i < 0$ become equal). Due to fluctuations, such a system is reconstructed and can not store the information any longer. That corresponds to the loss of memory.

Further on, the neuron graph is considered, for simplicity, in the form of two-dimensional regular lattice. Assume, within that graph the domain has been formed (the region, inside which all neurons are in the state $u_i > 0$, and outside – in the state $u_i < 0$). Let the perimeter length of that domain be L (in units being equal the size of the lattice cell). Then, at its boundary L pairs of neutrons in opposite states arise, that enlarges the system energy by $2LJ$, where $J \sim \langle w_{ij} \rangle$. To calculate the corresponding entropy, one needs to estimate what is the number of variants to draw the line of the length L, which does not get through network sites. If in each site the boundary can "choose" one of z directions, then the number of variants to draw such a boundary equals approximately $G = z^L$. Thus, the variation of the free energy equals $\Delta F \approx 2LJ - k_B T \ln G = L(2J - k_B T \ln z)$. The initial state (with no domain boundary) is stable when that variation is positive, i.e. at sufficiently low "temperature" $T < T_C$, where $k_B T_C = 2J/\ln z$.

Let us suggest now, that apart from considered "short" links, there are long-range connections of much more length (carrying through axons, for instance). They bring straightforward connections between neuron clusters that favors their synchronization. In terms of Ising model, these links encourage parallel magnetisations of those clusters, which is more energetically favorable then antiparallel ones. That should lead to enhancing stability of such a state (formally – to elevating Curie temperature). Corresponding estimate could be obtained, noticing that long-range links diminish the number of variants to draw the domain boundary considered: those boundaries, which intersect long links become "forbidden" (since every such an intersection results in the additional increase of the energy). Therefore, in every such a case, out of z ways to draw the domain boundary there remains $(z-1)$ ways only, that leads (in the next step of elongating the boundary) to reducing the number of possible boundaries by ~z.

If $p \ll 1$ is the fraction of long-range links, then the number of those special cases equals $\sim pL \ll L$, so that the total number of excluded variants for the boundary of the length L is $\sim pL \cdot z$. Hence, the total number of possible variants reduces to

$$G_p \sim z^{L-pLz}(z-1)^{pL}, \tag{5}$$

and the variation of the free energy becomes

$$\Delta F \approx 2LJ - kT \ln G_p = L\{2J - kT[(1-pz)\ln z + p\ln(z-1)]\}. \tag{6}$$

It is positive at $T < T_C(p)$, where

$$kT_C(p) = \frac{2J}{\ln\left[(z-1)^p \, z^{1-pz}\right]}. \tag{7}$$

It follows from (7) (see also Fig. 1), that $T_C(p) > T_C(0)$, i.e. appearing long-range links results in elevating the critical temperature. The physical reasons are those limitations which lead to diminishing the number of possible boundaries and, eventually, to the entropy decreasing. Thermodynamical approach, based on these considerations, results in the analytic dependence (7) of the critical temperature $T_C(p)$ on the fraction p of long links. (The estimate obtained is valid under the condition $p << 1$.)

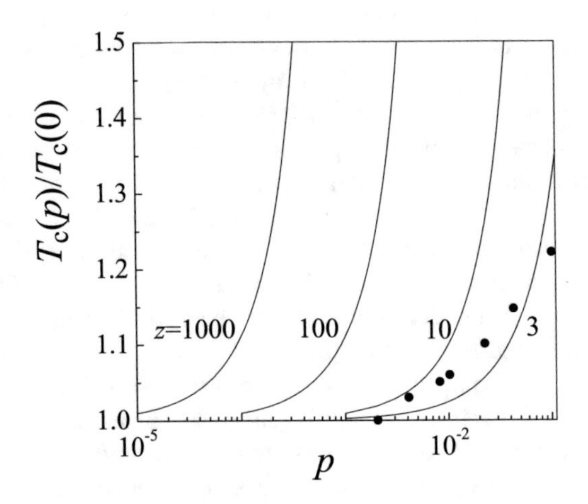

Fig. 1. Analytic dependencies (7) of elevating network Curie temperature $\Delta T_C(p)$ on the fraction p of long-range connections for the Ising problem. Points are numerical results of Monte-Carlo calculations for the lattice with $z = 3$.

The number z of variants to lengthen the domain boundary (in each lattice cell) depends on the structure of the neuron graph, and in order of magnitude coincides with the mean value of effective inter-neuron links (per one neuron). In the real network, that contact number could be very large due to branching and overlapping dendrites, starting from neurons, and building multiple dendrite connections. Often, they say of $10^2 - 10^4$ contacts. However, the mean value of *effective* inter-neuron links is, possibly, much less, that could be connected, for instance, with some physiological and/or geometrical factors, which limit the effective number of dendrites for each neuron, and also complicate allocation of dendrites and axons of the finite thickness within the finite volume of the three-dimensional space, where neurons are placed. Below we assume $z \sim 10 - 100$.

Multiplicity of dendrite links is by no means taken into account in the Hopfield model where all links are direct connections between neurons only (without any "intermediators" as in-between dendrite-dendrite contacts). Therefore, one should distinguish the *structural* and the *functional* neuron networks. The first comprises all, without exclusion, network contacts and has the structure of the real network, while the second one is the coarse (averaged) network structure, reflecting average characteristics of a real network, and, in part, the life-time of its existence (that is, time of memory storing). It is not, however, excluded, that multiplicity of inter-neuron links (including those ones that incorporate a lot of dendrite-dendrite contacts) is important for other functions of neuron networks (not connected with the memory functionality).

In Fig. 1 shown is the plot of the dependency (7) of the critical temperature on the fraction p of long-range links for $z = 10, 100, 1000$. It is seen, that in any case even appearing a small fraction of long links in the neuron network (about, for instance, $\sim$0.3% of the total link number for $z = 100$) results in elevating the critical temperature by half. In the same figure, presented are results (points) of the numerical Monte-Carlo simulation (Herrero 2002) for two-dimensional square lattice (with $z = 3$). It is seen that analytical and numerical results agree qualitatively, that allows to believe in the adequacy of the applied analytic model.

Recall, that according to the above-obtained estimate, transforming the memory from the short-term form into the long-term one (memory consolidation) requires enlarging the energy barrier hight (defining the life-time of a neuron graph) by about 50%. According to our estimates, requirable elevating the critical temperature, which corresponds to the needed enhancing of the neuron graph stability ("shaking" by chemical fluctuations in synapses), could be indeed provided by the mechanism associated with originating long-range (hypothetically, axon) links.

Finally, notice that the isomorphism of Hopfield and Ising models allows to predict the characteristic time of the neurodegenerative process resulting in the Alzheimer disease (AD). The latter is associated with the massive loss of neurons, that is accompanied primarily by the memory impairment and then – by the total its destruction. In this context, the question is raised "What is the fraction of destructed neurons for which the time τ_A of storing the rest part of the neuron engram drops, for example, by a factor of 100, in comparison with the "normal" time $\tau_L \sim 10^9$ s?" From the relation $\tau_A = \tau_0 \exp(\Delta_A/\langle\Phi\rangle) = 10^7$ s, analogous to (2), it follows $\Delta_A/\langle\Phi\rangle) \approx 16$. This corresponds to decreasing the activation barrier by $\Delta_L/\Delta_A \approx 23/16$ times only, i.e. by $\approx$40%. In the framework of the isomorphism hypothesis, that corresponds to lowering the Curie temperature by 40% in the Ising problem, where some fraction q of magnetic moments is removed randomly out of the lattice. The solution of that problem is known. It has been shown (Meilikhov and Farzetdinova 2014) that T_C lowering by 40% occurs when $q \approx 0.4$. Hence, the ruin of approximately 40% neurons should lead to radical memory degradation. That estimate agrees with known clinic data according to which $\approx$5% of neurons is lost each 10 years (Mukhin et al. 2017). Thus, the critical fraction (40%) of lost neurons is attained in 80 years. That perfectly corresponds to the typical age of persons suffering AD.

5 Conclusion

We have considered the hypothetical mechanism of consolidating memory from the short-term into the long-term form, being connected not with the biochemical structure of single neurons or synapses, but – with general topological properties of that region of brain neuronal network which is associated with the stored image. We think, there is the special type of the topological network reconstruction, which could provide the memory consolidation. Considering this phenomenon in the framework of the Hopfield model and the isomorphic Ising model shows that it is, in principle, possible. For that to happen, the neuronal network should not be neither regular, nor the random one, but have some properties of the SW-network.

To obtain numerical results, we need to clear the physical sense and estimate values of all relevant parameters of real neuronal networks, such as the energy of inter-neuron links, the effective (fluctuational) temperature, the network energy, energy barriers determining the rate of the network creep in the energy landscape, etc. It is shown that the memory consolidation could proceed by means of creating long-range inter-neuron connections, for instance – by organizing new axon-neuron synaptic contacts.

Besides, we estimate, that the natural ruin of about 40% neurons could trigger the radical memory degradation in cases of neurodegenerative disease (AD and etc.).

References

Albert, R., Barabási, A.-L.: Statistical mechanics of complex networks. Rev. Mod. Phys. **74**, 47–94 (2002)
Bagnard, D. (ed.): Axon Growth and Guidance. Springer, New York (2007)
Bartlett, F.C.: Remembering: A Study in Experimental and Social Psychology. Cambridge University Press, Cambridge (1932)
Baxter, R.J.: Exactly Solved Models in Statistical Mechanics. Academic Press, London (2007)
Burns, B.D.: The Uncertain Nervous System. Edward Arrnold (Publishers) Ltd., London (1968)
Ebbinghaus, H.: Memory; a contribution to experimental psychology. Teachers college, Columbia university, New York (1913)
Hebb, D.O.: The Organization of Behavior. A Neuropsychlogical Theory. Wiley, New York (1949)
Herrero, C.P.: Ising model in small-world networks. Phys. Rev. E **65**, 066110 (2002)
Hertz, J.A., Palmer, R.G., Krogh, A.: Introduction to the Theory of Neural Computation. Santa Fe Institute Series. Addison-Wesley, Boston (1991)
Huang, D., Pipa, G.: Achieving sinchronization of networks by an auxiliarly hub. Europhys. Lett. **77**(5), 50010 (2007)
McCulloch, W.S., Pitts, W.: A logical calculus of the ideas immanent in neurons activity. Bull. Math. Biophys. **5**, 115–133 (1943)
Meilikhov, E.Z., Farzetdinova, R.M.: Effective Field Theory for Disordered Magnetic Alloys. Phys. Sol. St. **56**, 707–714 (2014)
Mukhin, V.N., Pavlov, K.I., Klimenko, V.M.: Mechanisms of neuron loss in Alzheimer's disease. Neurosci. Behav. Physiol. **47**, 508–516 (2017)
Nicholls, J.C., Martin, A.R., Wallace, B.C., Fuchs, P.A.: From Neuron to Brain, 4th edn. Sinauer Associates Inc. Publishers, Sunderland (2001)
Shneidman, E., Berry, M.J., Segev, R., Bialek, W.: Weak pairwise corellations imply strongly correlated network states in a neural population. Nature **440**(7087), 1007–1012 (2006)

Tinkham, M.: Introduction to Superconductivity. Dover Publications (2004)
Vedenov, A.A.: Modelling Elements of Thinking. Nauka, Moscow (1988). (in Russian)
Verveen, A.A., DeFelice, L.J.: Membrane noise. Prog. Biophys. Mol. Biol. **28**, 189–265 (1974)
Watts, D.J., Strogatz, S.H.: Collective dynamics of "small-world" networks. Nature (London) **393**, 440 (1998)

Functional Neural Networks in Behavioral Motivations

Vyacheslav A. Orlov[1(✉)], Vadim L. Ushakov[1,2], Sergey I. Kartashov[1,2], Denis G. Malakhov[1], Anastasia N. Korosteleva[1,2], Lyudmila I. Skiteva[1], and Alexei V. Samsonovich[2]

[1] National Research Center "Kurchatov Institute", Moscow, Russia
ptica89@bk.ru
[2] National Research Nuclear University MEPhI (Moscow Engineering Physics Institute), Kashirskoe shosse 31, Moscow 115409, Russia

Abstract. Functional magnetic resonance imaging (fMRI) is an effective non-invasive tool for exploration and analysis of brain functions. Here functional neural networks involved in behavioral motivations are studied using fMRI. It was found that behavioral conditions producing different motivations for action can be associated with different patterns of functional network activity. At the same time, connection can be made to dynamics of socio-emotional cognition, decision making and action control, described by the Virtual Actor model based on the eBICA cognitive architecture. These preliminary observations encourage further fMRI-based study of human social-emotional cognition. The impact is expected on the emergent technology of humanlike collaborative robots (cobots) and creative cognitive assistants.

Keywords: fMRI · Functional neural networks · Behavioral motivations Humanlike AI · Cognitive architectures · Social-emotional cognition

1 Introduction

Near-future collaborative robots (cobots) and virtual actors will work side-by-side with humans, and therefore need to be humanlike for their successful functioning [18]. The challenge that becomes critical now is to implement human-level social-emotional intelligence in a machine. To solve this challenge, it is necessary to understand how social-emotional cognition works in the human brain. A proof of this understanding would be a computational model controlling an avatar in a heterogeneous team (including humans and automata), validated in an experimental paradigm. This work presents a study of this sort, in which brain imaging is used to compare internal model dynamics with cognitive brain dynamics. It continues the previous study [1].

Functional magnetic resonance imaging (fMRI) is an effective tool used for exploration and analysis of brain cognitive functions. It is a precise, non-invasive neurovisualization modality which allows tracking of functional brain responses using the Blood Oxygen Level Dependent (BOLD) contrast. This work presents preliminary

B. Kryzhanovsky et al. (Eds.): NEUROINFORMATICS 2018, SCI 799, pp. 274–283, 2019.
https://doi.org/10.1007/978-3-030-01328-8_33

results of the study of brain functional neural networks associated with behavioral motivations of actions, using the fMRI method. In particular, it was possible to compare brain activity associated with different motivations of behavior, that also correspond to distinct model dynamics.

2 Materials and Methods

2.1 Subjects

MRI data were obtained from 5 healthy subjects, mean age 22 (range from 20 to 25 years old). A signed Consent was provided by each participant. Permission to undertake this experiment has been granted by the Ethics Committee of the NRC "Kurchatov Institute". During the scanning, the participants were instructed to play the game "Teleport", specially developed for this study.

2.2 Experimental Paradigm

The experimental paradigm (see also [2, 3, 17]) was based on a virtual environment (VE), where participants engaged in a videogame. The VE setup and the game paradigm are illustrated in Fig. 1.

The VE consists of two platforms: the main platform (the action platform: the big circle in the middle of Fig. 1, top panel) and the escape platform (located to the left of the main platform, Fig. 1). The game session consists of a sequence of similar rounds. At the beginning of each round, three avatars labeled by letters A, B, C are placed at random locations on the main platform. These avatars are controlled by three *actors*: the test subject and two Virtual Actors (automata). Assignment of avatars to actors is random and persistent through the session. The objective of each actor in each round is to reach the escape platform. This can be achieved by means of teleportation: via teleports or by "saving", as explained below.

The main platform has two teleporters: these are small circles inside the big round platform (Fig. 1). Initially, actors need to use teleporters to get to the escape platform. When two actors occupy two teleporters, one of them can activate the other teleporter, so that the other actor can take off and land on the escape platform. This action therefore requires cooperation of two actors: one cannot escape on its own. When on the escape platform, the actor can (but does not have to) initiate teleportation of another selected actor from the main platform to the escape platform. This is done with pressing the Save button and clicking the mouse on the target avatar. Alternatively, the two actors remaining on the main platform still can use teleporters. Shortly after any two avatars have reached the escape platform, they become protected by a shield (blue ellipsoid in Fig. 1, bottom panel), and an explosion occurs that kills the remaining on the main platform avatar. For example, in the situation shown in Fig. 1, bottom panel, two avatars B, C located on the escape platform win, while the third avatar A located on the main platform loses. Alternatively, the first actor that reaches the escape platform could end the round with the Escape button, forcing the other two to lose. The explosion terminates the round either after two minutes from start – or when the Escape

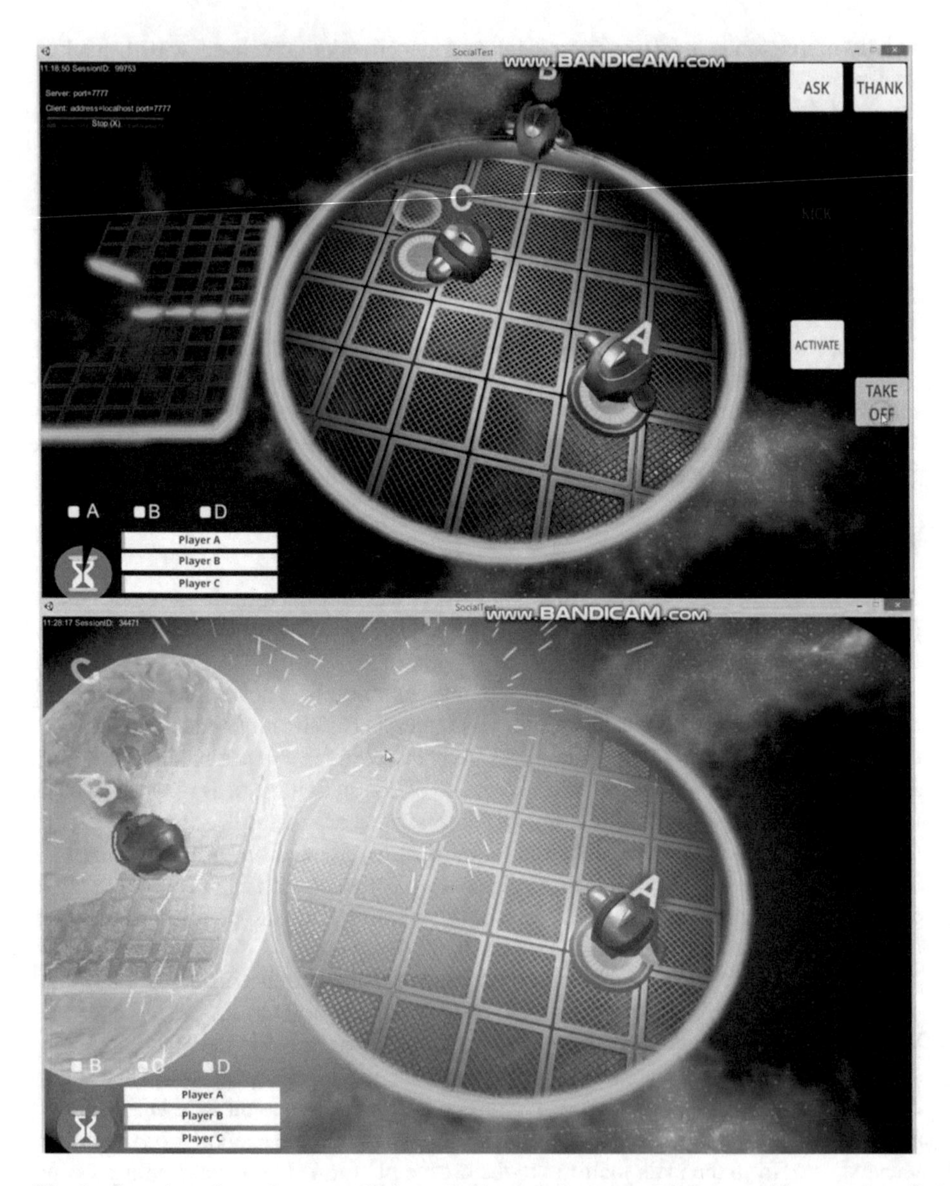

Fig. 1. Snapshots from the game "Teleport". Top panel: avatar C is in a position to take off. Bottom panel: the round ends with victory of B and C, while A loses.

button is pressed, or when two avatars reach the escape platform – whichever comes first. Avatars (if any) located at this moment on the escape platform win, and the rest lose. Then, after a short pause, the next round starts. The session lasts 10 min. The actors in each session in this study were one human participant and two Virtual Actors.

In this paradigm, the metric of effectiveness is the number of times the actor reached the escape platform during the session. The winner is determined by the highest score. The participant was told that the other two avatars may be controlled by humans or by automata. One of the tasks for the participant, except winning the game, was to identify the avatar that, according to participant's belief, was controlled by a human. This was done by marking a checkbox in the bottom-left corner (Fig. 1, top and bottom panels).

The complete list of game actions available to each actor includes: moving around the platform (mouse clicks), greeting another actor (Greet button), asking another actor for help (Ask button), thanking another actor (Thank button), kicking another avatar (Kick button), activating or deactivating a teleporter (Activate button), using an activated teleporter to jump to the goal platform (TakeOff button), saving another avatar by fetching it, while located on the escape platform (Save button), and terminating the round (Escape button). The last two actions are available only on the goal platform. Most actions require proximity to the target and clicking on the target. Essential buttons and associated with them behaviors are summarized in Table 1.

Table 1. Essential buttons and associated behaviors in the VE game paradigm.

Button	Behavior
Kick	Kick another actor - to push another actor, it is necessary to approach this actor and click on it, then, under the condition of the KICK button's activity, the actor, on which the action was committed, will push away for a long distance by a direction opposite to the location of the actor striking a given moment of time
Activate	Activate the teleport - the activation of the teleport can be performed by any actor by pressing the ACTIVATE button, provided that its location corresponds to the location of one of the teleports. This action takes another teleport to the active state
TakeOff	Make your own teleportation - your own teleportation can only be done if the teleport on which the actor is standing is activated. This action moves the actor from the zone of action to the safe zone
Save	To save one of the actors who are not in the rescue zone is possible only when you are in the rescue zone. Allows you to transfer any actor in the zone of action to the rescue zone. It's just enough to press the button SAVE, and then on the actor, located in the action area
Escape	Save yourself alone - is possible only when in the rescue zone. Allows the player in the rescue zone to complete the round without saving anyone in the zone of action. This is done by pressing the ESCAPE button

These actions are available to all actors, including test subjects and Virtual Actors. Corresponding behaviors of avatars would be identical, no matter who or what initiated them.

In each fMRI experiment, 3,000 time points were acquired (with a repetition period of 1 s), which resulted in a 50 min scan for each experimental session. The whole time was broken into four blocks of 10 min each, plus 2 min of rest between them. After the end of each time interval, the subject gave a verbal answer to the question whether he was playing with a person or with a computer.

2.3 Virtual Actors

The implementation of Virtual Actors used in this study was based on the emotional Biologically Inspired Cognitive Architecture (eBICA: [4, 5]). Details of implementation were described in previous works [2, 3]. Two versions of the Virtual Actor model were used in this study to control two avatars in each session: one with elements of social-emotional intelligence, and another without (purely "rational"). In particular, the "rational" Virtual Actor did not establish lasting social relationships.

2.4 fMRI Recording and Data Processing

The experiment was performed on the 3T MRI scanner Magnetom Verio installed in NRC "Kurchatov institute". Details of the MRI recordings are described in related works [7–10].

fMRI data were obtained using SIEMENS Magnetom Verio 3T (Germany) using a 32-channel head coil. T1-weighted sagittal three-dimensional magnetization with a rapid gradient echo was obtained with the following image parameters: 176 slices, TR = 1900 ms, TE = 2.19 ms, cut thickness = 1 mm, rotation angle = 9°, inversion time = 900 ms and FOV = 250 × 218 mm^2. The fMRI data were obtained using the ultrafast sequence with the following parameters: 52 slices, TR = 1000 ms, TE = 25 ms, cut thickness = 2 mm, rotation angle = 90° and FOV = 192 × 192 mm^2.

Maps of the inhomogeneity of the magnetic field during the experiment were also obtained. The ultrafast fMRI protocol was obtained from the Minnesota Center for Magnetic Resonance Research University. fMRI and anatomical data were preprocessed using SPM8 [11] based on Matlab. The center of anatomical and functional data were adducted to the anterior commissure and corrected for magnetic inhomogeneity using field mapping protocol. Slice-timing correction for fMRI data was performed (the correction of hemodynamic response in space and then in time to avoid pronounced motion artifacts) [1].

To exclude motion artifacts, the images were pre-corrected, using a least squares approach and a 6 parameter (rigid body) spatial transformation. Then, spatial normalization were performed to bring them to the coordinates of the MNI (Montreal Neurological Institute) atlas in a coordinate system.

Anatomical data were segmented into 3 possible tissues (grey matter, white matter, cerebrospinal fluid). After that, for the functional data, a smoothing procedure was performed using a Gaussian filter with a 6 × 6 × 6 mm^3 core. Statistical analysis was performed using Student's T-statistics ($p < 0.05$, with correction for multiple comparisons).

In this study, the following behavioral conditions associated with different motives were compared:

- act_actor_comp (Activated, actor, competitor) - activation (salvation) of the opponent playing (the opponent is a playing second person or a computer whose strategy is preformed in the game - not to help you to be saved);
- act_actor_partner (Activated, actor, partner) - Activation (rescue) by the playing partner (the partner is the second person or the computer whose strategy is preformed in the game - to help you to be saved);

- kick_actor - hit on the player;
- kick_target - hit the chosen target (any other avatar);
- saved_actor - saving any other avatar;
- rest - is the resting state of the subject.

3 Results and Discussion

The statistical maps obtained as a result of calculation of the general linear model were plotted on axial patterns for clarity. The resulting slice groups for each contrast of interest are shown in Fig. 2 A–H.

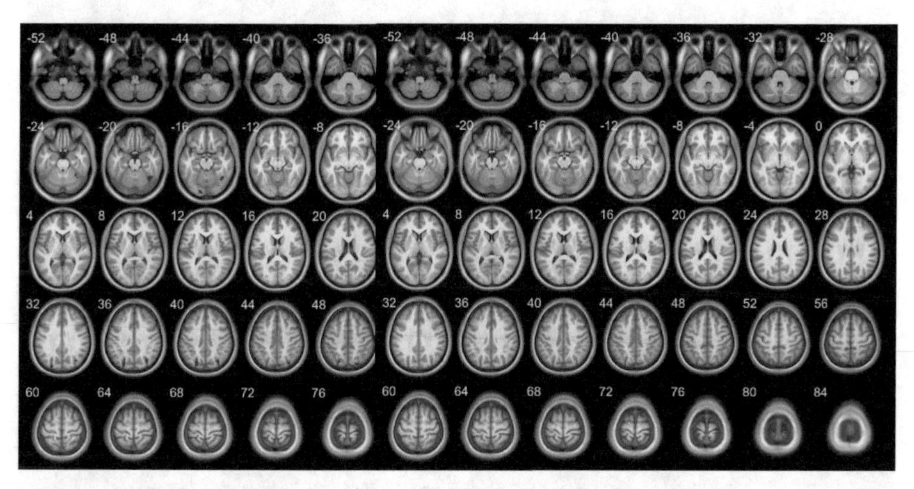

A: act_actor_comp-rest B: act_actor_partner-act_actor_comp

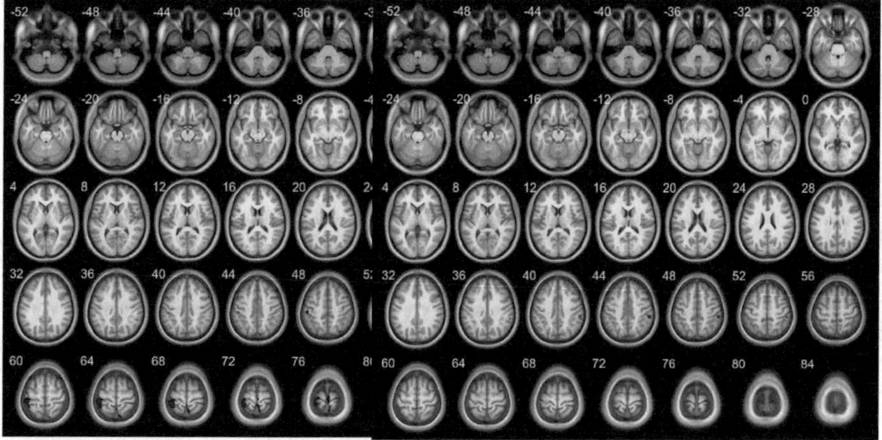

C: act_actor_partner-rest D: kick_actor-kick_target

Fig. 2. (*continued*)

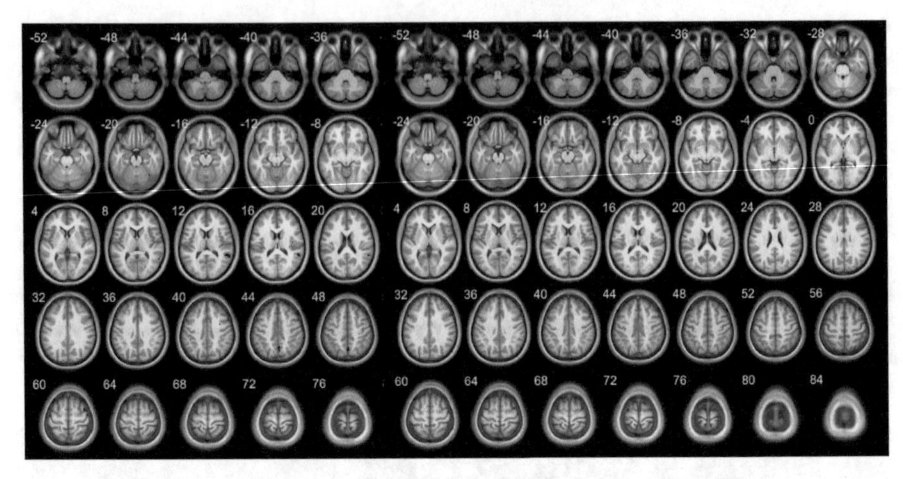

E: kick_target-rest F: kick_actor-rest

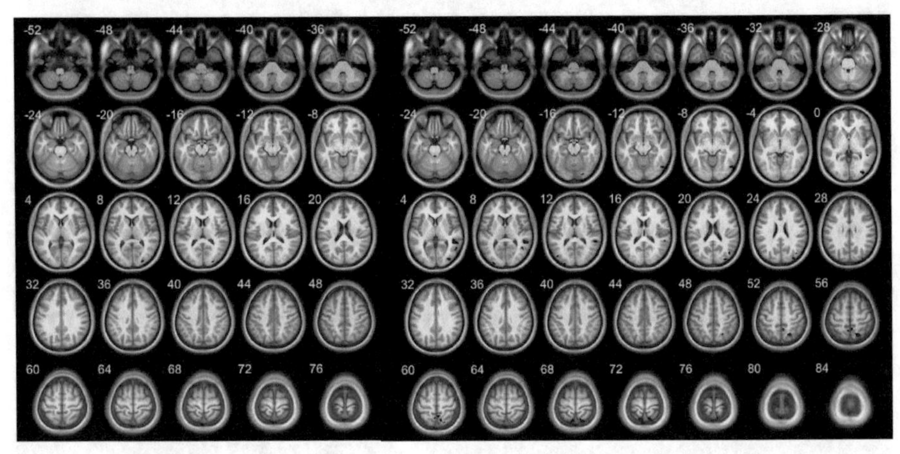

G: saved_actor-rest H: saved-kick_actors

Fig. 2. T-test maps imposed to axial sections, corresponding to different behavioral conditions.

These results (Fig. 2 A–H) show that the change in BOLD signal ($p < 0.05$ (FWE), L-left, R-right) when comparing the "hit" motive executed by the subject with respect to the resting state is observed in neural networks of Temporal_Mid_R area, visual Occipital_Mid_L area, sensorimotor Paracentral_Lobule_L area, in areas related to working memory, making decisions and getting reward - Frontal_Inf_Orb_L, Frontal_Mid_Orb_L. In the case of execution of the motive "hit" against the subject, changes in BOLD signal with respect to the resting state occur in Cerebelum_6_R area, in Temporal_Sup_R, Temporal_Mid_R, in areas of analysis of other people's intentions Fusiform_R, motor Precentral_R area, in frontal areas of attention, motivation, planning and short term memory - Frontal_Mid_RL, in systems related to empathy to other people - SupraMarginal_R, in systems related to processes of self-awareness, conscious integration and information processing - Precuneus_RL were observed ($p < 0.05$, FWE). A direct comparison of the two "hit" conditions differing in authorship showed a difference in the neural networks of empathy - SupraMarginal_R and of interpretation of sensory information - Parietal_Inf_R ($p < 0.05$, FDR). In the behavioral "safekeeping" motive with respect to the resting state, changes in neural network activity in Lingual_R cortex, Occipital_Inf_R, Occipital_Mid_LR, Occipital_Sup_R, Cerebelum_Crus1_R, Fusiform_R, Temporal_Sup_R, Temporal_Mid_R, in the area of observation of actions and strategies - Temporal_Inf_R, in Precuneus_RL, in the area related to the processes of attention and theories of consciousness - Angular_R, in Parietal_Sup_RL cortex, in somatosensory Postcentral_R area were observed ($p < 0.05$, FWE). In the motive called "activation of the teleport for a partner" comparing to the resting state, changes in activity in Cerebelum_Crus1_L, visual Occipital_Inf_L, Occipital_Sup_R, in Parietal_Inf_L, Parietal_Sup_LR cortex, temporal Fusiform_L areas, sensorimotor complex - Postcentral_L, Precentral_L, Supp_Motor_Area_R, integrating areas - Precuneus_RL - were observed ($p < 0.05$, FDR). In the motive called "activation of the teleport for an enemy" comparing to the resting state, changes in activity of neural networks in Cerebelum_6_R, in visual Occipital_Inf_R, Occipital_Mid_LR, Occipital_Sup_LR, Lingual_L, in dorsal flow of visual processing - Cuneus_R, in Fusiform_R, Temporal_Mid_LR, Temporal_Sup_R, in Parietal_Sup_LR, sensorimotor Precentral_R, frontal areas related to speech - Frontal_Inf_Oper_R, integrative areas - Precuneus_L - were observed ($p < 0.05$, FWE).

In general, the data obtained show, for various behavioral motives, the branchy organization of brain neural networks that participate in cognitive processes of attention, information integration, empathy, planning, understanding of the intentions of other people, etc. To compare different behavioral motives, a pair contrast is planned for all motives on a large group of subjects.

4 Concluding Remarks

In order to be successful in heterogeneous teams, virtual cobots need to be humanlike in their social emotionality. The challenge of finding the right computational model for the design of such cobots is a very hot topic today [12–14]. The popular approach based on statistical machine learning (e.g., [15]) has limitations, because it ignores actual mechanisms of human emotional cognition and does not lead to their understanding. These mechanisms can only be discovered empirically in studies where human brain dynamics recorded by neuroimaging techniques like fMRI and human behavior recorded by the virtual environment simulator are compared to those of a computer model of the human mind, based on the currently emerging scientific framework [16], specifically, the cognitive architecture eBICA [4, 5] embedded in the virtual environment [17]. The present study made one step toward this direction.

Acknowledgements. This study was partially supported by the Russian Science Foundation Grant # 18-11-00336 (Virtual Actors – intelligent agents based on the cognitive architecture eBICA), by the Russian Foundation for Basic Research Grant ofi-m 17-29-02518 (the study of thinking levels) and by the NRC "Kurchatov Institute" (11.07.2018 № 1649, MR compatible polygraphy). Authors are grateful to the MEPhI Academic Excellence Project for providing computing resources and facilities to perform experimental data processing.

References

1. Arinchekhina, J.A., Orlov, V., Samsonovich, A.V., Ushakov, V.L.: Comparative study of semantic mapping of images. Procedia Comput. Sci. **123**, 47–56 (2018). https://doi.org/10.1016/j.procs.2018.01.009
2. Azarnov, D.A., Chubarov, A.A., Samsonovich, A.V.: Virtual actor with social-emotional intelligence. Procedia Comput. Sci. **123**, 76–85 (2018). https://doi.org/10.1016/j.procs.2018.01.013
3. Chubarov, A., Azarnov, D.: Modeling behavior of virtual actors: a limited turing test for social-emotional intelligence. In: Advances in Intelligent Systems and Computing, 636, pp. 34–40. Springer Nature, Cham (2017). ISBN 978-3-319-63939-0
4. Samsonovich, A.V.: Emotional biologically inspired cognitive architecture. Biol. Inspired Cogn. Archit. **6**, 109–125 (2013). https://doi.org/10.1016/j.bica.2013.07.009
5. Samsonovich, A.V.: On semantic map as a key component in socially-emotional BICA. Biol. Inspired Cogn. Archit. **23**, 1–6 (2018). https://doi.org/10.1016/j.bica.2017.12.002
6. Ushakov, V.L., Samsonovich, A.V.: Toward a BICA-model-based study of cognition using brain imaging techniques. Procedia Comput. Sci. **71**, 254–264 (2015). https://doi.org/10.1016/j.procs.2015.12.222
7. Orlov, V.A., Kartashov, S.I., Ushakov, V.L., Korosteleva, A.N., Roik, A.O., Velichkovsky, B.M., Ivanitsky, G.A.: "Cognovisor" for the human brain: towards mapping of thought processes by a combination of fMRI and eye-tracking. Book Advances in Intelligent Systems and Computing. Springer Link, vol. 449. Biologically Inspired Cognitive Architectures (BICA) for Young Scientists Proceedings of the First International Early Research Career Enhancement School (FIERCES 2016), pp. 151–157 (2016). https://doi.org/10.1007/978-3-319-32554-5_20
8. Sharaev, M.G., Zavyalova, V.V., Ushakov, V.L., Kartashov, S.I., Velichkovsky, B.M.: Effective connectivity within the default mode network: dynamic causal modeling of resting-state fMRI data. In: Frontiers in Human Neuroscience, vol. 10, Article 14, pp. 1–9, February 2016. https://doi.org/10.3389/fnhum.2016.00014. WOS: 000368983100001
9. Ushakov, V.L., Sharaev, M.G., Kartashov, S.I., Zavyalova, V.V., Verkhlyutov, V.M., Velichkovsky, B.M.: Dynamic causal modeling of hippocampal links within the human default mode network: lateralization and computational stability of effective connections. In: Frontiers in Human Neuroscience, 25 October 2016. https://doi.org/10.3389/fnhum.2016.00528. WOS: 000385957500001
10. Zavyalova, V., Knyazeva, I.S., Ushakov, V.L., Poida, A., Makarenko, N.G., Malakhov, D.G., Velichkovsky, B.M.: Dynamic clustering of connections between fMRI resting state networks: a comparison of two methods of data analysis. Book Advances in Intelligent Systems and Computing, vol. 449. Springer Link (2016). Biologically Inspired Cognitive Architectures (BICA) for Young Scientists Proceedings of the First International Early Research Career Enhancement School (FIERCES 2016), pp. 265–271. https://doi.org/10.1007/978-3-319-32554-5_34
11. URL: http://www.fil.ion.ucl.ac.uk/spm/software/spm8/

12. Gratch, J., Marsella, S.: A domain-independent framework for modeling emotion. Cogn. Syst. Res. **5**, 269–306 (2004)
13. Hudlicka, E.: Affective game engines: motivation and requirements. In: Proceedings of the 4th International Conference on Foundations of Digital Games, pp. 299–306. ACM (2009)
14. Marsella, S.C., Gratch, J.: EMA: a process model of appraisal dynamics. Cogn. Syst. Res. **10**(1), 70–90 (2009)
15. Shum, H.-Y., He, X.-D., Li, D.: From Eliza to XiaoIce: challenges and opportunities with social chatbots. Front. Inf. Technol. Electron. Eng. **19**(1), 10–26 (2018)
16. Laird, J.E., Lebiere, C., Rosenbloom, P.S.: A standard model of the mind: toward a common computational framework across artificial intelligence, cognitive science, neuroscience, and robotics. AI Mag. **38**(4), 13–26 (2017)
17. Bortnikov, P.A., Samsonovich, A.V.: A simple virtual actor model supporting believable character reasoning in virtual environments. Advances in Intelligent Systems and Computing, vol. 636, pp. 17–26. Springer Nature, Cham (2017)
18. Parker, L.: Creation of the national artificial intelligence research and development strategic plan. AI Mag. **39**(2), 25–32 (2018)

Contrasting Human Brain Responses to Literature Descriptions of Nature and to Technical Instructions

Vadim L. Ushakov[1,2(✉)], Vyacheslav A. Orlov[1], Sergey I. Kartashov[1,2], Denis G. Malakhov[1], Anastasia N. Korosteleva[1,2], Lyudmila I. Skiteva[1], Lyudmila Ya. Zaidelman[1], Anna A. Zinina[1], Vera I. Zabotkina[3], Boris M. Velichkovsky[1,3], and Artemy A. Kotov[1]

[1] National Research Center "Kurchatov Institute", Moscow, Russia
ushakov_vl@nrcki.ru
[2] National Research Nuclear University "MEPhI", Moscow, Russia
[3] Russian State University for the Humanities, Moscow, Russia

Abstract. The problem of semantic mapping of the brain is one of the urgent problems in human neurocognitive studies. At the present time there are only few studies reported in the world literature, all of which are made on the material and with the participation of English language native speakers. Russian language can thus become the second language for which this kind of research will be carried out, namely, finding out a correspondence between the semantic classes of the Russian vocabulary and the cortical areas responsible for processing these semantic classes when the text is orally presented. To solve this problem, it is necessary to develop techniques that allow us to investigate cognitive and neurolinguistic mechanisms of perception and understanding of the continuous text segments in natural language. In this paper, we present data on the comparative mapping of the human brain structures involved in the perception of meaningful texts containing technical instructions and literature descriptions of nature.

Keywords: Semantic mapping
Functional magnetic resonance imaging (fMRI) · Semantic categorization
Localization of higher mental functions · Russian language

1 Introduction

The work of the brain is closely related to the categorization of objects in the external environment, the formation of concepts and signs of surrounding objects in memory and the manipulation of these categories in cognitive and speech processes. Certain areas of the human brain are associated in a specific way with the categories of words, as well as with understanding and generation of speech. For example, the Broca area located in the left premotor part of the frontal cortex is associated with the generation of speech—with "motor patterns of words", while the Wernicke area located in the upper part of the posterior temporal cortex, the left hemisphere as well, is responsible for the

B. Kryzhanovsky et al. (Eds.): NEUROINFORMATICS 2018, SCI 799, pp. 284–290, 2019.
https://doi.org/10.1007/978-3-030-01328-8_34

ability to understand oral speech—with "sensory patterns of words" [1]. Other areas of the brain that participate in speech and categorical processes are also found: the "center of concepts" in the lower parietal zone of the left hemisphere and the "center of writing" in the anterior part of the middle frontal gyrus of the left hemisphere, etc. Luria believed that in the processes of encoding and decoding messages, different brain regions work together as a coordinated functional system [1].

In 2016, Huth et al. [2] developed a special experimental procedure to create a semantic atlas of English language corpus in the human brain. The research was carried out on the basis of 12 semantic categories: 'Tactile' (a cluster containing words like 'fingers'), 'Visual' (words like 'yellow'), 'Numeric' ('four'), 'Spatial' ('Stadium'), 'Abstract' ('natural'), 'Temporary' ('minute'), 'Professional' ('meetings'), 'Forcible' ('deadly'), 'Public' ('schools'), 'Mental' ('awake'), 'Emotional' ('despised') 'Social' ('child'). As a result of the work, on a small sample of subjects, the authors showed that different semantic groups of the stimulus material have different representations in the brain. Remarkably, no left hemisphere dominance so obvious in clinical observations has been discovered in the study.

In this paper, we present the first results of a study on the representation of the semantics of technical instructions and literature descriptions of nature in the human brain, performed on the material of the Russian language corpus.

2 Materials and Methods

MRI data were obtained from 10 healthy subjects, mean age 24 (range from 20 to 35 years). Permission to undertake this experiment has been granted by the Ethics Committee of the NRC "Kurchatov Institute". Eleven texts were selected as the stimuli: 6 extracts from classic literature with the descriptions of nature (prose by I. Turgenev, K. Paustovsky, M. Prishvin and A. Ivanov) and 5 texts with the popular description of diverse technical devices, like *door lock*, *refrigerator*, *steam engine*. The total number of words in all the texts was 2064. The texts from two groups contained overlapping concepts – like 'sun', 'reflect' and 'beams' for nature and the description of *a telescope*. Text order within each group was randomized, and then the texts from the two domains were interchanged for presentation. This kind of data preparation allows us to study the relative perception of texts from different thematic domains. In order to evaluate a familiarity of words, the relative frequency of each word was assessed on the basis of a frequency dictionary. Texts were recorded as orally articulated by a professional broadcaster.

The stimuli were linguistically annotated. Each word was aligned with presentation time manually with the help of ELAN annotation software [3]. Figure 1 shows an example of an interface when working with audio material. In the center of the screen is a sonogram of the audio file being processed. Working with the sonogram allows the expert to navigate faster in the audio signal and successfully determine the boundaries of words and pauses. To accelerate the work of the expert, stimulus texts were pre-divided into words, for each word, its relative expected duration was calculated, after which the prepared markup file was saved in the EAF format of the ELAN program. Thus, as an initial material, the expert received pre-selected words with an approximate time reference. The time-aligned word marking line is shown at the bottom of Fig. 1. The task of the expert, therefore, was to check the established word boundaries and, if necessary, move the word boundaries to more accurately match the audio signal.

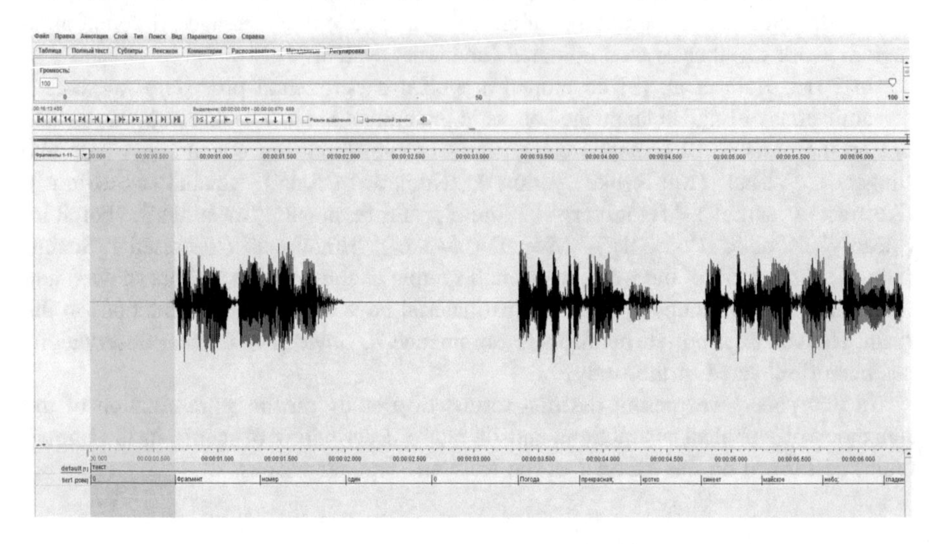

Fig. 1. Markup of an audio file in the ELAN program.

The consistency of the resulting markup was evaluated and, if necessary, corrected by two other experts. We have tagged 2350 annotations for words and pauses. It was assumed that the end of the previous word coincides with the beginning of the next word if the pause is less than 110 ms, otherwise a pause was annotated between the two words. A specially developed automatic parser [4] has been used to assign grammatical characteristics (part of speech, case, number, voice, etc.) to each word. Each word was lemmatized to its lexeme: from *koshki* ('cats') to *koshka* ('cat'). Semantic tags were assigned to nouns and verbs following the Russian semantic dictionary [5], lexical homonymy has been manually resolved. Each semantic tag is a triple $< x, y, z >$, where x is the general lexical class (part of speech), y is the general semantic category and z is the semantic series. For example, a semantic tagging for the word *konyuh* (a groom, a stable boy) is a triple (1, 38, 231), where 1 is a non-abstract noun, 38 is the semantic category 'something corresponding to transport or transportation' and 231 is the lexical semantic series 'person in relation to manual, horse-drawn and other non-automatic means of transportation'. Grammatical and semantic annotation was used as the general data to verify the processing of lexical semantic categories by means of correlation to Blood oxygenation level dependent (BOLD).

The MRI data was acquired using a 3 T SIEMENS Magnetom Verio MR tomograph. The T1-weighted sagittal three-dimensional magnetization-prepared rapid gradientecho sequence was acquired with the following imaging parameters: 176 slices, TR = 1900 ms, TE = 2.19 ms, slice thickness = 1 mm, flip angle = 9°, inversion time = 900 ms, FOV = 250 mm × 218 mm^2. fMRI data was acquired with the following parameters: 30 slices, TR = 2000 ms, TE = 25 ms, slice thickness = 3 mm, flip angle = 90°, FOV = 192 × 192 mm^2. The fMRI and structural MR data were preprocessed using SPM8 (available free at http://www.fil.ion.ucl.ac.uk/spm/software/spm8/). After Siemens DICOM files into SPM NIFTI format all images were manually centered at the anterior commissure. EPI images were corrected for magnetic field inhomogeneity using FieldMap toolbox for SPM8. Next, slice-timing correction for fMRI data was performed. Both anatomical and functional data was normalized into the ICBM stereotactic reference frame. T1 images were segmented into 3 tissue maps (gray/white matter and CSF). Functional data was smoothed using Gaussian filter with a kernel of 6 mm FWHM. Statistical analysis was performed using Student's T-statistics ($p < 0.05$, with correction for multiple comparisons (FWE).

3 Results

In total, 10 people participated in the experiment. For a preliminary analysis, we selected data of five participants who have a more pronounced activity in the auditory area ($p < 0.05$, FWE), when presenting auditory stimuli. The other five subjects also had a change in the BOLD signal in the auditory areas at a significant albeit lesser level of $p = 0.05$ (without correction for multiple comparisons). Their data were then included only in the group analysis. For the selected subgroup of subjects, three contrast comparisons were made: perception of the technical text against the resting state, perception of the natural text against the resting state and perception of the technical text against the natural text. In a contrast comparison of perception of the natural text with respect to the resting state ($p < 0.001$), a change in the BOLD signal was observed in the Temporal_Sup_LR, Temporal_Mid_L, Temporal_Inf_R, Precentral_L, Frontal_Sup_2_L, Frontal_Mid_2_L, Frontal_Inf_Tri_L, Frontal_Sup_Medial_L areas (L - left, R - right). In a contrast comparison of perception of the technical text with respect to the resting state ($p < 0.001$), a change in the BOLD signal in the Temporal_Sup_LR, Temporal_Mid_L, Frontal_Mid_2_L, Frontal_Sup_2_L, Frontal_Inf_Tri_L areas was observed in four subjects out of five. It should be noted that in all these contrasts, there was an individual specificity of the activity of the brain zones in the perception of the text, affecting a large number of zones in cortical and subcortical structures.

An individual specificity of text perception was also noted for cortical structures by Huth et al. [2]. The same individual variability is observed when comparing the perception of the technical instructions in comparison to the descriptions of nature, given in Table 1 ($p = 0.001$). Like Huth et al. [2], we were unable finding left hemisphere dominance for lexical semantic processing in healthy subjects.

Group analysis for 10 subjects, in technical-natural text contrast, showed statistically significant ($p < 0.001$) differences in voxel activity in Temporal_Mid_LR, Calcarine_R, Frontal_Mid_L, Frontal_Inf_Tri_R, SupraMarginal_LR, Thalamus_LR, Parietal_Inf_L, Parietal_Sup_R, Frontal_Inf_Orb_L, Hippocampus_L, Caudate_LR, Pallidum_L, Occipital_Mid_L, Occipital_Sup_L, Lingual_R, Precentral_R, Precuneus_LR, Rolandic_Oper_R, Postcentral_R, Parietal_Sup_L, Putamen_LR, Calcarine_L (L-left, R-right). In natural-technical text contrast, neural network activity was observed in

Insula_L, Cingulum_Ant_LR, Cingulum_Mid_R, Temporal_Mid_L, Temporal_Sup_L, Frontal_Inf_Orb_L, Frontal_Mid_L, Supp_Motor_Area_R, Cerebelum_4_5_R areas. Figure 2 shows the activity of neural networks based on the results of group analysis ($p < 0.001$) applied to the developed map of cerebral cortex obtained in contrasting comparisons of perception of the technical text with respect to the resting state, perception of the natural text with respect to the resting state, and perception of the technical text relative to the natural text.

4 Discussion

The obtained data show that the brain network activation undelaying perception of technical instructions and literature descriptions of nature has a pronounced interindividual specificity. This fact has been noted earlier, in the pioneering study by Huth and colleagues [2]. In a difference to this earlier investigation, we found statistically significant changes not only in neocortex but in subcortical structures as well. In particular, such structure of the limbic system as the hippocampus and amygdala were involved (see Table 1). In general, neural network activity in frontal and temporal areas is predominant in perception of two types of texts. When the descriptions of nature are perceived in comparison to the technical instruction, the activities in the left temporal and frontal regions, in the emotional part of limbic system, in the posterior cingular cortex and insula are registered. This could be described as a mostly sensual and emotional pattern of activation.

In the perception of the technical text in relation to the natural, the activity prevails in the temporal, frontal, parietal (precuneus), and occipital (visual brain) regions. Two latter localizations can be related to self-relational (egocentic) processing [6] with elements of visualizations [7]. At the same time, activity in the subcortical areas—basal ganglia, thalamic structures of the right hemisphere, left hippocampal regions—was observed. Thus, the perception of the technical text in comparison with the description of nature relies on the activation of much broader network architecture including the cortical-subcortical loops of interaction.

Table 1. Contrast comparison of technical text perception with respect to natural text ($p = 0.001$)

Subj.1	Subj.2	Subj.3	Subj.4	Subj.5
Positive				
Frontal_Inf_Oper_R	OFCmed_R	Cuneus_R	Postcentral_R	Frontal_Inf_Tri_L
Paracentral_Lobule_R	Frontal_Inf_Orb_2_R	Frontal_Inf_Tri_L	Rolandic_Oper_R	Frontal_Mid_2_L
Postcentral_R	Temporal_Pole_Sup_R	Frontal_Mid_2_L		Frontal_Sup_2_L
Temporal_Pole_Sup_R		Frontal_Sup_Medial_L		Fusiform_L
		Fusiform_L		Occipital_Mid_L
		Lingual_L		Occipital_Mid_R
		Parietal_Inf_L		Parietal_Inf_L
		Precuneus_L		Parietal_Sup_L
		Supp_Motor_Area_L		Parietal_Sup_R
		Temporal_Inf_L		Precentral_L
		Temporal_Mid_L		Temporal_Mid_L

(*continued*)

Table 1. (*continued*)

Subj.1	Subj.2	Subj.3	Subj.4	Subj.5
Negative				
Cerebelum_6_L	Angular_R	SupraMarginal_L	Frontal_Inf_Oper_R	Amygdala_R
Cingulate_Mid_L	Frontal_Mid_2_R	Temporal_Mid_R	Frontal_Mid_2_R	Cuneus_R
Cingulate_Mid_R	Frontal_Sup_2_R	Temporal_Pole_Mid_R	Frontal_Sup_2_R	Frontal_Inf_Oper_R
Cuneus_L	Frontal_Sup_Medial_L	Temporal_Pole_Sup_L	OFCpost_R	Frontal_Inf_Tri_L
Frontal_Inf_Orb_2_R	Occipital_Mid_L		Postcentral_L	Frontal_Inf_Tri_R
Frontal_Mid_2_L	Parietal_Inf_L		Postcentral_R	Frontal_Med_Orb_R
Frontal_Mid_2_R	Postcentral_R		Precentral_L	Frontal_Mid_2_L
Frontal_Sup_2_L	Precuneus_L		Rolandic_Oper_R	Frontal_Sup_Medial_R
Frontal_Sup_2_R	Precuneus_R			Fusiform_R
Frontal_Sup_Medial_L	Temporal_Mid_R			Hippocampus_R
Frontal_Sup_Medial_R	Temporal_Pole_Mid_R			Lingual_L
Fusiform_R				Lingual_R
Occipital_Mid_L				Occipital_Inf_R
Occipital_Mid_R				Occipital_Mid_L
OFCmed_L				Occipital_Mid_R
Parietal_Inf_L				OFCant_R
Parietal_Inf_R				OFCpost_L
Parietal_Sup_R				OFCpost_R
Precentral_L				Precentral_L
Precuneus_L				Temporal_Inf_L
Supp_Motor_Area_R				Temporal_Pole_Mid_L
Temporal_Mid_L				Temporal_Pole_Mid_R
				Temporal_Pole_Sup_R
				Temporal_Sup_L
				Temporal_Sup_R

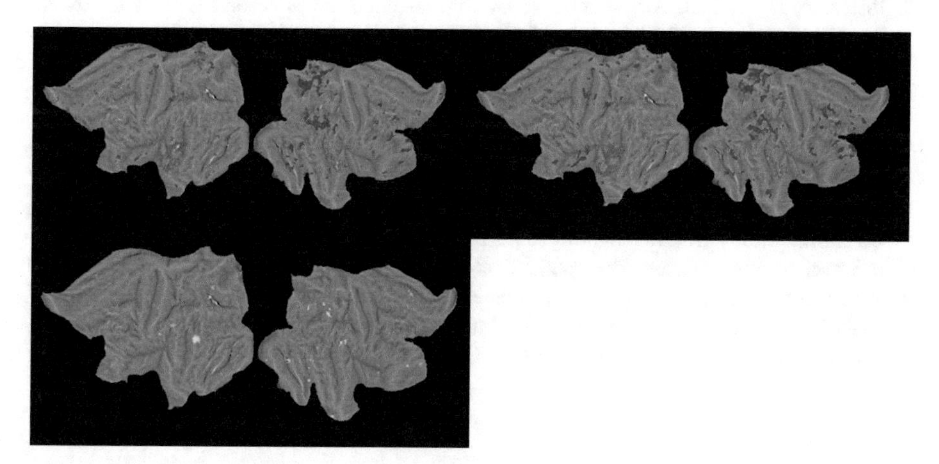

Fig. 2. Obtained activities of neural networks in contrasting comparisons are shown. Perception of the natural text with respect to the resting state (the upper left figure), perception of the technical text regarding the resting state (the upper right figure), and perception of the technical text relative to the natural text (lower left figure) are mapped using group analysis ($p < 0.001$). Warm color indicates an increase in activity, a cold color—a decrease in activity.

These results suggest that processing of semantic information demonstrates a high degree of individual variability. The signs of such processing can be found not only in different areas of the neocortex but also in the deeper subcortical structures. With respect to the discussion of methods suited for such innovative research, we can conclude that the particular approach used in the work allows us to visualize the architecture of the neural networks of the brain involved in the formation and extraction of semantic maps underlying perception and understanding of continuous segments of meaningful texts in Russian language.

Acknowledgements. This study was initiated and partially supported by the RSF (grant 17-78-30029, the brain mapping of semantic categories). In preparation of experimental material, we also used algorithms developed in framework of two ofi-m projects supported by the RFBR: grant 17-29-02518 on the cognitive-affective structures of the human brain and grant 16-29-09601 on automatic detection of emotional expression in continuous texts. The authors are grateful to the MEPhI Academic Excellence Project for Providing computing resources and facilities to perform experimental data processing.

References

1. Luria, A.R.: The main problems of neurolinguistics, 3rd edn. The LIBROKOM Book House, Moscow (2009). [In Russian]
2. Huth, A.G., de Heer, W.A., Griffiths, T.L., Theunissen, F.E., Gallant, J.L.: Natural speech reveals the semantic maps that tile human cerebral cortex. Nature **532**(7600), 453–458 (2016)
3. Brugman, H., Russel, A.: Annotating multimedia multi-modal resources with ELAN. In: Proceedings of the 4th International Conference on Language Resources and Language Evaluation (LREC 2004), pp. 2065–2068 (2004)
4. Kotov, A., Zinina, A., Filatov, A.: Semantic parser for sentiment analysis and the emotional computer agents. In: Proceedings of the AINL-ISMW FRUCT 2015, pp. 167–170 (2015)
5. Shvedova, N.Y.: The Russian Semantic Dictionary. Azbukovnik, Moscow (1998). [In Russian]
6. Soch, J., Deserno, L., Assmann, A., Barman, A., Walter, H., Richardson-Klavehn, A., Schott, B.H.: Inhibition of information flow to the default mode network during self-reference versus reference to others. Cereb. Cortex **27**, 3930–3942 (2017)
7. Verkhlyutov, V.M., Ushakov, V.L., Sokolov, P.A., Velichkovsky, B.M.: Large-scale network analysis of imagination reveals extended but limited top-down components in human visual cognition. Psychol. Russ.: State-of-the-Art **7**(4), 4–19 (2014)

fMRI and Tractographic Studies of Cognitive Systems in the Human Brain at the Norm and the Paranoid Schizophrenia

Vadim L. Ushakov[1,2(✉)], Vyacheslav A. Orlov[1], Denis G. Malakhov[1], Sergey I. Kartashov[1,2], Alexandra V. Maslennikova[3], Andrey Yu. Arkhipov[3], Valeria B. Strelez[3], Maria Arsalidou[4], Alexandr V. Vartanov[5], Georgy P. Kostyuk[6], and Natalia V. Zakharova[6,7]

[1] National Research Center "Kurchatov Institute", Moscow, Russia
ushakov_vl@nrcki.ru
[2] National Research Nuclear University "MEPhI", Moscow, Russia
[3] Institute of Higher Nervous Activity and Neurophysiology of RAS, Moscow, Russia
[4] National Research University Higher School of Economics, Moscow, Russia
[5] Lomonosov Moscow State University, Moscow, Russia
[6] Alekseyev Psychiatric Hospital No. 1, Moscow, Russia
[7] Pirogov Russian National Research Medical University, Moscow, Russia

Abstract. This study is aimed at a systematic study of the work of neural networks of the human brain and their architecture in norm and in schizophrenia. To obtain the neurophysiological data, a unique complex of experimental equipment for world-class neurocognitive studies was used. The data obtained showed a significant decrease in the structural connectivity relationships for the rich club coefficient for a group of schizophrenic patients compared with the norm. Perception of emotionally negative visual and audio stimuli related to delusions in patients with schizophrenia does not lead to a significant decrease in BOLD signal as compared with the norm in Calcarine_L, Cerebelum_4_5_R, ParaHippocampal_LR, Precuneus_L, Temporal_Sup_R areas. The differences found in the structural and functional patterns of cognitive-affective disorders can serve as prognostic biomarkers in patients with schizophrenia and will make a significant contribution to the development of high-tech diagnostics in the early stages of mental illness.

Keywords: fMRI · Rich-club · Cognitive architecture · Connectome
Schizophrenia · Hallucinatory-paranoid syndrome

1 Introduction

With the help of modern methods of non-invasive neuroimaging, in particular, functional magnetic resonance imaging (fMRI), it was shown that the pathogenesis of schizophrenia includes a neurodegenerative process (possibly starting at the earliest stages of development) and pathophysiological changes. There is evidence of structural

B. Kryzhanovsky et al. (Eds.): NEUROINFORMATICS 2018, SCI 799, pp. 291–299, 2019.
https://doi.org/10.1007/978-3-030-01328-8_35

changes in the limbic system, thalamus, basal ganglia, and prefrontal cortex [1, 2], both as a result of a decrease in the number of neurons, a decrease in the volume of gray matter, and a violation of the integrity of the white matter [3]. The pathophysiological process is caused by abnormal neuronal activity in various cognitive tasks which is associated with morbid distortion of working and long-term memory, decision making and emotional response. These changes are observed at all stages of schizophrenia-spectrum disorders [4–6].

Functionality of the brain depends on effective neural communication and neural integration in different areas. This exchange of information is facilitated by a "connectome" - a complex network of all neural elements and neural connections of the organism, which provides an anatomical basis for the emerging functional dynamics. One of the basic elements of the macro-scale connection architecture associated with global integration is the concept of "neural hubs" - areas of the brain that have many connections and, therefore, occupy a topologically central position in the common network. In addition to the individual rich connectivity, hubs in neural systems, as a rule, are also closely interrelated with each other, forming a central "rich-club" [7]. The paranoid syndrome in schizophrenia has been studied in several works performed with the fMRI method and tractography. Primarily, they focus on clarifying the pathophysiology of disturbance of separate psychological functions - sensory perception and motion behavior, visual perception and hallucinations, and cognitions, including executive functions, working memory, emotional and social cognition, cognitive control. In these studies, methods of experimental presentation of relevant stimuli were used to clarify the localization of the pathological process. Differences in effective neural communication and integrative capacity are particularly noticeable in constructing dynamic connectomes at resting state when comparing patients with schizophrenia to healthy controls [8, 9].

Patients with schizophrenia exhibit reduced activation in the ventral striatum in the evaluation of reward, a violation of the activation pattern in the primary information processing phase in the thalamus, the lower prefrontal cortex, and the posterior part of the parietal lobe which indicates a violation of the ability to multimodal integration, etiopathogenetically associated with the development of paranoid syndrome and the errors of source monitoring, responsible for the development of hallucinations.

Dysfunctional activity in reward and cognition networks remain the main characteristics of schizophrenia, both from the point of view of brain activity and from the point of view of interrelations of brain structures [10]. In particular, in patients with more severe symptom expression, there is a lower activation in the ventral striatum in the evaluation of reward, and the connection between the ventral striatum and the prefrontal regions is significantly reduced [10].

In the design of experimental neuroimaging studies of schizophrenia, it is advisable to include emotionally significant stimuli due to characteristic for this disease pathognomonic ambivalence of the information and the instability of assessing the significance of the information presented. The experiments show a decrease in the activation of the areas of the network, when emotionally significant stimuli are presented [11] and in selection of rare deviating stimuli [12, 13]. At the present stage, specific stimuli in experiments are presented in both active and passive paradigms. The works devoted to the study of visualization of the whole complex of paranoid syndrome (with systematized delirium, symptoms of mental automatism and pseudo-hallucinations). Despite the completeness of the data presented, no studies combining tractography and functional MRI with the presentation of specific stimuli in the period of acute or subacute paranoid syndrome have been carried out so far.

The purpose of this study was to register, in patients with schizophrenia with hallucinatory-paranoid syndrome, with the help of functional MRI, localization of key brain activity areas based on the analysis of the brain response to the presentation of subjectively emotionally significant images (i.e., corresponding to delirium) specially selected with individual assessment of psychological cognitive space. Using the tractography mode in the MRI study, the task was to visualize the rich-club areas of the brain of patients with schizophrenia.

2 Materials and Methods

The study involved 20 patients with paranoid schizophrenia (12 men, 8 women, under 30 y.o.) from among those placed in Alekseyev Psychiatric Hospital No. 1, whose condition met the criteria for the diagnosis of paranoid schizophrenia (F20.0 for ICD-10). The control group for patients with schizophrenia included volunteers without cognitive and affective disorders, comparable to the experimental sample by sex and age – 10 men, 10 women, under 30 y.o. This study was approved by the ethics committee of the NRC Kurchatov Institute, ref. no. 5 (from 05.04.2017). All subjects signed an informed consent for participation in the study. For the patients with paranoid schizophrenia it was confirmed that they were able to follow instruction and understand the consent process. The mental state at the time of the experiment was characterized by the stage of decay of the paranoid syndrome with individual elements of psychic automatism (the influx of thoughts, mentism), criticality to the past psychosis while preserving the memory of the past psychotic state with the possibility to describe a subjective history of the disease (with details of delusions of delirium and the assessment of the affective state in that period). In most cases, the disease developed according to the "classical pathway": the debut of the disease was preceded by a nonspecific prodromal stage (asthenia, disturbances of thinking, individual supervalued ideas) followed by a transformation into a paranoid form (in two cases subcatatonic phenomena were noted at the height of the psychosis; in 12 observations, psychosis was transformed into paraphrenic syndrome). Lucid and onyroid forms of psychosis are excluded, all patients were treated with antipsychotics in therapeutical doses and were in a state of reconvalescence - in the stage of remission formation.

To conduct a functional study for each patient, specific stimuli were selected - cards with words (40 pcs.) and photographs (40 pcs.) associated with the plot of the hallucinatory-delusional syndrome (for example, "video cameras", "persecution", "apocalypse", "psychic", etc.) and the same number of neutral stimuli ("lake", "sky", "coffee", etc.). Two days before the fMRI experiments were conducted, psychological testing of patients was carried out for three tasks: categorization and ranking of stimuli for subjectively emotionally significant groups, categorization and ranking of pictures for subjectively emotionally significant groups, and a comparative scaling of 20 pictures to build a cognitive-psychological space of perception of emotionally significant stimuli. 5 of that pictures were associated with the history of the hallucinatory-delusional syndrome and 15 pictures were taken from the IAPS annotated database (for three pictures on the extreme positions along the axes of Valence, Arousal, Dominance and 3 neutral ones). Presentation of stimuli was carried out in the NBS Presentation software with custom scripts.

Based on this psychological testing, a selection of stimulant material in audio and visual modality for the fMRI study was carried out. Audio stimuli were recorded in two versions – with a male and female voice. Presentation of stimuli occurred in pairs for patients and healthy subjects. The block paradigm of presentation of stimuli was used. Before the presentation of stimuli, the subjects were given 20 s of rest for each fMRI study. For each fMRI study, there were three blocks lasting 64 s. Each block consisted of 16 s of presentation of emotional stimuli, 16 s of rest and 16 s of presentation of neutral stimuli, 16 s of rest.

The study was performed on an MR scanner Magnetom Verio 3T (Siemens, Germany) using a 32-channel head MR coil. For each subject, high-resolution anatomical data was obtained based on the T1-weighted sequence (TR = 1900 ms, TE = 2.21 ms, 176 slices, voxel size $1 \times 1 \times 1$ mm^3).

The fMRI experiment consisted of 2 different sessions:

(1) Recording of functional data for 3 block paradigms using EPI-sequences (TR = 2000 ms, TE = 20 ms, 42 slices, voxel size is $2 \times 2 \times 2$ mm^3).
(2) Recording of functional data for 10 min of resting state using ultrafast (multi-band, MCMRR) sequence (TR = 720 ms, TE = 25 ms, 42 slices, voxel size is $2 \times 2 \times 2$ mm^3).

For all modes, the inhomogeneity of the magnetic field was measured using Siemens gre_field_mapping sequence. To objectively assess the response of the subject to emotionally significant stimuli, the parameters of skin conduction response, photoplethysmogram, upper and lower respiration of the subject were recorded using MRI-compatible polygraphic sensors.

After recording the MRI functional data, the subjects passed an assessment of the resting state using Amsterdam test. Further, a survey was conducted on the effect of emotionally significant and control stimuli of each modality and presentation.

The data obtained during the experiments were processed in the software package SPM8 (Statistical Parameter Mapping http://www.fil.ion.ucl.ac.uk/spm/). Structural and functional data were brought to the center in the forward commissure with normalization to template images with a voxel size of $2 \times 2 \times 2$ mm^3. Functional data were coregistered with structural data for their subsequent three-dimensional alignment. After the registration to remove random emissions for functional data, a Gaussian smoothing procedure with a core of $8 \times 8 \times 8$ mm^3 was performed. For ultrafast sequences, the time offset of the data acquisition was also corrected. Based on Student's statistics (p = 0.001), two conditions were compared statistically - activation of the brain when emotionally significant stimuli were perceived with respect to neutral stimuli for visual and auditory modality.

In the second part of the experiment, the tractographic data were recorded. Diffusion images were acquired using a diffusion sequence with TE = 101 ms, and TR = 13700 ms. A DTI diffusion scheme was used, and a total of 64 diffusion sampling operations were acquired. The b-value was 1500 s/mm^2. The in-plane resolution was 2 mm, the slice thickness was 2 mm. The analysis of diffusion data was conducted using DSI Studio. Each matrix contains positive integers in range from 0 to 10000, that describe the number of white matter tracts between two regions of interests. Connectivity matrices were converted to graph format for data analysis purposes. For rich-club analysis, we used a method described in [7], and we calculated two types of rich-club coefficients: unweighted and weighted.

3 Results

Table 1 (p = 0.001) shows the data of the group analysis based on the BOLD signal change for the groups of tested norm and patients with schizophrenia obtained in the perception paradigms of visual and audio stimuli in contrast to the comparison of emotional stimuli with neutral.

Table 1. Contrast comparison of emotional stimuli in relation to neutral for groups of tested norms and patients with schizophrenia (p = 0.001)

Visual			
norm, positive	norm, negative	patients with schizophrenia, positive	patients with schizophrenia, negative
Fusiform_L Fusiform_R Occipital_Inf_L Occipital_Inf_R Temporal_Mid_R	Calcarine_L Cerebelum_4_5_R Cingulate_Mid_R Cuneus_L Cuneus_R Frontal_Mid_2_R Frontal_Sup_2_L Frontal_Sup_2_R Hippocampus_L Hippocampus_R Occipital_Mid_L ParaHippocampal_L Parietal_Inf_L Parietal_Inf_R Parietal_Sup_L Postcentral_R Precentral_L Precuneus_L Rolandic_Oper_R Temporal_Sup_R Vermis_6	Occipital_Inf_L Temporal_Inf_R Temporal_Mid_L	Cingulate_Mid_R Frontal_Mid_2_R Insula_R Parietal_Inf_R Parietal_Sup_R Precuneus_R
Audio, male voice			
norm, positive	norm, negative	patients with schizophrenia, positive	patients with schizophrenia, negative
Caudate_L Frontal_Inf_Tri_L	Amygdala_R Angular_R	–	Angular_R Cerebelum_3_L

(*continued*)

Table 1. (*continued*)

Visual			
Supp_Motor_Area_L Temporal_Mid_L Temporal_Mid_R Temporal_Pole_Sup_L	Calcarine_L Calcarine_R Caudate_L Cerebelum_4_5_R Cingulate_Ant_L Cingulate_Mid_R Frontal_Inf_Tri_L Frontal_Inf_Tri_R Frontal_Sup_Medial_L Fusiform_R Insula_R ParaHippocampal_R Precuneus_L Precuneus_R Putamen_L Putamen_R Rolandic_Oper_R SupraMarginal_L SupraMarginal_R Temporal_Mid_R Temporal_Sup_L Temporal_Sup_R Thalamus_L Vermis_4_5		Cerebelum_6_L Cerebelum_6_R Cingulate_Mid_R Frontal_Inf_Oper_R Frontal_Inf_Orb_2_R Frontal_Inf_Tri_R Frontal_Mid_2_L Frontal_Mid_2_R Frontal_Sup_2_R Fusiform_L Fusiform_R Insula_L Lingual_L Parietal_Inf_L Parietal_Inf_R Parietal_Sup_R Precentral_L Precuneus_R Putamen_L Rolandic_Oper_L Temporal_Inf_L Temporal_Inf_R Temporal_Mid_R Thalamus_R Vermis_4_5 Vermis_6
Audio, female voice			
norm, positive	norm, negative	patients with schizophrenia, positive	patients with schizophrenia, negative
Caudate_L Caudate_R Cingulate_Ant_L Frontal_Inf_Tri_L Frontal_Med_Orb_L Frontal_Sup_Medial_L Frontal_Sup_Medial_R Hippocampus_R Occipital_Mid_R OFCpost_L Olfactory_R Temporal_Mid_L Temporal_Pole_Sup_R Temporal_Sup_L Thalamus_L	Angular_R Calcarine_L Calcarine_R Cerebelum_4_5_R Cingulate_Mid_L Cingulate_Mid_R Cuneus_R Frontal_Inf_Tri_L Frontal_Mid_2_R Frontal_Sup_2_L Frontal_Sup_2_R Frontal_Sup_Medial_L Fusiform_L Fusiform_R Hippocampus_R Insula_L Insula_R Lingual_R Occipital_Mid_L Occipital_Mid_R ParaHippocampal_L	Frontal_Inf_Tri_L OFCpost_L Rolandic_Oper_R Temporal_Mid_L Temporal_Pole_Sup_R	Occipital_Mid_L

(*continued*)

Table 1. (*continued*)

Visual			
	ParaHippocampal_R		
	Parietal_Inf_L		
	Parietal_Sup_R		
	Precuneus_L		
	Precuneus_R		
	Supp_Motor_Area_L		
	Supp_Motor_Area_R		
	SupraMarginal_L		
	SupraMarginal_R		
	Temporal_Inf_L		
	Temporal_Inf_R		
	Temporal_Mid_L		
	Temporal_Mid_R		
	Temporal_Sup_R		
	Vermis_4_5		

Figure 1 shows comparison data of the rich-club for a control subject (left) and for a patient with schizophrenia (right).

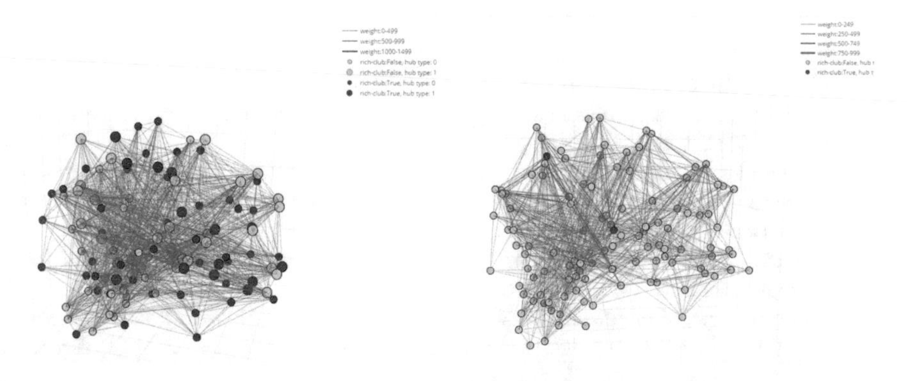

Fig. 1. 3D-graphs calculated from connectivity. Red nodes are in the weighted rich-club versus yellow ones, which aren't. Big nodes are connector hubs, while the small nodes are provincial. Data for a control subject (left) and a patient with schizophrenia (right) are shown.

4 Discussion

The data obtained (Table 1) show that perception of emotionally negative visual and audio stimuli related to delusions in patients with schizophrenia does not lead to a significant decrease in BOLD signal as compared with the norm in Calcarine_L, Cerebelum_4_5_R, ParaHippocampal_LR, Precuneus_L, Temporal_Sup_R areas. At the same time, depending on the modality of the stimuli presented, there is an increase in the BOLD signal for the limbic system structures in a group of healthy subjects, which is not observed for a group of patients. In general, when the emotional stimuli of different modalities are perceived, the number of active areas of the brain changing the activity in the group of patients is less than in the group of healthy subjects. In behavioral reactions in patients, the displacement of emotional estimates of perceived stimuli is observed. Rich club structure for a group of patients contains a small number of zones, while for a group of healthy subjects it contains multiple sites (see Fig. 1). Since normal operation of the brain requires synchronization of neural networks that can be provided by branched rich club structures, their absence in patients leads to violations of labile synchronization and can be the basis of the hallucinatory-delusional state. The obtained data show the prospects of these methods for understanding the mechanisms of the development of cognitive-affective disorders in patients and for detection of prognostic neurophysiological markers in patients with schizophrenia in the development of personalized therapy.

Acknowledgements. This study was partially supported by RFBR Grant ofi-m 17-29-02518 (the cognitive-effective structures of the human brain) and by the NRC "Kurchatov Institute" (11.07.2018 № 1649, MR compatible polygraphy). The authors are grateful to the MEPhI Academic Excellence Project for providing computing resources and facilities to perform experimental data processing.

References

1. Bogerts, B.: The temporolimbic system theory of positive schizophrenic symptoms. Schizophr. Bull. **23**, 423–436 (1997)
2. McKenna, P.J.: What works in schizophrenia: cognitive behaviour therapy is not effective. BMJ Br. Med. J. **333**, 353 (2006)
3. Romme, I.A.C., de Reus, M.A., Ophoff, R.A., Kahn, R.S., van den Heuvel, M.P.: Connectome disconnectivity and cortical gene expression in patients with schizophrenia. Biol. Psychiatry **81**, 495–502 (2017)
4. Karlsgodt, K.H., Sun, D., Cannon, T.D.: Structural and functional brain abnormalities in schizophrenia. Curr. Dir. Psychol. Sci. **19**, 226–231 (2010)
5. van den Heuvel, M.P., Mandl, R.C.W., Stam, C.J., Kahn, R.S., Hulshoff Pol, H.E.: Aberrant frontal and temporal complex network structure in schizophrenia: a graph theoretical analysis. J. Neurosci. **30**, 15915–15926 (2010)
6. Bohlken, M.M., Brouwer, R.M., Mandl, R.C.W., Van den Heuvel, M.P., Hedman, A.M., De Hert, M., et al.: Structural brain connectivity as a genetic marker for schizophrenia. JAMA Psychiatry **73**, 1–9 (2015)
7. van den Heuvel, M.P., Sporns O.: Rich-club organization of the human connectome. J. Neurosci. **31**(44), 15775–15786 (2011)
8. Allen, E.A., Erhardt, E.B., Damaraju, E., Gruner, W., Segall, J.M., Silva, R.F., et al.: A baseline for the multivariate comparison of resting-state networks. Front. Syst. Neurosci. **5**, 2 (2011)
9. Damaraju, E., Allen, E.A., Belger, A., Ford, J.M., McEwen, S., Mathalon, D.H., et al.: Dynamic functional connectivity analysis reveals transient states of dysconnectivity in schizophrenia. NeuroImage Clin. **5**, 298–308 (2014)
10. Simon, J.J., Cordeiro, S.A., Weber, M.A., Friederich, H.C., Wolf, R.C., Weisbrod, M., et al.: Reward system dysfunction as a neural substrate of symptom expression across the general population and patients with schizophrenia. Schizophr. Bull. **41**, 1370–1378 (2015)

11. Lee, S.-K., Chun, J.W., Lee, J.S., Park, H.-J., Jung, Y.-C., Seok, J.-H., et al.: Abnormal neural processing during emotional salience attribution of affective asymmetry in patients with schizophrenia. PLoS One **9**, e90792 (2014)
12. Galderisi, S., Mucci, A., Volpe, U., Boutros, N.: Evidence-based medicine and electrophysiology in schizophrenia. Clin. EEG Neurosci. **40**, 62–77 (2009)
13. Adrian, E.D, Matthews, B.H.C.: The interpretation of potential waves in the cortex. J. Physiol. **81**, 440–471 (1934)

The Principle of Implementing an Assistant Composer

Timofei I. Voznenko(✉), Alexei V. Samsonovich, Alexander A. Gridnev, and Aliona I. Petrova

National Research Nuclear University "MEPhI" (Moscow Engineering Physics Institute), Kashirskoe shosse 31, Moscow 115409, Russia
snaipervti@gmail.com

Abstract. The task of developing an assistant composer is a very interesting task, as the composer's work is complex and creative. Lately there has been done a lot of research, devoted both to systems that help composers to write melodies and systems that generate melodies. In the case of a melody generation, the system offers a composition based on some rules embedded in it according to the musics theory, or obtained by analyzing a large number of compositions with the help of neural networks, while the assistant's task is to offer its ideas that can help composer with creating the work of music. In this article the basic principle of interaction between the composer and the assistant, who helps with writing the melody, is considered. The melody itself can be conditionally divided into two components: the main melody (the main theme, the main idea of the work), and its accompaniment (the arrangement of the main melody). Accordingly, the composer's assistant can help with composing both the main melody and its accompaniment. In this article we consider possible problems that the composer may encounter while composing melodies, as well as possible ways of implementing an assistant which will be able to solve them.

Keywords: Assistant composer · Musical melodies · Emotional preference

1 Introduction

The task of composing a melody is a difficult creative task. In order to compose musical works, the composer must have such qualities as knowledge of harmony, polyphony, as well as knowledge of various musical directions, styles, genres. To the moment, many attempts have been made to implement both the composer's assistant and systems capable to generate melodies. For example, in the paper [1] authors consider the library "Bach: automated composer's helper" for automated composition in real time and musical notation. In the work [2] the cognitive model of music creation based on a semantic map of musical sounds is considered. To achieve this goal, the authors have analyzed the basic theory of music and connected it with models of emotional perception of music. In the paper [3] authors constructed a stochastic musical model that, while analyzing the frequencies of N-grams of a musical theme, does not independently compose a melody, but helps the composer by recommending the following notes for a melody. In the paper [4] authors propose an approach to the creation of polyphonic

B. Kryzhanovsky et al. (Eds.): NEUROINFORMATICS 2018, SCI 799, pp. 300–304, 2019.
https://doi.org/10.1007/978-3-030-01328-8_36

music using LSTM, based on two stages. At the first stage, the LSTM predicts the next chord sequence. At the second stage LSTM generates polyphonic music from the predicted chord sequence. In article [5] authors introduced a model with the goal of learning musical style using deep neural networks and successfully demonstrated a method of using a distributed representation of style to influence the model to generate music with a given mixture of artist styles. In article [6] authors describe MorpheuS: automatic music generation with recurrent pattern constraints. MorpheuS uses state-of-the-art pattern detection techniques to find repeated patterns in a template piece. These patterns are then used to constrain the generation process for a new polyphonic composition. In article [7] authors describe music generation method using recognition of human music conductor's gestures. All these works were aimed at implementing the system being either an assistant or a melody generator. In this paper we consider a possible way to implement namely the composer's assistant.

2 Interaction Scheme

The main purpose of the assistant is to help the composer. The task of the assistant is to offer its ideas (melodies), which will help composer with creating the musical work. Figure 1 shows the scheme of interaction between the composer and the assistant. Let's consider this scheme in more detail.

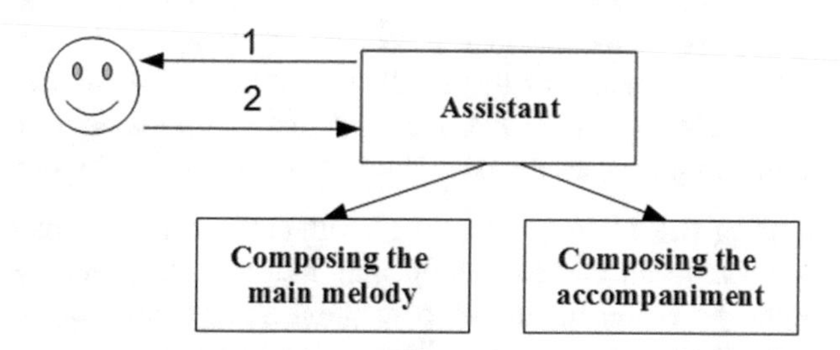

Fig. 1. The scheme of interaction between the assistant and the composer.

2.1 Selecting the Time for Offering Help Automatically

If the assistant will constantly offer its melody variations (ideas), then it will rather obstruct than help the composer with composing the melody. Therefore, it is necessary to determine the time moments at which the assistant will propose its variations. The best option for proposing the ideas would be the moments (detected behaviorally or via psychometrics) when the composer has no idea how to start or continue this or that melody. At these moments the assistant is able to offer its ideas.

2.2 Selecting the Time for Offering Help Manually

The assistant can be given some manual commands to initiate and control interaction with it. For example, there can be the following basic commands:

- Your ideas? – to get assistant's ideas
- Wait – if the composer has not played anything yet, but he ponders the melody at the moment
- Next idea – to propose next idea (if the composer doesn't like the current one)
- Repeat once more – to repeat the idea (if it seems interesting to the composer).

This scheme shows the basic interaction between the composer and the assistant. In the following section, several ways to implement the assistant are considered.

3 Assistant Implementation

Melody can be conditionally divided into two components: the main melody (the main theme, the main idea of the work), and its accompaniment (the decoration of the main melody). Accordingly, the composer's assistant can help with composing both the main melody and its accompaniment (as shown in Fig. 1). Let's consider possible ways of implementing such an assistant.

3.1 Composing the Main Melody

Consider an assistant helping with the basic melody. One way is to train a neural network to compose music, but this can lead to a slightly different task (not to develop an assistant, but to develop a system capable of generating a melody, as, for example, shown in [4]). The task of the assistant is to help the composer to create a melody, which literally means to help, not to get all the work done for the composer. The assistant can help the composer in two ways: (i) help with starting the melody and (ii) help with continuing a given melody. We consider them one by one.

Starting Composing the Melody. One of the difficult tasks is to come up with the beginning of the melody. In this case an opening music phrase can be generated, and then the composer, based on his or her perception and experience, tries to use this phrase to develop the main melody. It is also possible to generate a melody accounting for the composer's emotional mood and preferences changing over time.

Emotional preferences of the composer imply the sequence of composer's preferred emotional states (for example, when the composer is sad, (s)he likes one type of melody, when (s)he feel joyfully – the other one would be preferable, as shown in Fig. 2). Accordingly, there are two tasks: (1) identification of the composer's emotional states, and (2) identification of the generating types of musical compositions that match the composer's emotional states.

The first problem can be solved in several ways, one of which is based on psychometrics: e.g., the analysis of electroencephalogram (EEG) data obtained from brain-computer interface (BCI) to recognize the current emotional state of the composer.

The second task is to develop a system that will check whether this generated melody excerpt is harmonious compared to the type that the composer likes. Examples of the implementation of such a system are provided in [3, 4, 6].

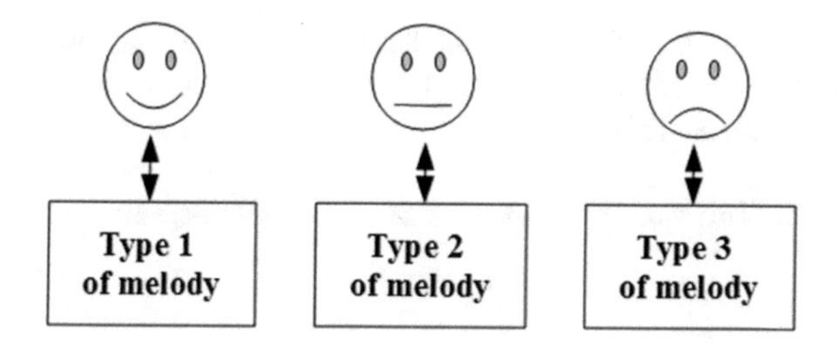

Fig. 2. Composer's emotional preference of various melodies.

Continuing Composing the Melody. One can try to generate random sequences, but with high probability they will not work, as they will not be consistent with the other part of the melody. One possible approach is to choose among previously generated melodies that will be consistent with the current main melody. However, how can one determine whether the generated part of the melody is consistent with the already composed part of the main melody? To this end, as for the previous case, it is necessary to develop a system that takes into account the previous part of the melody and determines whether this generated fragment is harmonious in the context of the current melody. To implement this system, one can use various approaches, including a stochastic musical model [3], LSTM [4], deep neural networks [5], etc. It is also possible to consider the main melody as a combination of frequencies (given that each note corresponds to a certain frequency). By feeding the sequence of frequencies of the main melody to a neural network, one can obtain reasonable choices for its continuation. A disadvantage of this approach is that it is essentially a black box.

3.2 Composing the Accompaniment

To develop an assistant that will help one with creating the accompaniment, it is necessary to take into account the music theory to make sure that the accompaniment does not contradict the main melody. This can be achieved either by integration of the principles of the musical theory (harmony, polyphony, as well as knowledge of various musical directions, styles, genres) into the system, or by using a neural network to analyze a large number of compositions.

4 Conclusions

In this article the basic interaction between the composer and the assistant, which helps with writing the melody, was considered. The melody itself can be conditionally divided into two components: the main melody (the main tune, the main idea of the work), and its accompaniment. Accordingly, the composer's assistant can help with composing the main melody and/or the accompaniment. In this work we considered several approaches and associated problems that the composer assistant designer may encounter, and discussed possible ways of their solution. It was not possible to give here a comprehensive review of the state of the art. One essential original idea that can be taken as the bottom line of this work is that the work of a composer assistant can be based on interpretation of the sequence of emotional states of the composer, in conjunction with a convenient formalization of the music theory.

Acknowledgments. This work was supported by the Russian Science Foundation Grant #18-11-00336.

References

1. Agostini, A., Ghisi, D.: A max library for musical notation and computer-aided composition. Comput. Music J. **39**(2), 11–27 (2015)
2. Buravenkova, Y., Yakupov, R., Samsonovich, A., Stepanskaya, E.: Toward a virtual composer assistant. Procedia Comput. Sci. **123**, 553–561 (2018)
3. Isowa, Y., Iino, N., Okino, S., Iizuka, Y.: A musical composition assistant system using a stochastic musical model. In: 6th IIAI International Congress on Advanced Applied Informatics (IIAI-AAI) (2017)
4. Brunner, G., Wang, Y., Wattenhofer, R., Wiesendanger, J.: JamBot: music theory aware chord based generation of polyphonic music with LSTMs. In: IEEE 29th International Conference on Tools with Artificial Intelligence (ICTAI) (2017)
5. Mao, H., Shin, T., Cottrell, G.: DeepJ: style-specific music generation. In: IEEE 12th International Conference on Semantic Computing (ICSC) (2018)
6. Herremans, D., Chew, E.: MorpheuS: automatic music generation with recurrent pattern constraints and tension profiles. In: IEEE Region 10 Conference (TENCON) (2016)
7. Chen, S., Maeda, Y., Takahashi, Y.: Melody oriented interactive chaotic sound generation system using music conductor gesture. In: IEEE International Conference on Fuzzy Systems (FUZZ-IEEE) (2014)

Neurobiology

Forecasting of Influenza-like Illness Incidence in Amur Region with Neural Networks

A. V. Burdakov[1(✉)], A. O. Ukharov[2], M. P. Myalkin[1], and V. I. Terekhov[1]

[1] Bauman Moscow State Technical University, Moscow, Russia
aleksey.burdakov@gmail.com, maxmyalkin@gmail.com, terekchow@bmstu.ru
[2] PH Informatics, Corp., New York, USA
ukharov@phinformatics.net

Abstract. Influenza-like illness (ILI) incidence forecasting strengthens disease control and prevention. The virus, host and host behavior factors influencing ILI outbreaks have been studied for decades. A range of statistical and machine learning forecasting methods was developed. These novel machine learning methods require inclusion of a proper factor set based on a systematic research. The conventional forecast evaluation metrics such as Mean Absolute Error (MAE) do not adequately reflect the epidemiological requirements and shall be replaced with tailored evaluation criteria. This paper discusses selection of the main influencing factors based on the recent epidemiological research, and proposes new epidemiological forecast evaluation criteria to asses early-warning power of the short-term forecasting model. It describes development of a prediction model based on a Long-Short Term Memory (LSTM) neural network. The model was implemented, trained, validated and tested on the 2007–2018 historical data set and compared to Local Autoregressive Models, Autoregressive Integrated Moving Average, and Multivariate Regression methods.

Keywords: Influenza-like-Illness · Forecast · Neural networks
Deep learning · Machine learning · LSTM

1 Introduction

The Influenza-like illness (ILI) is a human respiratory system disease, which is considerably infectious and mainly spreads through coughs and sneezes of an infected person. Influenza spreads rapidly and creates a considerable economic and social burden. Around 10–15% of the total population especially influencing the school and preschool age groups with 20–30% incidence.

"Rospotrebnadzor", the Public Health Office (PHO) of the Amur Region, performs ILI monitoring, and coordinates preventive and response measure to outbreaks. The monitoring is performed post-factum and does not allow preventive and response measures advance planning. Thus, the ILI disease incidence forecasting and outbreaks early warning capabilities shall strengthen the ILI disease control and prevention.

B. Kryzhanovsky et al. (Eds.): NEUROINFORMATICS 2018, SCI 799, pp. 307–314, 2019.
https://doi.org/10.1007/978-3-030-01328-8_37

The ILI spread is influenced by a number of factors which can be attributed to the following three groups: virus, host and host behavior [5]. The virus survival and transmission rate are attributed to a particular strain, temperature and relative humidity of the air. In particular, the virus transmission rate is very low for relative humidity between 40% and 60%, going up for higher humidity as well as dry conditions below 40%. The latest research [14] though does not address the ultra-dry conditions below 10–15%, which are very common in certain regions of the Russian Federation due to very low winter temperatures below −15°C. It is believed that the transmission rate may drop again for these conditions. The host factors include Vitamin D and Melatonin level [1] which is gained mostly from the UV exposure influencing the immune system of the population with 8…10-week lag. The virus is also prevalent within the school-age and preschool-age groups. The behavioral factors are attributed to hosts proximity due to certain behavior patterns, such as going to work, taking public transport, going to kindergarten or school [6].

There are no precise formulas to infer the variety of the factors to the resulting disease incidence in Amur Region, however, there is a historical set of weekly disease incidence rates and a set of additional factors for 2007–18 that can be used to build a predictive model based on Machine Learning techniques. Due to a short ILI incubation period and its sensitivity to weather factor fluctuations we focus on a short-term forecast from one to two weeks.

2 Existing Forecasting Models Overview

Time series analysis and forecasting is a well-developed domain with a range of methods. Univariate approaches mostly focus on various autoregressive variations. Based on our observations these models do not accurately reflect seasonality due to the weather, holiday and school break fluctuations from year to year [12]. Multivariate models are built with different regression models accompanied by various data transformations, e.g. differentiation. These models perform better due to accounting for additional factors like weather. However, it is challenging to find an optimal model in practice due to the manual feature engineering requirement [16].

Machine learning techniques address nonlinear dependencies and interactions and are considered to be the most promising. Artificial Neural Networks (ANN) were applied to ILI forecasting based on weekly data [9, 11, 13]. Authors in [10] added some climatic and spatial factors. However, the authors did not systematically research the range of factors influencing the ILI spread process. So, a significant improvement can be gained by including a proper set of factors into a forecasting model.

3 Input Features Selection

Based on the ILI spread factors research [1, 5, 6, 14] we have evaluated a range of factors, their reliability, availability, and influence on the model (see Table 1 for details). Due to the absence of Vitamin D and Melatonin level data we built a model to estimate UV exposure based on sunrise, sunset data, solar angle, snow coverage and cloud density. These factors showed a weak influence on the model, so they were excluded from the set. The vaccination data appeared to be non-reliable, while the population by age groups did not show any significant changes over time, so these factors were also excluded from the input data set.

Table 1. Input features selection

Factor group	Factors	Model input
Virus	Temperature	8 per day
	Humidity	8 per day
	Strain	N/A
Host	Weekly case counts	1 per day
	Vitamin D level	N/A
	Melatonin level	N/A
	Sunlight, Snow coverage, Clouds	Not included
	Vaccination	Not included
	Population by age groups	Not included
Behavioral	Holidays	1 per day
	School Breaks	1 per day
	Days-off	1 per day

The selected factors were collected from the PHO, weather archives, government resources and the Internet for a period of 11¼ years. The weekly counts and weather-related measurements were transformed into a 4100 days set.

4 Model Evaluation Criteria

Despite extensive research on epidemic forecasting as well as a number of forecasting competitions held by domain experts, including the Centers for Disease Control and Prevention (CDC) and National Institutes of Health (NIH), there are no unified evaluation criteria [8].

The conventional error metrics, such as Mean Absolute Error (MAE) or Mean Squared Error (MSE), while minimizing the distance between the actual and predicted epidemic curves [7], favor models which provide forecast by repeating the observed values, i.e. performing the "right shift". The value of these "right shift" forecasts is virtually non-existing.

Authors in [2] suggested a multi-criteria framework for evaluating epidemic forecasts. We selected the following Virginia Tech-based metrics (VTM) to asses early-warning power of the short-term forecasting model: (1) Peak Time - time when peak value is attained; (2) First-Take-off-Time - the start time of a sudden increase in the number of new infected case counts; (3) Start-Time of disease season - time at which the fraction of infected individuals exceeds a specific threshold.

Since the ILI data is collected on the weekly basis we measured the difference between prediction and actual data in weeks for each of the VTM-based criterion. The closer the difference to zero, the better the prediction. The positive and negative values designate the "right shift" and "left shift" of the forecast respectively.

5 Neural Network Architecture

Artificial Neural Network (ANN) models are robust to noise in data and allow identifying hidden nonlinear dependencies [15]. Recurrent neural networks (RNN) eliminate the temporal dependence problem, in particular, the Long Short-Term Memory (LSTM) ANNs capture long-term dependencies in a sequence [4].

Figure 1 shows the LSTM unit structure. Forget gate (f_t) layer decides which information will be removed from the cell state. Input gate layer (i_t) decides which values will be updated. Candidate values (c_t') could be added to the new cell state. New cell state (c_t) is calculated by forgetting information from the old cell state and adding scaled candidate values. Finally, the output gate (o_t) and hidden state (h_t) decide what information is going to output [17].

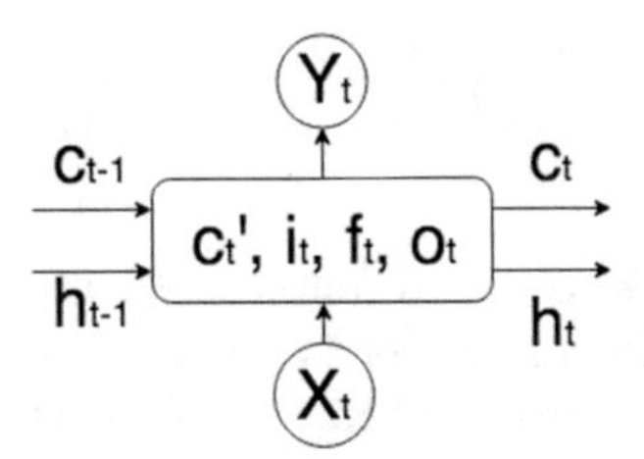

Fig. 1. LSTM unit structure.

Based on the evaluation of ~ 400 model architectures with 1…5 layers, 5…256 neurons, 0.0…0.5 dropouts, we selected a model architecture that consists of 2 stacked LSTM layers with 20 and 15 neurons respectively, lookback of 14 days. Dropout regularization in both layers prevents overfitting. The output dense layer has one neuron predicting disease count and receives input from the second LSTM layer. Figure 2 shows the architecture of the ANN with layer tensor dimensions.

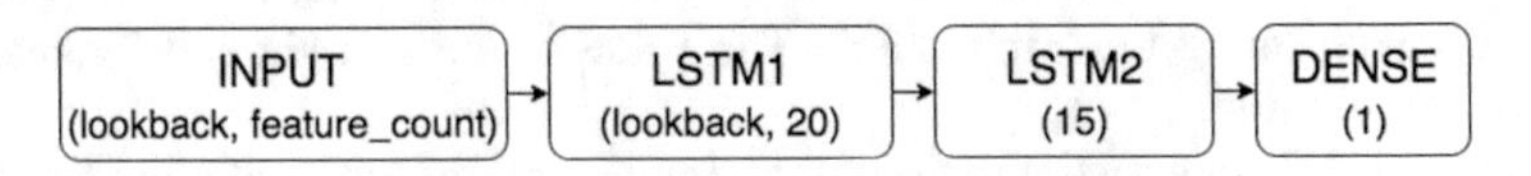

Fig. 2. Stacked ANN architecture.

LSTM ANNs are sensitive to the input data scale, so the MinMaxScaler was used to normalize data features within the (0, 1) interval:

$$X_{t,scaled} = lb + (ub - lb) * (X_t - min(X))/(max(X) - min(X)), \quad (1)$$

where t is time point, X_t is a value to normalize, ub and lb are upper and lower bounds of normalizing interval.

The model was trained for 500 epochs with early stopping. The model used Adam optimizer with the default parameters [3] and MAE loss function.

$$MAE = \Sigma_{i=1}^{n} |y_i - x_i|/n, \quad (2)$$

where y_i is the actual value, x_i is the predicted value, n is the predictions count.

6 Model Training and Evaluation

The dataset was split into the training, validation and test sets. The test data set with ILI counts and thresholds is shown in Fig. 3a. Thresholds are calculated by the PHO based on statistical averaging of the previous 5 years and are used for outbreak detection. Figure 3b present LSTM model predictions for 2015–16.

We used the following univariate models: Local Autoregressive Models (LOESS), Autoregressive Integrated Moving Average (ARIMA), and multivariate regression (MVR) model as the forecasting baseline depicted in Fig. 3c–e.

The univariate models provide weekly predictions based on a rolling window over the complete data set. The multivariate model was trained on the training data set, including weather factors, and evaluated over the test data set, providing a daily forecast for the next 7 days.

Each year in the observed dataset has 4 major peaks. We applied the evaluation criteria to all test data major peaks and averaged the results (see Table 2).

The LSTM model showed the best early prediction and season start capturing accuracy. LOESS demonstrated similar performance in peak prediction, but significantly worse case counts prediction due to the noticeable tendency for the "right shift". ARIMA suffered from the "right shift" which resulted in poor forecasting power. The multivariate regression showed high volatility despite the fact that original output was slightly smoothed with moving average.

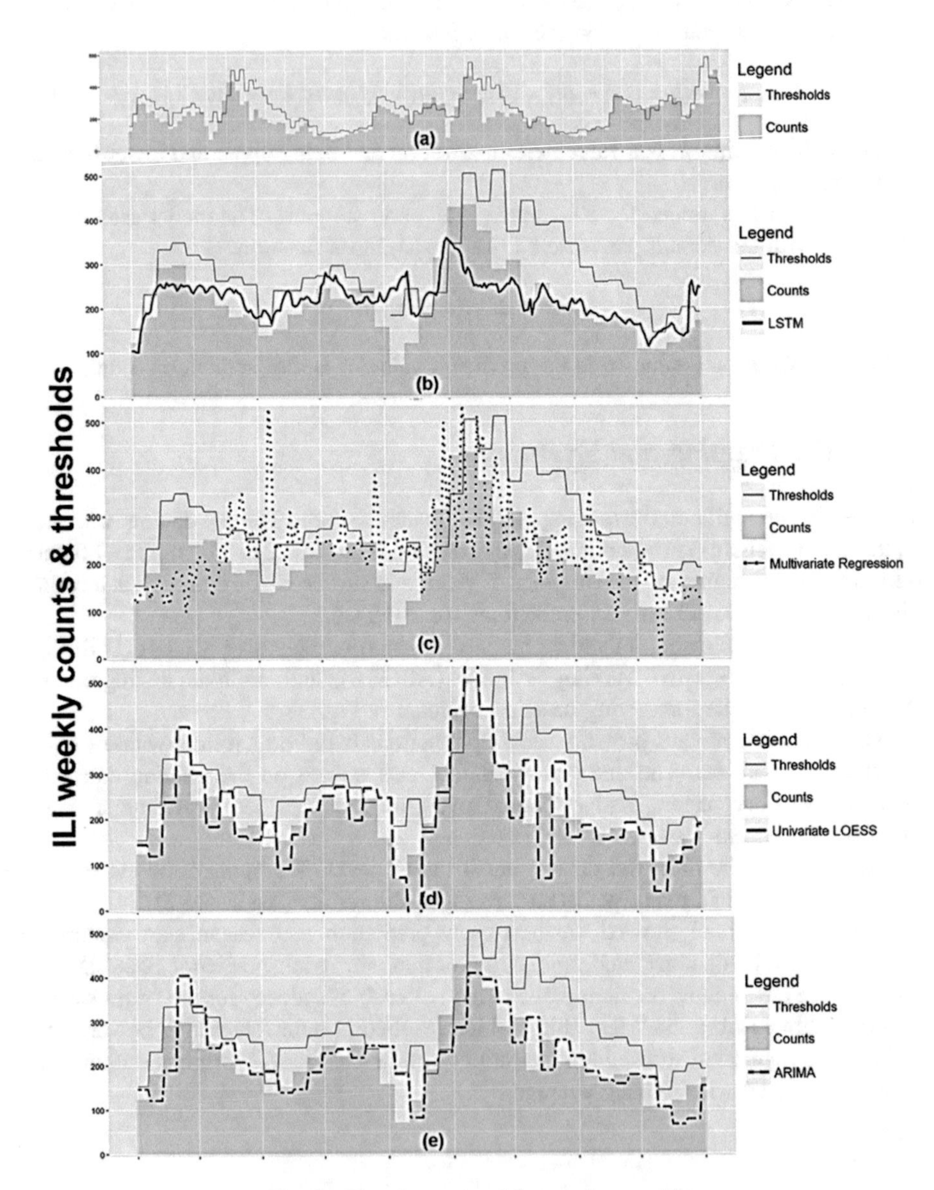

Fig. 3. Test data set and forecasting models.

Table 2. Input features selection.

Criterion	LSTM	MVR	LOESS	ARIMA
Average peak time difference, weeks	0.64	2.64	0.73	1.18
Average first-take-off-time, weeks	0.55	2.36	1.09	1.45
Average start-time of disease season difference, weeks	0.66	1.00	1.33	3.5
Start seasons captured	3 of 3	1 of 3	3 of 3	2 of 3

7 Conclusions

The value of the forecast provided by the existing ILI forecasting models is diminished by their tendency to repeat the current observation as the forecast (the "right shift" effect). The ILI incidence depends on a wide range of virus, host and host behavior factors, thus calling for a multivariate forecasting model. The LSTM ANN fits this tasks nicely due to its robustness to data noise and ability to identify hidden nonlinear dependencies.

Based on the epidemiological studies of the influence factors we developed a forecasting model, which included case counts, temperature, humidity, school breaks, holidays and weekend calendars. The 4100 days (11¼ years) archive was used for training, validation and testing purposes.

The loss functions such as MAE or MSE provide a poor metric for ILI forecasting evaluation allowing the "right shift" effect. We have applied an evaluation method based on the assessment of the peak time, first-take-off-time and start-time of the disease season to adequately assess the forecasting quality.

The Developed LSTM Model Showed Better Prediction Quality Scoring the Best Results on All Four Selected Criteria and Virtually no "Right Shift" Effect.

Further improvement for the proposed model can be achieved by predicting on daily vs. weekly ILI case counts. We also plan to increase the depth of the ILI forecasting with the help of the 5-day weather forecast. Further improvement in accuracy might be achieved by application of ensemble methods coupling the ANN models with other machine learning and statistical models.

References

1. Pittaway, J.K., et al.: Make vitamin D while the sun shines, take supplements when it doesn't: a longitudinal, observational study of older adults in Tasmania, Australia. PLoS One **8**(3), e59063 (2013)
2. Tabataba, F.S., et al.: A framework for evaluating epidemic forecasts. BMC Infect. Dis. **17**(1), 345 (2017)
3. Kingma, D.P., Adam, B.J.: A method for stochastic optimization (2014). arXiv preprint arXiv:1412.6980
4. Hochreiter, S., Schmidhuber, J.: Long short-term memory. Neural Comput. **9**(8), 1735–1780 (1997)
5. Lipsitch, M., Viboud, C.: Influenza seasonality: lifting the fog. Proc. Natl. Acad. Sci. **106**(10), 3645–3646 (2009)

6. Garza, R.C., et al.: Effect of winter school breaks on influenza-like illness, Argentina, 2005–2008. Emerg. Infect. Dis. **19**(6), 938 (2013)
7. Shcherbakov, M.V., et al.: A survey of forecast error measures. World Appl. Sci. J. **24**, 171–176 (2013)
8. https://predict.phiresearchlab.org
9. Xue, H., et al.: Influenza activity surveillance based on multiple regression model and artificial neural network. IEEE Access **6**, 563–575 (2018)
10. Venna, S.R. et al.: A novel data-driven model for real-time influenza forecasting. bioRxiv, 185512 (2017)
11. Zhang, J., Nawata, K.: A comparative study on predicting influenza outbreaks. Biosci. Trends **11**(5), 533–541 (2017)
12. Altizer, S., et al.: Seasonality and the dynamics of infectious diseases. Ecol. lett. **9**(4), 467–484 (2006)
13. Leonenko, V.N., Bochenina, K.O., Kesarev, S.A.: Influenza peaks forecasting in Russia: assessing the applicability of statistical methods. Procedia Comput. Sci. **108**, 2363–2367 (2017)
14. Lowen, A.C., Steel, J.: Roles of humidity and temperature in shaping influenza seasonality. J. Virol. **88**(14), 7692–7695 (2014)
15. Dorffner, G.: Neural networks for time series processing. Neural Network World (1996)
16. Chollet, F.: Deep Learning with Python, 386 p. Manning Publications Co., Shelter Island (2018)
17. Zhang, L., Wang, S., Liu, B.: Deep learning for sentiment analysis: a survey. Wiley Interdisciplinary Reviews: Data Mining and Knowledge Discovery (2018)

Epileptic Seizure Propagation Across Cortical Tissue: Simple Model Based on Potassium Diffusion

Anton V. Chizhov[1,2](✉)

[1] Ioffe Institute, Saint-Petersburg, Russia
anton.chizhov@mail.ioffe.ru
[2] Sechenov Institute of Evolutionary Physiology and Biochemistry of the Russian Academy of Sciences, Saint-Petersburg, Russia

Abstract. Mechanisms of epileptic discharge generation and spread are not well known. Interictal and ictal discharges (IIDs and IDs) are determined by neuronal interactions and ionic dynamics. In order to reproduce the discharges in a simplest way, we have recently proposed a minimal mathematical model that is alternative to the known model Epileptor. Our model is of similar complexity, but in contrast to the Epileptor formulated in terms of abstract variables, it attributes physical meaning to the main variables. The model is expressed in ordinary differential equations for four principal variables, extracellular potassium and intracellular sodium concentrations, a mean membrane potential and a short-term depressing synaptic resource. Our model reproduces IIDs as bursts of spikes, and IDs as clusters of spike bursts. Potassium accumulation governs the transition to IDs. Here we generalize the model to the case of spatial propagation. Diffusion of the extracellular potassium concentration is assumed to govern the spatial spread of spiking activity across cortical tissue. Simulations are consistent with experimental registrations of waves in pro-epileptic conditions, propagating at a speed of about 0.5 mm/s.

Keywords: Epilepsy · Biophysical model · Potassium diffusion

1 Introduction

The spread of activity through cortical circuits has been studied in experiments by means of electrical registrations and optical imaging [1–3]. Experiments show slow propagation of an ictal wavefront and fast spread of discharges behind the front [3]. A speed of the ictal wavefront is similar in different electrophysiological [1, 2] and imaging studies [4], it is about tenths of millimeters per second (0.6 mm/s in [4]). The mechanism is still an open question [3]. Some models that considers spatial propagation suggest that the excitability of the tissue surrounding the seizure core may play a determining role in the seizure onset pattern [5]. Whereas the generation of interictal discharges is modelled in the conditions of impaired but fixed ionic concentrations [6], the dynamics of ictal discharges and the excitability of the cortical tissue is hypothesized to be governed by the ionic dynamics. Generally, a computational approach to this issue requires a biophysical consideration of the neuronal population interactions in

B. Kryzhanovsky et al. (Eds.): NEUROINFORMATICS 2018, SCI 799, pp. 315–320, 2019.
https://doi.org/10.1007/978-3-030-01328-8_38

the conditions of changing ionic concentrations of sodium, potassium, chloride and calcium ions inside and outside the neurons and glial cells. This problem is quite complex and computationally expensive. The most well-elaborated biophysical models considers either a single neuron [7] or a network [8, 9] without a spatial structure. Thus, the consideration of spatial propagation requires a reduced but biophysically detailed model able to reproduce ictal discharges. Recently, we have proposed a spatially concentrated biophysical model of ictal and interictal discharges [10], called "Epileptor-2" after the known abstract model "Epileptor" [11]. Our model might be extended to the spatially distributed case. As shown, the major role in excitability belongs to the extracellular potassium concentration. Recently, the spatial patterns of the extracellular potassium distribution have been registered by means of nanoparticle based technique [12]. The wavefront of potassium elevation from a seizure in 4-AP (4-apinopyridyne that strengthens synaptic connections) based model of cortical epilepsy spreads with a speed about tenths of millimeters per second. In the present work, the Epileptor-2 model is extended by introducing the diffusion equation for the potassium concentration.

2 Methods

2.1 Governing Equations of the Epileptor-2

The Epileptor-2 model [10] is based on the previous biophysically detailed considerations of ionic dynamics during the pathological states of brain activity [8, 13, 14]. The ionic dynamics were implemented in a rate-based model for recurrently connected excitatory and inhibitory neuronal populations, where the inhibitory population has been accounted for implicitly, and the firing rate was assumed to be proportional to that of the excitatory population. The firing rate has been described as a rectified sigmoid function of a membrane potential. The membrane potential has been described by Kirchoff's current conservation law, which was written for a one-compartment neuron. The expressions for the excitatory and inhibitory synaptic currents, the input-output-function, the rate-based equations for the ionic dynamics, etc., are justified in [10]. The short-term synaptic depression is described according to the Tsodyks-Markram model. An adaptive quadratic integrate-and-fire model was used as a model for a representative neuron. Thus, the proposed model consists of three subsystems that describe: (i) the ionic dynamics, (ii) the neuronal excitability, and (iii) a neuron-observer. The equations are as follows:

$$\frac{d[K]_o}{dt} = \frac{[K]_{bath} - [K]_o}{\tau_K} - 2\gamma I_{pump} + \delta [K]_o \nu(t), \tag{1}$$

$$\frac{d[Na]_i}{dt} = \frac{[Na]_i^0 - [Na]_i}{\tau_{Na}} - 3 I_{pump} + \delta [Na]_i \nu(t), \tag{2}$$

$$C\frac{dV}{dt} = -g_L V + u(t), \tag{3}$$

$$\frac{dx^D}{dt} = \frac{1 - x^D}{\tau_D} - \delta x^D x^D v(t). \tag{4}$$

The main variables are as follows: $[K]_o$ and $[Na]_i$ represent extracellular potassium and intraneuronal sodium concentrations, respectively; $V(t)$ is the membrane depolarization; $x^D(t)$ is the synaptic resource; $v(t)$ is the firing rate of an excitatory population; and an inhibitory population firing rate is assumed to be proportional to $v(t)$. The dynamics described by these equations is driven $v(t)$, which is calculated with a sigmoidal input-output function:

$$v(V(t)) = v_{\max} \left[\frac{2}{1 + \exp[-2(V(t) - V_{th})/k_v]} - 1 \right]_+, \tag{5}$$

where $[x]_+$ is equal to x for the positive argument and 0 otherwise. The input current $u(t)$ includes the potassium depolarizing current, the synaptic drive, and the noise $\xi(t)$, respectively:

$$u(t) = g_{K,leak}\left(V_K(t) - V_K^0\right) + G_{syn} v(t)\left(x^D(t) - 0.5\right) + \sigma\xi(t). \tag{6}$$

The potassium reversal potential is obtained from the ion concentrations via the Nernst equation:

$$V_K = 26.6\,\text{mV}\ 1\text{n}\left([K]_o/130\,\text{mM}\right). \tag{7}$$

The Na^+/K^+ pump current is taken from [14] in the form:

$$I_{pump} = \frac{\rho}{\left(1 + \exp(3.5 - [K]_o)\right)\left(1 + \exp((25 - [Na]_i)/3)\right)} \tag{8}$$

The parameters are as follows. The time constants for ionic dynamics, membrane polarization, and synaptic depression are $\tau_K = 100\,\text{s}$, $\tau_{Na} = 20\,\text{s}$, $\tau_m = C/g_L = 10\,\text{ms}$, and $\tau_D = 2\,\text{s}$; the concentration increments at spike are $\delta[K]_o = 0.02\,\text{mM}$, $\delta[Na]_i = 0.03\,\text{mM}$, $\delta x^D = 0.01$; the Gaussian white noise $\xi(t)$ has zero mean and unity dispersion, $\langle \xi(t)\xi(t') \rangle = \tau_m \delta(t - t')$; the noise amplitude is $\sigma/g_L = 25\,\text{mV}$; the maximum pump flux is $\rho = 0.2\,\text{mM/s}$; the volume ration is $\gamma = 10$ (close to 7 in [13]); the postsynaptic charge is $G_{syn}/g_L = 5\,\text{mV} \cdot \text{s}$; the potassium leak conductance is $g_{K,leak}/g_L = 0.5$; the initial extracellular and elevated bath potassium concentrations are $[K]_o^0 = 3\,\text{mM}$ and $[K]_{bath} = 8.5\,\text{mM}$; $V_K^0 = 26.6\,\text{mV}\ 1\text{n}\left([K]_o^0/130\,\text{mM}\right)$; initial and resting intracellular sodium concentrations are the same, $[Na]_i^0 = 10\,\text{mM}$. The parameters of the input-output function are the maximal rate $v_{\max} = 100\,\text{Hz}$, the threshold potential $v_{th} = 25\,\text{mV}$, and the gain $k_v = 20\,\text{mV}$.

A representative neuron is modeled with an adaptive quadratic integrate-and-fire neuron [15]. The equations for the membrane potential $U(t)$ and the adaptation current $w(t)$ are as follows:

$$C\frac{dU}{dt} = g_U(U - U_1)(U - U_2) - w + u + I_a, \qquad (9)$$

$$\tau_w \frac{dw}{dt} = -w, \qquad (10)$$

$$\textit{if } U > V^T \textit{ then } U = V_{reset},\ w = w + \delta w, \qquad (11)$$

with the following parameters: $g_U = 1.5\,nS/mV$, $C = 1nF$, $\tau_w = 200\,\text{ms}$, $V^T = 25\,\text{mV}$, $V_{reset} = -40\,\text{mV}$, $\delta_w = 100\,pA$, $U_1 = -60\,\text{mV}$, $U_2 = -40\,\text{mV}$; initial conditions are $U = -70\,\text{mV}$; $w = 0$; the tonic current is $I_a = 116\,pA$.

2.2 Equation for Extracellular Potassium Diffusion

In order to take into account the diffusion of extracellular potassium, we generalize Eq. (1) in the following way:

$$\frac{\partial[K]_o}{\partial t} = D\left(\frac{\partial[K]_o}{\partial x^2} + \frac{\partial^2[K]_o}{\partial y^2}\right) + \frac{[K]_{bath} - [K]_o}{\tau_K} - 2\gamma I_{pump} + \delta[K]_o v(t), \qquad (12)$$

where D is the diffusion coefficient, set to be $D\Delta x^2 = 0.001\,\text{s}^{-1}$. The calculation domain of modelled cortical area is discretized with 40×40 mesh. The conditions of 4-AP application are modelled with $G_{syn}/g_L = 5\,\text{mV}\cdot\text{s}$ set in a small circle and $G_{syn}/g_L = 1\,\text{mV}\cdot\text{s}$ in the periphery.

3 Results

Simulations (Fig. 1) reproduce the most characteristic features of the discharges. The experiment with optical visualization of potassium distribution after application of 4-AP in the center of the slice (supporting material in [12]) shows the spread of the wavefront, the decay of average potassium concentration and the decay of the concentration below normal level in the center of the slice. Similar scenario is seen in the simulation (Fig. 1A). The speed of the wavefront is comparable with the experimental estimations, tenths of millimeters per second. The mechanism of ictal discharges (IDs) is similar to that in the concentrated, spatially homogeneous model described in [10]. Briefly, the IDs occur quasi-periodically (Fig. 1B,C). Each ID consists of a cluster of short bursts (SBs) that remind interictal discharges (Fig. 1D,E). A single neuron generates a few spikes during each SB (Fig. 1F). Each SB terminates because of synaptic depression. Each ID terminates because of activation of Na-P pump at high sodium levels.

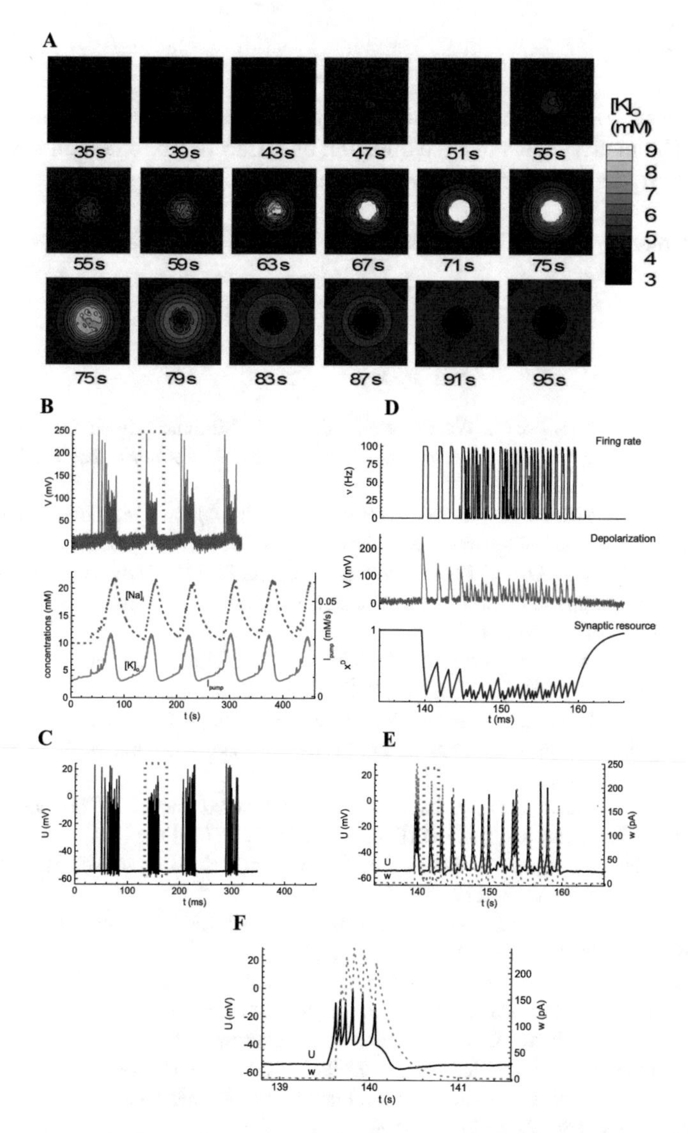

Fig. 1. Simulations of IDs, each consisting of separate series of short bursts (SBs). **A**, 2D-patterns during the first ID. **B–F**, Signals in the focus of the discharges. **B**, the nominal depolarization (top plot), the intracellular sodium and extracellular potassium concentrations (bottom), and the ionic flux through the Na-K-pump (bottom plot) during six IDs. **C**, Simulations of a single neuron activity during the IDs. IDs are the bursts of clustered short bursts (SBs) seen in the membrane voltage. **D**, the population firing rate, the nominal depolarization, and the synaptic resource during a single ID consisting of some SBs. Note the decrease of $[K]_o$ at the high level of the peaks of I_{pump} and following termination of the ID. **E**, a single ID containing a few SBs. Membrane voltage and the adaptation current. **F**, a single SB.

4 Conclusion

A simple biophysical model of ictal discharge generation and propagation has been proposed. The model shows that the diffusion of extracellular potassium ions may play a major role in the spread of the wavefront of the ictal discharges.

Acknowledgments. This work was supported by the Russian Science Foundation (project 16-15-10201).

References

1. Trevelyan, A.J., Sussillo, D., Watson, B.O., Yuste, R.: Modular propagation of epileptiform activity: evidence for an inhibitory veto in neocortex. J. Neurosci. **26**(48), 12447–12455 (2006)
2. Trevelyan, A.J., Sussillo, D., Yuste, R.: Feedforward inhibition contributes to the control of epileptiform propagation speed. J. Neurosci. **27**(13), 3383–3387 (2007)
3. Smith, E.H., Liou, J., Davis, T.S., Merricks, E.M., Kellis, S.S., Weiss, S.A., et al.: The ictal wavefront is the spatiotemporal source of discharges during spontaneous human seizures. Nat. Commun. **7**, 11098 (2016)
4. Wenzel, M., Hamm, J.P., Peterka, D.S., Yuste, R.: Reliable and elastic propagation of cortical seizures in vivo. Cell Rep. **19**, 2681–2693 (2017)
5. Wang, Y., Trevelyan, A.J., Valentin, A., Alarcon, G., Taylor, P.N., Kaiser, M.: Mechanisms underlying different onset patterns of focal seizures. PLoS Comput. Biol. **13**(5), e1005475 (2017)
6. Chizhov, A., Amakhin, D., Zaitsev, A.: Computational model of interictal discharges triggered by interneurons. PLoS ONE **12**(10), e0185752 (2017)
7. Wei, Y., Ullah, G., Ingram, J., Schiff, S.J.: Oxygen and seizure dynamics: II. Computational modeling. J. Neurophysiol. **112**, 213–223 (2014)
8. Bazhenov, M.: Potassium model for slow (2–3 Hz) in vivo neocortical paroxysmal oscillations. J. Neurophysiol. **92**, 1116–1132 (2004)
9. Krishnan, G.P., Bazhenov, M.: Ionic dynamics mediate spontaneous termination of seizures and postictal depression state. J Neurosci. **31**(24), 8870–8882 (2011)
10. Chizhov, A.V., Zefirov, A.V., Amakhin, D.V., Smirnova, EYu., Zaitsev, A.V.: Minimal model of interictal and ictal discharges "Epileptor-2". PLoS CB **14**(5), e1006186 (2018)
11. Jirsa, V.K., Stacey, W.C., Quilichini, P.P., Ivanov, A.I., Bernard, C.: On the nature of seizure dynamics. Brain **137**, 2210–2230 (2014)
12. Müller, B.J., Zhdanov, A.V., Borisov, S.M., Foley, T., Okkelman, I.A., Tsytsarev, V., et al.: Nanoparticle-based fluoroionophore for analysis of potassium ion dynamics in 3D tissue models and in vivo. Adv. Funct. Mater. (2018). 1704598
13. Kager, H., Wadman, W.J., Somjen, G.G.: Simulated seizures and spreading depression in a neuron model incorporating interstitial space and ion concentrations. J. Neurophysiol. **84**(1), 495–512 (2000)
14. Cressman, J.R., Ullah, G., Ziburkus, J., Schiff, S.J., Barreto, E.: The influence of sodium and potassium dynamics on excitability, seizures, and the stability of persistent states: I. Single neuron dynamics. J. Comput. Neurosci. **26**(2), 159–170 (2009)
15. Izhikevich, E.M.: Simple model of spiking neurons. IEEE Trans. Neural Netw. **14**, 1569–1572 (2003)

Spatial and Temporal Dynamics of EEG Parameters During Performance of Tasks with Dominance of Mental and Sensory Attention

Irina Knyazeva[1,2,4(✉)], Boytsova Yulia[3], Sergey Danko[3], and Nikolay Makarenko[1,2]

[1] Pulkovo Observatory, Saint-Petersburg, Russia
iknyazeva@gmail.com
[2] Saint-Petersburg State University, Saint-Petersburg, Russia
[3] Institute of the Human Brain, Russian Academy of Sciences, Saint-Petersburg, Russia
[4] Institute of Information and Computational Technologies, Almaty, Kazakhstan

Abstract. The purpose of this work was to analyze the spatial and temporal dynamics of EEG parameters during performance of tasks with dominance of mental and sensory attention. In addition, an attempt to identify differences in dynamics of EEG parameters in such close mentally oriented tasks as a productive and reproductive imagination was made. EEG wavelet spectrum and phase coherence (relationships between EEG-channels) were studied for θ, α and β frequency ranges of EEG. Analysis of the data showed significant numerous differences of EEG wavelet spectrum and phase coherence between tasks with the dominance of mental and sensory attention and also between such close mental states as productive and reproductive imagination. Time dynamics of EEG differences, obtained in different frequency ranges, allows us to trace which processes were consequentially involved in realization of investigated states on different time intervals and provide new information for understanding of brain mechanisms of these states.

Keywords: Mental and sensory attention
Reproductive and productive imagination · EEG
Analysis of non-stationary signals

1 Introduction

In psychophysiological literature are usually used terms attention to internal vs. external information. Also can be found attention to the environment or to internal processing [10,12]. We consider sensory attention as externally directed to the incoming sensory information, mental attention we consider as internally directed to operating with the information already in the brain. Despite the large

B. Kryzhanovsky et al. (Eds.): NEUROINFORMATICS 2018, SCI 799, pp. 321–327, 2019.
https://doi.org/10.1007/978-3-030-01328-8_39

number of publications, the brain mechanisms of sensory and mental attention have not been studied in equal measure.

Earlier studies of brain mechanisms of mental attention were conducted in a stationary regime of tasks performing, with prolonged (up to several minutes) stay in a state of mental attention [3,6,7,10,12]. In studies of sensory attention is usually used the regime of experiment in which short-term, event-related changes in EEG spectruml power (ERD/ERS) are analyzed. In recent works, the same regime of experiment used to study processes in which mental attention play a dominant role [1,2,8]. Such regime allows comparing correlates of mental and sensory attention and creates opportunities for considering correlates of mental attention at different stages of its realization in time. In present work, we use the above-mentioned regime of experiment for research of spatial and temporal dynamics of EEG parameters during performance of tasks with dominance of mental and sensory attention. We also compared dynamics of EEG parameters in such close mentally oriented tasks as visual imagination and retrieval of visual representations from memory.

2 Experiment Description

28 healthy volunteers (average age 29, 17 women) all right-handed participated in the study. The study described herein was approved by the Ethics Committee of Institute of the Human Brain. All subjects signed written informed consent, in accordance with the ethical standards laid down in the Declaration of Helsinki (1964), prior to their participation in the study. In this paper, we used tasks in which mental attention (MA) or sensory attention (SA) dominates. Tasks for mental attention were also of two types: the dominance of visual imagination (Im) and the dominance of retrieval of visual representations from memory (Mm). Tasks were presented in blocks, 80 trials in each block.

In the block of SA task, the trials consisted of pair of stimuli: word (for example, apple) and corresponding color image, subjects memorized the image. In the block of retrieval of visual representations from memory the trials consisted of pair of stimuli: word and white screen, the words used in the previous SA task were presented. Here, the subjects were asked to recall and visualize on a white background an image, corresponding to presented word. In the block of visual imagination the trials consisted of pair of stimuli: 2 words and a white screen. Here, after simultaneous presentation of 2 words (for example: apple, machine), the subjects were asked to invent and visualize a chimera image (for example, an apple-shaped machine, seeds are pour out when the doors are opened). The duration of verbal stimulus presentation is 400 ms, the duration of second stimulus (color image or white screen) - 5 s, the interval between stimuli - 800 ms.

3 Methods

19 channels of EEG were recorded using standard 10–20 electrode placement on the scalp by computer electroencephalograph “Mitsar-202”. EEG recording and the subsequent removal of eye artifacts were conducted by a software

package WinEEG, version 2.83 (copyright V.A. Ponomarev, J.D. Kropotov, RF2001610516, 08.05.2001). Independent Component Analysis (ICA) was used in the package to correct artifacts due to vertical and horizontal eye movements and blinks. All trials with outliers and trends were also excluded.

Next, for each subject for each type of task, the wavelet power spectrum density (PSD), the square of the wavelet transformation amplitude, and the phase coherence (PC) between each pair of channels were calculated. To calculate phase wavelet coherence and wavelet spectrum, algorithms adapted for EEG analysis, described in the book [5] and in our previous work [8] were applied. The code implemented in Python with application examples is available at the https://github.com/iknyazeva/EEGprocessing repository. The wavelet transformation was done for a grid of 10 frequencies from the 4–30 Hz range, the minimum sample length in the experiment was 3,480 counts or 6.96 s. For the subsequent analysis, a PSD normalized to the average power value during the prestimul interval was used to describe each channel. The dynamics of PC over time was also calculated for a grid of 10 frequencies, for each pair of channels, that is, only 171 dynamics episodes for each of the 10 frequencies. Thus, for each subject, samples of 10 normalized PSD were obtained for each channel and 10 dynamics of phase coherence for each pair of channels. The power spectrum for each channel can be represented as an image by the dimension of the number of frequencies per time, and by the same image each pair of channels for phase coherence. Figure 1 shows examples of wavelet transforms for one channel and phase coherence for one pair of channels for one subject.

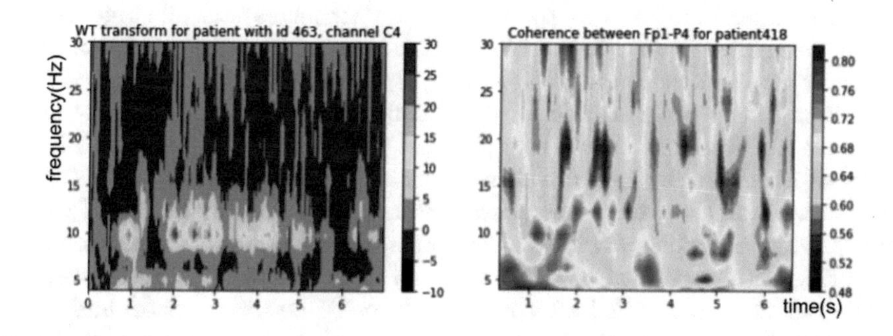

Fig. 1. Wavelet transform and phase coherence example for one subject

Further, these dynamics were averaged over the frequency ranges of θ (4–7 Hz), $\alpha 1$ (7–10 Hz), $\alpha 2$ (10–13 Hz), $\beta 1$ (13–18 Hz), $\beta 2$ (18–30 Hz). Thus, there were 5 dynamics, one for each frequency range. To construct a comparative picture, we considered the averaged parameters at intervals of 1–2, 2–3, 3–4, 4–5, 5–6 s from the beginning of the experiment. Further, the differences in the parameters of the wavelet spectrum and interchannel coherence were compared separately for each time interval. The identification of differences between the groups for each parameter was carried out using the F-test of Fisher.

4 Results and Discussion

4.1 Differences in Tasks with Mental and Sensory Attention

Statistically significant differences in wavelet power spectrum and coherence between mental (MA) and sensory attention (SA) tasks are presented in the Fig. 2.

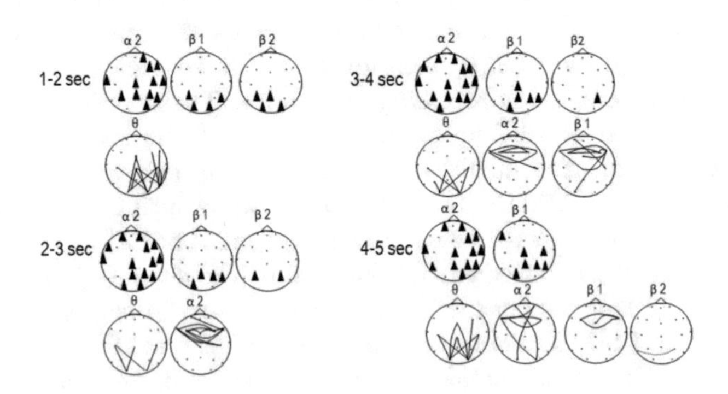

Fig. 2. Differences in the wavelet PSD and PC in different EEG frequency bands for different time intervals during MA with SA task. A triangle up - larger power spectrum (larger PC), a triangle down - smaller PSD (PC)in the MA (first task).

MA is differ from SA task by $\alpha2$ wavelet PSD increase at all stages of execution. α-synchronization is considered as an indicator of the activation of the internally-directed, mental attention [6,10,12]. The fact that $\alpha2$ increase is localized mainly in the regions of the right hemisphere is also consistent with the literature [2]. Observed increase in $\beta1$- and $\beta2$-amplitude in MA tasks is also observed in the parietal regions of right hemisphere at all stages of tasks performance. These changes can be considered as a result of suppressing of ventral attention system, which prevents attention switching to irrelevant sensory stimuli during performance of MA tasks. At all stages of tasks performance, MA tasks differ from SA task by θ-coherence increase between central and occipital zones. Dynamics of EEG parameters in θ range is often associated with memory [11]. In our case, MA tasks were related to extraction and retention of visual information before the mind's eye. These processes can be associated with θ-coherence increase. α-activity usually considered according to processes of suppression of irrelevant information [11]. So, observed in MA tasks local increase of a2-coherence in frontal regions in time interval 2–4 s after start of the tasks, can be considered as a temporary disabling of the regions from general context of activity. Further (3–5 s) to these differences between tasks, large $\beta1$-coherence in frontal regions in MA tasks are added. This $\beta1$-coherence increase can be interpreted as a result of attention concentration on internal mental context during MA tasks. At the final stage of tasks performance (4–5 s), local $\alpha2$-coherence in MA tasks is replaced by $\alpha2$-coherent interactions between frontal and occipital regions. Simultaneous involvement of these regions was shown for attention and memory processes [4,14] So, the fronto-parietal interactions at the final stages of MA tasks may indicate completion of mental image creation and monitoring of obtained results.

4.2 Differences in Tasks of Imagination and Retrieval from Memory

Statistically significant differences in wavelet spectrum amplitude and coherence between tasks of imagination (Im) and retrieval information from memory (Mm) are presented in the Fig. 3.

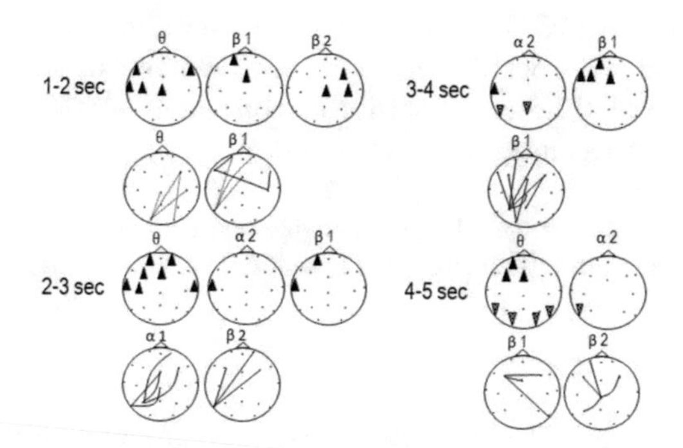

Fig. 3. Differences in the PSD amplitude and interchannel PC in different EEG frequency bands for certain time intervals during performance of Im with Mm task.

Im task at the initial stages of execution (1–3 s) differs from Mm task by PSD amplitude increases in the anterior and parietal regions in θ, β- and $\beta2$-frequency ranges. At the same time, Im task is characterized by the decrease of θ- and $\beta1$-coherence between the front and parietal regions. The detected differences in PSD amplitude and coherence may indicate a greater involvement of the anterior and parietal cortical regions in the process of imagination, and also indicate a more isolated work of these departments in the process of imagination at the initial stages. The differences between Im and Mm tasks may be explained by the fact that process of imagination, in contrast to retrieval, was accompanied not only by the retrieval of visual representations from memory but also by retaining of representations before the inner eye and by process of recombination for creation new chimeric object. At later stages of execution (2–4 s), Im task differs from Mm task by coherence increase in $\beta1$-, $\beta2$- and $\alpha1$-bands between the frontal and parietal regions. Increased interaction between the frontal and parietal regions was shown when performing creative tasks [9,13]. Thus, our results reinforce the existing understanding of the imagination as a more creative process than the process of remembering. At the final stage of tasks performance (4–5 s) Im task differs from Mm task by θ-amplitude increase in the frontal regions but by the decrease of θ - and $\alpha2$-amplitude in the occipital and parietal regions. Differences between tasks may be related to the fact that final stage of retrieval information from memory is associated with comparing the emerging visual image with original, whereas imagination is more connected with the free flow of associations.

5 Conclusion

The obtained data demonstrate not only significant differences in the dynamics of EEG parameters in the performance of tasks with a predominance of mental and sensory attention, but allow us to trace the differences between such close mental states as visual imagination and retrieval visual representations from memory. The dynamics of the EEG differences obtained in different frequency ranges allows us to trace which processes are involved in the execution of the investigated states at different instants of time and provide new material for understanding brain mechanisms of these states.

Acknowledgments. We gratefully acknowledge financial support of Institute of Information and Computational Technologies (Grant AR05134227, Kazakhstan).

References

1. Bartsch, F., Hamuni, G., Miskovic, V., Lang, P.J., Keil, A.: Oscillatory brain activity in the alpha range is modulated by the content of word-prompted mental imagery. Psychophysiology **52**(6), 727–735 (2015). https://doi.org/10.1111/psyp.12405
2. Benedek, M., Schickel, R.J., Jauk, E., Fink, A., Neubauer, A.C.: Alpha power increases in right parietal cortex reflects focused internal attention. Neuropsychologia **56**(1), 393–400 (2014). https://doi.org/10.1016/j.neuropsychologia.2014.02.010
3. Boytsova, Y.A., Danko, S.G.: EEG differences between resting states with eyes open and closed in darkness. Hum. Physiol. **36**(3), 367–369 (2010). https://doi.org/10.1134/S0362119710030199
4. Bressler, S.L., Tang, W., Sylvester, C.M., Shulman, G.L., Corbetta, M.: Top-down control of human visual cortex by frontal and parietal cortex in anticipatory visual spatial attention. J. Neurosci. **28**(40), 10056–10061 (2008). https://doi.org/10.1523/JNEUROSCI.1776-08.2008
5. Cohen, M.: Analyzing Time Series Data. MIT Press, Cambridge (2014)
6. Cooper, N.R., Croft, R.J., Dominey, S.J., Burgess, A.P., Gruzelier, J.H.: Paradox lost? Exploring the role of alpha oscillations during externally vs. internally directed attention and the implications for idling and inhibition hypotheses. Int. J. Psychophysiol. **47**(1), 65–74 (2003). https://doi.org/10.1016/S0167-8760(02)00107-1
7. Danko, S.: The reflection of different aspects of brain activation in the electroencephalogram: quantitative electroencephalography of the states of rest with the eyes open and closed. Hum. Physiol. **32**(4), 377–388 (2006). https://doi.org/10.1134/S0362119706040013
8. Efitorov, A., Knyazeva, I., Boytsova, Y., Danko, S.: GPU-based high-performance computing of multichannel EEG phase wavelet synchronization. Procedia Comput. Sci. **123**, 128–133 (2018). https://doi.org/10.1016/J.PROCS.2018.01.021

9. Fink, A., Neubauer, A.C.: EEG alpha oscillations during the performance of verbal creativity tasks: differential effects of sex and verbal intelligence. Int. J. Psychophysiol. **62**(1), 46–53 (2006). https://doi.org/10.1016/j.ijpsycho.2006.01.001
10. Harmony, T., et al.: EEG delta activity: an indicator of attention to internal processing during performance of mental tasks. Int. J. Psychophysiol. **24**(1–2), 161–171 (1996). https://doi.org/10.1016/S0167-8760(96)00053-0
11. Klimesch, W., Doppelmayr, M., Schwaiger, J., Auinger, P., Winkler, T.: 'Paradoxical' alpha synchronization in a memory task. Cogn. Brain Res. **7**(4), 493–501 (1999). https://doi.org/10.1016/S0167-8760(02)00107-1
12. Ray, W.J., Cole, H.W.: EEG alpha activity reflects attentional demands, and beta activity reflects emotional and cognitive processes. Science (New York, N.Y.) **228**(4700), 750–752 (1985). https://doi.org/10.1126/science.3992243
13. Razumnikova, O.M.: Creativity related cortex activity in the remote associates task. Brain Res. Bull. **73**(1–3), 96–102 (2007). https://doi.org/10.1016/j.brainresbull.2007.02.008
14. Sauseng, P., Klimesch, W., Schabus, M., Dopplemayr, M.: Fronto-parietal EEG coherence in theta and upper alpha reflect central executive functions of working memory. Int. J. Psychophysiol. **57**(2), 97–103 (2005). https://doi.org/10.1016/j.ijpsycho.2005.03.018

Bursting in a System of Two Coupled Pulsed Neurons with Delay

Elena A. Marushkina(✉)

P.G. Demidov Yaroslavl State University, Yaroslavl, Russia
marushkina-ea@yandex.ru

Abstract. In this paper, a mathematical model of synaptic interaction between two pulsed neuron elements is considered. Each of the neurons is represented by a singularly perturbed difference-differential equation with delay. The connection between elements is assumed to be at the threshold, taking into account the time delay. The problems of existence and stability of relaxation cycles are studied. In the framework of the task an algorithm for finding periodic solutions with bursting behavior is proposed. As the delay in the coupling link between the oscillators grows, it is shown that the system exhibits a lot of pulse regimes with different number of spikes over a period length interval. Of particular importance is the fact that the system can have a large number of coexisting relaxation oscillations. The appearance of such bursting effect in the system is a consequence of delay in the coupling link between the oscillators. The obtained result has a natural neurobiological meaning, which is that the growth in the delay leads to an increase in the capacity of this dynamic system as a memory device.

Keywords: Neural models · Synaptic interaction
Relaxation oscillations · Asymptotics · Stability · Bursting effect

1 Introduction

Self-oscillating processes in neural systems have one characteristic feature, which has received a special name: bursting effect. This feature consists in the alternation of pulse packets (sets of several consecutive intense spikes) with relatively calm areas of membrane potential changes. The study of this effect is devoted to a quite large number of publications (see, for example, works [1–3] and the bibliography available in them). Most often, for mathematical modeling of the bursting behaviour, singularly perturbed systems of ordinary differential equations with two fast and one slow variable are selected. It is well known that in such systems stable bursting cycles can exist at appropriate parameter values. However, another approach to solving this problem is possible, related to the time delay (see, for example, [4–6]). The book [6] shows that for the differential-difference equation with two delays modeling a single pulsed neuron, it is possible to choose the values of the parameters in an appropriate way to realize the bursting behaviour.

B. Kryzhanovsky et al. (Eds.): NEUROINFORMATICS 2018, SCI 799, pp. 328–333, 2019.
https://doi.org/10.1007/978-3-030-01328-8_40

In [7], a new approach to simulation of chemical synapses is proposed. It is based on a modification of the idea of fast threshold modulation (FTM). This phenomenon, which was described for the first time in [8,9], is a specific way of coupling dynamic systems. It is characterized by abrupt changes in the right hand sides of the corresponding differential equations when certain control variables exceed their thresholds.

2 Problem Statement

As a mathematical model of the association of two neurons with synaptic interaction (see [7]) take a system

$$\begin{aligned} \dot{u}_1 &= [\lambda f(u_1(t-1)) + bg(u_2(t-h))\log(u_*/u_1)]u_1, \\ \dot{u}_2 &= [\lambda f(u_2(t-1)) + bg(u_1(t-h))\log(u_*/u_2)]u_2, \end{aligned} \tag{1}$$

where $b = const > 0$, $u_* = \exp(c\lambda)$, $c = const \in \mathbb{R}$, the parameter $h > 1$ determines the delay in the coupling element, and the functions $f(u)$, $g(u) \in C^2(\mathbb{R}_+)$ are such that

$$\begin{aligned} &f(0) = 1, \quad g(0) = 0, \quad g(u) > 0 \quad \forall u > 0; \\ &uf'(u), f(u) + a, g(u) - 1, ug'(u) = O(1/u) \quad as \quad u \to +\infty. \end{aligned} \tag{2}$$

One of the reasons for choosing a system (1) is that the corresponding coupling summands $bg(u_2(t-h))u_1 \log(u_*/u_1)$, $bg(u_1(t-h))u_2 \log(u_*/u_2)$ change sign with "+" on "−" as the potentials u_1, u_2 grow and exceed the critical value u_*. This behaviour is typical for synaptic interaction. In addition, we have managed to correctly identify the limit object for system (1), which proves to be a certain relay system with delay.

Indeed, upon transferring to new variables $x_j = (1/\lambda)\log u_j$, $j = 1, 2$ the system (1) is rewritten as

$$\begin{aligned} \dot{x}_1 &= F(x_1(t-1), \varepsilon) + b(c - x_1)G(x_2(t-h), \varepsilon), \\ \dot{x}_2 &= F(x_2(t-1), \varepsilon) + b(c - x_2)G(x_1(t-h), \varepsilon), \end{aligned} \tag{3}$$

where $\varepsilon = 1/\lambda \ll 1$, $F(x, \varepsilon) = f(\exp(x/\varepsilon))$, $G(x, \varepsilon) = g(\exp(x/\varepsilon))$. Note that the following limit equalities hold:

$$\lim_{\varepsilon \to 0} F(x, \varepsilon) = R(x) = \begin{cases} 1 & for \quad x < 0, \\ -a & for \quad x > 0, \end{cases} \quad \lim_{\varepsilon \to 0} G(x, \varepsilon) = H(x), \tag{4}$$

where $H(x) = \begin{cases} 0 & for \quad x < 0, \\ 1 & for \quad x > 0. \end{cases}$

It follows, in turn, that the system (3) goes into a relay system

$$\begin{aligned} \dot{x}_1 &= R(x_1(t-1)) + b(c - x_1)H(x_2(t-h)), \\ \dot{x}_2 &= R(x_2(t-1)) + b(c - x_2)H(x_1(t-h)), \end{aligned} \tag{5}$$

as $\varepsilon \to 0$.

The presence of a limit object (5) significantly facilitates the problem of finding the attractors of the system (3) and allows, in particular, to apply to it the general results of [10] on the correspondence between the stable cycles of the relay and relaxation systems.

Only the simplest regimes of the system (3) can be found by analytical asymptotic methods. Among them is a homogeneous regime found and described in [11,12], the regime of oscillations in the antiphase and the regime in which one of the oscillators is in the oscillatory mode, and the second remains immune to external action (refractory state) [13].

The system (3), apparently, has a large number of different stable regimes [14]. This idea is suggested by the results obtained in the article [15]. It proves the existence of an arbitrarily large number of solutions close to harmonic of a singularly perturbed equation with two delays, similar to the equation presented in the system (5). At the same time, the result of the article [15] is obtained for the values of parameters close to the critical case of infinite-dimensional degeneration. The oscillations obtained in this work, as mentioned earlier, are close to harmonic and have a small amplitude. But the real oscillations arising in the systems modeling the electrical activity of nerve cells have the usual relaxation character with high exponential bursts. Below we consider the method of numerical finding such solutions.

3 The Numerical Determination of Coexisting Regimes with Several High-Amplitude Spikes on the Period

Let us first consider the solution of the system (3) in the form $x_1 = x(t)$, $x_2 = x(t+\Delta)$. After this replacement, the system (3) becomes a pair of the following equations:

$$\begin{aligned} &\dot{x} = F(x(t-1),\varepsilon) + b(c-x)G(x(t+\Delta-h),\varepsilon), \\ &\dot{x}(t+\Delta) = F(x(t+\Delta-1),\varepsilon) + b(c-x(t+\Delta))G(x(t-h),\varepsilon). \end{aligned} \tag{6}$$

Suppose that the first of the equations of the system (6) is executed. Then the second will also take place, provided that the value of the period T_* of the function $x(t)$ and the shift Δ are in the following ratio $T_*/\Delta = m/k$, what follows from equality $x_1 = x_2(t-\Delta)$.

Thus, the initial problem can be reduced to the search for a periodic solution of the following equation with two delays:

$$\dot{x} = F(x(t-1),\varepsilon) + b(c-x)G(x(t-\Delta-h),\varepsilon). \tag{7}$$

The task for the numerical computation, which can be supplied, is the following: it is necessary to choose the parameters b, c, Δ in Eq. (7) such that for all sufficiently small ε Eq. (7) has an exponentially orbitally stable cycle $x = x_*(t,\varepsilon)$ with period $T_*(\varepsilon)$, where $\lim\limits_{\varepsilon\to 0} T_*(\varepsilon) = T_*$. The corresponding asymptotics for $T_*(\varepsilon)$ was given in article [11]. Thus, choosing different m and k, we can select the value Δ of such that $T_*(\varepsilon)/\Delta = m/k$. Therefore, it is possible to simulate the solution $x_*(t,\varepsilon)$ of equation (7) such way that it is in the time interval of the period length had on one section the form of the periodic function, and the second was negative. The period for the oscillatory section should be chosen equal to $T_0 = 1 + a + 1/a$. This value is the period of solution of the Eq. (7) at $b = 0$.

Taking into account an exponential replacement this would mean that the components u_1 and u_2 of the solution in the system (1) will have the same number of high-amplitude spikes on the period with the phase difference Δ between solutions.

4 Numerical Analysis of a Pair of Synaptically Coupled relaxation Oscillators

For the system (5) by the method of a large parameter it is possible to prove that with a appropriate choice of values a, b, c and h in the phase space of this system, three simple regimes coexist: inphase oscillations, antiphase oscillations and pulse-refractory oscillations [11–13]. At the same time, with sufficiently large values of delay h the problem, in all probability, has not three, but a significantly greater number of coexisting solutions. In order to verify this, a numerical experiment was performed for the system (5). It was shown, that for sufficiently large values of the parameter h, for example, when $h = 10$ the system manages to find up to 16 different coexisting regimes.

To effectively search for bursting cycles, the following approach was used: when initial conditions were formed at the interval of the length of the bursting, one of the variables was chosen oscillating with respect to zero, and the second one was chosen negative at the whole interval. The periodic part of the initial condition was chosen to correspond to the relaxation cycle of the equation $\dot{x} = f(\exp(x(t-1)/\varepsilon))$. For this, the initial condition is convenient to choose a such kind function $x_c(t) = \alpha - \sin(2\pi(t+\gamma)/T_0)$, where $\alpha = \sin(\pi(a - 1/a)/2T_0)$.

To make the solution of the system (5) a set of fixed number of spikes on the period, as $x_{1,2}$ the following functions were selected:

$$\begin{aligned} x_1(t) &= x_c(t)\mathrm{sgn}(t_c - t) - \mathrm{sgn}(t - t_c), \\ x_2(t) &= -2. \end{aligned} \tag{8}$$

Here, the constant t_c determines the number of spikes occurs in the interval of length h. Further steps were as follows: the value t_c was selected so that the interval $[-h, 0]$ has $1, 2, \ldots, k$ spikes, where $k = [h/T_0]$. After this was performed the integration of the system (5) and look for stable oscillations, which converge to the solution of the system with these initial conditions. As it turned out, for almost every number of k spikes there is a separate stable regime of the system (5).

The Fig. 1 shows the graphics of several stable solutions, received at a delay value $h = 10$.

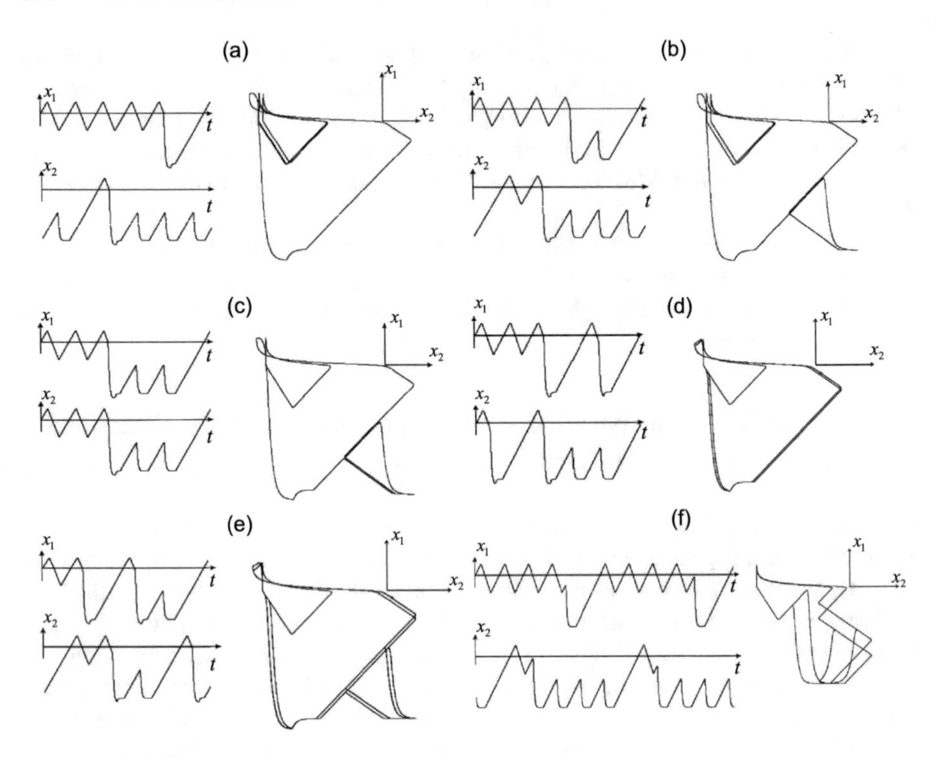

Fig. 1. Stable cycles and their phase portraits obtained by $a = 1.5$, $b = 7$, $c = -4.5$, $h = 10$: (a) $\gamma = 0.95$; (b) $\gamma = 0.7$; (c) $\gamma = 0.5$; (d) $\gamma = 1.1$; (e) $\gamma = 1.3$; (f) $\gamma = 0.8$

Note that each of these solutions corresponds to a symmetrical one. Hence, at least 16 oscillatory regimes of the initial problem were received together with those found earlier. Also we note the fact observed in the numerical experiment that the total number of spikes on the period for the values of $x_1(t)$ and $x_2(t)$ remains equal to the constant.

5 Conclusions

The combination of analytical asymptotic methods and numerical experiment allowed us to show that as the delay h in the coupling link between the oscillators grows, the system of two synaptically connected pulsed neurons demonstrates a lot of regimes with different number of spikes on the period. The total number of spikes for both components of the solution also increases with the growth h. The obtained result has a natural neurobiological meaning, which is that the increase in the delay of the connection elements leads to an increase in the capacity of this dynamic system as a memory device.

Acknowledgments. The reported study was funded by RFBR, according to the research project No. 16-31-60039 mol_a_dk.

References

1. Chay, T.R., Rinzel, J.: Bursting, beating, and chaos in an excitable membrane model. Biophys. J. **47**(3), 357–366 (1985)
2. Ermentrout, G.B., Kopell, N.: Parabolic bursting in an excitable system coupled with a slow oscillation. SIAM J. Appl. Math. **46**(2), 233–253 (1986)
3. Coombes, S., Bressloff, P.C.: Bursting: The Genesis of Rhythm in the Nervous System. World Scientific Publishing Company, Singapore (2005)
4. Glyzin, S.D., Kolesov, A.Y., Rozov, N.K.: Modeling the bursting effect in neuron systems. Math. Notes. **93**(5–6), 676–690 (2013)
5. Glyzin, S.D., Kolesov, A.Y., Rozov, N.K.: Self-excited relaxation oscillations in networks of impulse neurons. Russ. Math. Surv. **70**(3), 3–76 (2015)
6. Kashchenko, S.A., Mayorov, V.V.: Wave Memory Models. Librokom, Moscow (2009)
7. Glyzin, S.D., Kolesov, A.Y., Rozov, N.K.: On a method for mathematical modeling of chemical synapses. Differ. Equ. **49**(10), 1193–1210 (2013)
8. Somers, D., Kopell, N.: Anti-phase solutions in relaxation oscillators coupled through excitatory interactions. J. Math. Biol. **33**, 261–280 (1995)
9. Somers, D., Kopell, N.: Rapid synchronization through fast threshold modulation. Biol. Cybern. **68**, 393–407 (1993)
10. Kolesov, A.Y., Mishchenko, E.F., Rozov, N.Kh.: A relay with delay and its C^1-approximation. Proc. Steklov Inst. Math. **216**, 126–153 (1997)
11. Glyzin, S.D., Kolesov, A.Yu., Marushkina, E.A.: Relaxation oscillations in a system of two pulsed synaptically coupled neurons. Autom. Control Comput. Sci. **50**(7), 658–665 (2017)
12. Preobrazhenskaia, M.M.: Relaxation cycles in a model of synaptically interacting oscillators. Autom. Control Comput. Sci. **51**(7), 783–797 (2017)
13. Preobrazhenskaia, M.M.: The impulse-refractive mode in the neural network with ring synaptic interaction. Model. Anal. Inf. Syst. **24**(5), 550–566 (2017)
14. Glyzin, S.D., Kolesov, A.Y., Rozov, N.K.: Buffer phenomenon in neurodynamics. Dokl. Math. **85**(2), 297–300 (2012)
15. Glyzin, S.D., Kolesov, A.Y., Rozov, N.K.: Extremal dynamics of the generalized Hutchinson equation. Comput. Math. Math. Phys. **49**(1), 71–83 (2009)

Function and Molecular Design of the Synapse

Anna L. Proskura(✉) and Tatyana A. Zapara

The Institute of Computational Technologies of SB RAS, Novosibirsk, Russia
annleop@mail.ru

Abstract. The majority of excitatory synapses in the mammalian forebrain and the hippocampus terminate onto dendritic spines. Even though their intricate molecular composition remains obscure in many aspects, available evidence suggests that these structures are highly specialized to support the short- and long-term plasticity crucial for flexible information processing. One concept that is extensively used to describe synaptic function is the Hebb's postulate. However, the knowledge accumulated throughout following decades advocate for a broader scale in intercellular connection functions. Synaptic activity depends on interactions among sets of proteins (synaptic interactome) that assemble into complex supramolecular machines. Molecular biology, electrophysiology, and live-cell imaging studies have provided tantalizing glimpses into the inner workings of the synapse, yet fundamental questions regarding the functional organization of these "molecular nanomachines" remain to be answered. The presence of accessory receptors for secondary messengers in synapses along with receptors for primary mediator gave us the idea that these molecular constructs could be responsible for initial processing of the incoming signals. The purpose of this initial processing could be determined by analyzing and reconstructing this molecular informational machine that is essentially dendritic spine of hippocampal pyramidal neurons.

Keywords: Pyramidal synapse · Dendritic spine · Interactome
AMPA receptors

1 Introduction

The hippocampus is a brain structure involved in learning and memory [1]. Learning and memory are mediated by changes in the excitability of the cell and modulations of synaptic transmission, usually it takes place in proteome of the postsynaptic (dendritic) part of a synaptic contacts.

The dendritic spines are the postsynaptic part of exciting synapses. Dendritic spine interactome (set consisting of every protein interaction) represents dynamic multi-level and multi-component system [2].

Ionotropic glutamate receptors (NMDA (N-methyl-D-aspartate), AMPA (α-amino-3-hydroxy-5-methyl-4-isoxazolepropionic acid) receptors) are the key components of

In this paper the data obtained during the implementation of RAS IV.35.2.6 and the RFBR 17-04-01440a projects are used.

B. Kryzhanovsky et al. (Eds.): NEUROINFORMATICS 2018, SCI 799, pp. 334–338, 2019.
https://doi.org/10.1007/978-3-030-01328-8_41

excitatory synapses in brain. Through structural proteins of postsynaptic density, they form various protein macrocomplexes. These complexes mediate intermolecular rearrangements and form the functional regulatory protein-protein networks [3].

In addition, the receptors for various hormones, notably leptin, growth and neurotrophic factors, as well as the receptors of other mediator systems, in particular dopamine receptors are present in the synaptic contact zone. It is known that leptin and dopamine can modulate the regulation of synaptic plasticity in hippocampus [4, 5], but the exact molecular mechanisms behind these effects are unclear.

2 Molecular Functional System of the Dendritic Spine Mediated Synaptic Efficiency

Postsynaptic interactome provides the foundation for dynamic multi-level and multi-component system allowing synapse to change and maintain its efficiency depending on incoming signals. The phenomenon of long-term potentiation (LTP) (increasing the efficiency of synaptic transmission after an intense and short-term release of neurotransmitter) provides a commonly accepted cellular model of learning and memory [6]. Information on main proteins involved in modulating and maintaining the efficiency of synaptic transmission in the CA1 hippocampal field during LTP has been accumulated in the GeneNet database (Rospatent no. 990006 dated February 15, 1999) [7], which is available as a graphic diagram “AMPA Receptors Delivery Mechanisms in LTP” (http://wwwmgs.bionet.nsc.ru/mgs/gnw/genenet/viewer/AMPA.html) [8, 9].

After LTP induction occur a local short-term entry of calcium ions triggering the cascades of molecular reactions [10]. Functional AMPA receptors (AMPARs) are assembled from combinations of the four subunits (GluR1, GluR2, GluR3). The basic neurotransmission and LTP induction depend on AMPAR including GluR2 and GluR3 subunits (GluR2/3 AMPAR). After LTP is induced GluR2/3 internalize and AMPAR including the GluR1 subunit (GluR1/2 and/or GluR1/1 are GluR1-AMPARs) are introduced into the synaptic zone instead [11, 12].

A key event is the groundwork of second messengers' pools. P4 → PIP2 (PtdIns (4,5)P2) catalyzed Phosphoinositide 5-kinase (PI5K), PIP2 → PIP3 (PtdIns(3,4,5)P3) mediated Phosphoinositide 3-kinase (PI3K). Each of those initiates its own molecular pathway regulating various cellular events (endocytosis, receptor mobility on the plasmatic membrane, etc.). At the same time together, they form a functional dendritic spine interactome that changes the efficiency of synapse in response to the incoming stimulus and respond by changing its efficiency, affected by the previous synaptic activity, novelty of the signal, the metabolic state of the neuron itself and environment.

PtdIns(4,5)P2 is a substrate for hydrolysis by phospholipase C (PLC) therefore produced inositol 1,4,5-trisphosphate (InsP3; IP3) and diacylglycerol (DAG), both of them function as second messengers. DAG remains on the cell membrane and sends down signal cascade by activating protein kinase C (PKC). PKC on its own activates other cytosolic proteins including AMPAR by phosphorylating them.

After LTP induction GluR2/3 are move from the synaptic contacts after being phosphorylated by PKC. Simultaneously GluR1-AMPARs move from the perisynaptic membrane to synaptic zone [12]. Thus, the perisynaptic region may be considered a transition zone or reservoir for receptor trafficking into or out of the synapse contact [11]. Increase in DAG pool as we think would therefore prolong PKC activity and as a result simultaneously remove GluR2/3 from the synaptic zone and introduce GluR1-AMPAR by lateral diffusion on spine plasmatic membrane.

The GluR1 subunit is regulated by protein phosphorylation at two sites on its intracellular carboxy-terminal domain: serine 831 (S831) and serine 845 (S845). S831 is phosphorylated by CaMKII and PKC, what is necessary for expression of LTP. S845 is phosphorylated by protein kinase A (PKA), which is activated by cyclic AMP (cAMP). Phosphorylation of either of these sites potentiates AMPA receptor function through distinct biophysical mechanisms [13]. PKA colocalization with Adenyl cyclase (ADCY8) within perisynaptic region [14]. Stimulation of adenyl cyclase results in formation of cAMP which is released from the membrane into the dendritic cytoplasm. Evidence exists for phosphorylation in S845 GluR1-AMPAR being necessary for their anchoring in perisynaptic zone [15]. Receptors move in perisynaptic zone from extrasynaptic zone of a spine where they gate in from neuronal dendritic shaft and soma by exocytosis [11]. We've suggested the presence of mobile GluR1-AMPAR pool in the perisynaptic zone of the spine to be responsible for the high-speed changes in synaptic efficiency [16].

2.1 Dopamine and Leptin Regulate Functional System for Changing and Maintaining Synaptic Efficiency in Hippocampus

Dopamine belongs to one of factors of internal reinforcement. It is known that the activity of dopamine receptors provides the maintenance of higher AMPAR density in spine, possibly mediating the facilitation of long-term potentiation in CA1 hippocampal field; dopamine also contributes to the processes of learning and spatial memory development, acting as a novelty detection signal [5, 17]. Leptin, hormone associated with energy metabolism regulation, is secreted by adipocytes and can transported across the blood–brain barrier. The concentration of leptin is a physiological signal for energy resource sufficiency in the body [4].

Dopamine receptors in hippocampal CA1 field (D1/D5R types) are Gs-bound proteins. They trigger the cascade generation of cAMP that, as shown before, is needed for PKA activation. PKA phosphorylates GluR1-AMPAR providing, as we assume, certain pool of this receptor type. We suggest NMDAR can cooperate with D1/D5Rs on the molecular level: activate NMDAR → Ca^{2+}_{ex} → ADCY8 → cAMP → PKA → increase the perisynaptic GluR1-AMPAR density → introduce GluR1-AMPAR to the synapse. Thus, D1/D5R form a loop of positive regulation in this molecular cascade, supporting the increase in AMPAR density and subsequently increasing synaptic efficiency: Dopamine → D1/D5R → Gs → ADCY8 → cAMP → PKA → increase the perisynaptic GluR1-AMPAR density.

Subunit GluR1 is phosphorylated on S831 as shown before PKC. It is necessary for incorporation receptors on synaptic sites from perisynaptic zone of dendritic spine by lateral diffusion what is very important for LTP expression [13] and underlying increase synaptic AMPARs density. The activity of PKC is mediated by protein kinase PDK1 (3-phosphoinositide-dependent protein kinase 1), which binds to PtdIns(3,4,5)

P3. PDK1 is proposed to phosphorylate a key threonine residue within the catalytic domain of PKC that controls the stability and catalytic competence of these kinase.

Thereby, emergence into membrane PIP3 mediates a positive regulation loop promoting anchoring of GluR1-AMPARs into synaptic contacts after NMDAR-depending LTP induction: PIP2 → PIP3 → PDK1 → PKC → GluR1-AMPARs incorporation at the synaptic contact. PtdIns(3,4,5)P3 is dephosphorylated by the phosphatase PTEN on the 3 position, generating PIP2, which is observed when the basic synaptic activity takes place (when NDMAP is not active). Leptin contributes significantly to removal of PTEN from spines and promote of the activity PI3K after NMDAR activation and LTP induction. It will contribute generating PIP3 and, thus, support mediated increased of the GluR1-AMPARs expression at the synaptic contacts [18].

3 Conclusion

The dendritic spine interactome in hippocampal CA1 field neurons forms a functional system that changes synaptic efficiency in response to incoming stimuli. This process depends from the past synaptic activity, stimulus novelty and the metabolic state of the neuron itself.

Dopamine seemingly acts as a signal importance agent. In presence Dopamine is provides the pool of perisynaptic GluR1-AMPA receptors for the round of NMDAR-mediated activity in the corresponding synapse, therefore facilitating of LTP induction in hippocampal CA1. Leptin supposedly acts as an agent of an organism's resource adequacy to provide a current synapse efficiency through maintaining PKC activity.

References

1. Turgut, Y.B., Turgut, M.: A mysterious term hippocampus involved in learning and memory. Childs Nerv. Syst. **27**(12), 2023–2025 (2011). https://doi.org/10.1007/s00381-011-1513-y
2. Proskura, A.L., Ratushnyak, A.S., Vechkapova, S.O., Zapara, T.A.: Synapse as a multi-component and multi-level information system. In: Kryzhanovsky, B., Dunin-Barkowski, W., Redko, V. (eds.) Advances in Neural Computation, Machine Learning, and Cognitive Research, NEUROINFORMATICS 2017. Studies in Computational Intelligence, vol. 736, pp. 186–192. Springer, Cham (2018)
3. Proskura, A.L., Vechkapova, S.O., Zapara, T.A., Ratushnyak, A.S.: Reconstruction of the molecular interactome of glutamatergic synapses. Russ. J. Genet. Appl. Res. **5**(6), 616–625 (2015)
4. Harvey, J.: Leptin regulation of neuronal morphology and hippocampal synaptic function. Front. Synaptic Neurosci. **5**(3), 1–7 (2013). https://doi.org/10.3389/fnsyn.2013.00003
5. Li, S., Cullen, W.K., Anwyl, R., Rowan, M.J.: Dopamine-dependent facilitation of LTP induction in hippocampal CA1 by exposure to spatial novelty. Nat. Neurosci. **6**(5), 526–531 (2003). https://doi.org/10.1038/nn1049
6. Pastalkova, E., Serrano, P., Pinkhasova, D., Wallace, E., Fenton, A.A., Sacktor, T.C.: Storage of spatial information by the maintenance mechanism of LTP. Science **313**(5790), 1141–1144 (2006). https://doi.org/10.1126/science.1128657

7. Ananko, E.A., Podkolodny, N.L., Stepanenko, I.L., Podkolodnaya, O.A., Rasskazov, D.A., Miginsky, D.S., Likhoshvai, V.A., Ratushny, A.V., Podkolodnaya, N.N., Kolchanov, N.A.: GeneNet in 2005. Nucl. Acids Res. **33**, 425–427 (2005). https://doi.org/10.1093/nar/gki077
8. Proskura, A.L., Malakhin, I.A., Zapara, T.A., Turnaev, I.I., Suslov, V.V., Ratushnyak, A.S.: Intermolecular interactions in neuronal functional systems. Russ. J. Genet. Appl. Res. **17**, 620–628 (2013)
9. Proskura, A.L., Ratushnyak, A.S., Zapara, T.A.: The protein-protein interaction networks of dendritic spines in the early phase of long-term potentiation. J. Comput. Sci. Syst. Biol. **7**, 40–44 (2014). https://doi.org/10.4172/jcsb.1000136
10. Raghuram, V., Sharma, Y., Kreutz, M.R.: Ca2+ sensor proteins in dendritic spines: a race for Ca2+. Front. Mol. Neurosci. **5**, 61 (2012). https://doi.org/10.3389/fnmol.2012.00061
11. Newpher, T.M., Ehlers, M.D.: Glutamate receptor dynamics in dendritic microdomains. Neuron **58**(4), 472–497 (2008). https://doi.org/10.1016/j.neuron.2008.04.030
12. Petrini, E.M., Lu, J., Cognet, L., Lounis, B., Ehlers, M.D., Choquet, D.: Endocytic trafficking and recycling maintain a pool of mobile surface AMPA receptors required for synaptic potentiation. Neuron **63**(1), 92–105 (2009). https://doi.org/10.1016/j.neuron.2009.05.025
13. Lee, H.K., Barbarosie, M., Kameyama, K., Bear, M.F., Huganir, R.L.: Regulation of distinct AMPA receptor phosphorylation sites during bidirectional synaptic plasticity. Nature **405** (6789), 955–959 (2000). https://doi.org/10.1038/35016089
14. Kim, M., Park, A.J., Havekes, R., Chay, A., Guercio, L.A., Oliveira, R.F., Abel, T., Blackwell, K.T.: Colocalization of protein kinase A with adenylyl cyclase enhances protein kinase A activity during induction of long-lasting long-term-potentiation. PLoS Comput. Biol. **7**(6), e1002084 (2011). https://doi.org/10.1371/journal.pcbi.1002084
15. He, K., Song, L., Cummings, L.W., Goldman, J., Huganir, R.L., Lee, H.K.: Stabilization of Ca2+-permeable AMPA receptors at perisynaptic sites by GluR1-S845 phosphorylation. Proc. Natl. Acad. Sci. USA **106**(47), 20033–20038 (2009). https://doi.org/10.1073/pnas.0910338106
16. Zapara, T.A., Proskura, A.L., Malakhin, I.A., Vechkapova, S.O., Ratushnyak, A.S.: The mobility of AMPA-type glutamate receptors as a key factor in the expression and maintenance of synaptic potentiation. Neurosci. Behav. Physiol. **47**(5), 528–533 (2017). https://doi.org/10.1007/s11055-017-0430-2
17. Lemon, N., Manahan-Vaughan, D.: Dopamine D1/D5 receptors gate the acquisition of novel information through hippocampal long-term potentiation and long-term depression. J. Neurosci. **26**(29), 7723–7729 (2006). https://doi.org/10.1523/JNEUROSCI.1454-06.2006
18. Moult, P.R., Cross, A., Santos, S.D., Carvalho, A.L., Lindsay, Y., Connolly, C.N., Irving, A.J., Leslie, N.R., Harvey, J.: Leptin regulates AMPA receptor trafficking via PTEN inhibition. J. Neurosci. **30**(11), 4088–4101 (2010). https://doi.org/10.1523/JNEUROSCI.3614-09.2010

Evolutionary Origins and Principles of the Organization of Biological Information Systems

A. S. Ratushnyak(✉), E. D. Sorokoumov, and T. A. Zapara

The Institute of Computational Technologies of SB RAS, Novosibirsk, Russia
ratushniak.alex@gmail.com

Abstract. The main challenge of the present time vital for advancement in the fields of designing new generations of information systems and in the problems of neuroinduced pathologies correction is the need to understand the principles underlying brain activity. The conducted studies are vastly scattered, being carried out both in different fields and on different conceptual and technological levels. This disparity contributes to the fact that the concepts developed more than 50–100 years ago are still dominating modern neuroscience. The article attempts to form a new paradigm, based on an understanding of the physical essence and evolutionary roots of the formation of the basic properties of the brain as a biological information system. Since without the creation of a new integrative concept, efforts and resources aimed at identifying the principles of the functioning of the brain will continue to be of low productive.

Keywords: Neuroscience · Neurobiology
Information processes in neurons and neural systems · Bioinformatics
Computational biology · Learning · Memory · Neuron · Synapse

1 Introduction

It is necessary to understand the physical principles on which the existence of biological information systems (BIS) is based to create a new paradigm of the principles and mechanisms of their work [1]. One of the problems arising in the development of the conceptual foundations for the emergence and existence of living systems is the notion that the main functional property of such systems is self-replication (reproduction). Such a "genetic" theory of life, despite obvious contradictions, is quite common. The main problem in the framework of this concept is significant difficulties in trying to present a naturally scientific theory of the emergence of processes with complex system of self-reproduction. In addition, as a rule, it is not taken into account that not all biological systems are endowed with such a property. As a rule, reproduction is not used, for example, in neurons. It can be assumed that at the initial stage of evolution, when simple

The data presented in this paper were obtained during the implementation of the basic project of fundamental research of the Russian Academy of Sciences VI.35.2.6 and RFBR № 17-04-01440-a.

B. Kryzhanovsky et al. (Eds.): NEUROINFORMATICS 2018, SCI 799, pp. 339–342, 2019.
https://doi.org/10.1007/978-3-030-01328-8_42

molecular constructs appeared, such proto cells existed long enough in the absence of self-copying systems. The function of reproduction could arise on the basis of the simple merging and fragmentation that existed at these stages of evolution.

The ability to lower internal entropy is the main feature of biological systems. All known BIS functions arise on this basis. Such an opportunity, existing in information and thermodynamically open systems, can be realized only based on predicting future events (anticipating the display of reality based on knowledge). This allows such systems in accordance with the negentropic principle to increase internal ordering using information and energy received from the external environment [2–4].

The aspect of causality is also reflected in the framework of the Hebb principle [5]. It is emphasized that the generation of an action potential in a presynaptic cell must precede (outstrip) the initiation of a postsynaptic cell. Such a prior stimulus can be considered as a signal for the postsynaptic cell. Post-synaptic cells can use this signal in the future for prognostic purposes if its effectiveness is changed and remembered [6]. We can assume the possibility of spontaneous formation of the simplest molecular structures that can be organized into a system that reduces internal entropy on the basis of processes of "supromolecular training" [7, 8]. The use of information acquired with such "learning" for forecasting the future state of the environment allows increasing the orderliness of the system, enabling additional energy and/or avoiding the conditions for its loss. Molecular systems possessing such qualities create a zone of stability, a space of life. Merging and fragmenting such protobionts, aromorphosis under the control of feedbacks to the environment, probably led to the improvement of the interaction of receptor-effector systems. More effective and developed receptor-effector molecular ensembles took precedence. Homeostatic properties, "motivational" and self-replicating routines were formed on their basis.

2 Approaches and Methods of BIS Modeling

BIS modeling trends being developed are based primarily on extremely simplified, long-outdated notions about the physical principles, functions and molecular mechanisms of operating the brain basic units - nerve cells. One of such widespread and so far widely used concept is the postulate on the principle of the neuron synaptic sensor - the Hebb rule [5].

However, the knowledge existing at present exceed far beyond the limits of these concepts. It has been shown that within the synapse, besides receptors to traditional neurotransmitters, there are receptors to many other signaling molecules. It can be assumed that already within the framework of this receptor zone of cell, a simple molecular machine performs the primary processing of polysensory information, perhaps, the decision-making on the reaction to such a matrix of signals.

However, the known data on the principles of operating such molecular machines are often not taken into account in BIS modeling. This makes it difficult to move toward creating the information systems commensurable with the brain in terms of capabilities. Accordingly, the possibilities of solving all the problems based on this basis in the field of the complex of neurosciences, informatics, medicine, etc., are substantially limited.

When modeling such systems, it is necessary to create both a BIS model and a model of the environment in which they exist. As a rule, this is not taken into account. Modifications to BIS are possible when there is variability in the environment. Entropic and other processes can cause such variability. Important is the set of factors that can play the role of a signal for BIS. Of great importance is the anisotropy of the medium (transparency for sound, light, viscosity, electrical conductivity, thermal conductivity, etc.). Biosystems adapt to changing conditions based on feedback. In these conditions emergence both new properties of bioobjects and various forms occupying certain niches.

3 BIS Characteristics and Conditions of Origination

One of the main conditions and factors for the emergence of biological systems is the presence in the environment of gradients that can allow emerging molecular systems to concentrate energy in some internal space. This can be gradients of temperature, concentration, pressure, light, etc. Examples of existing of the biological systems under such conditions are found everywhere. Black smokers (as well as other places of the stable-gradients presence) can be considered as an obvious case of possible conditions for the emergence and existence of BIS communities.

At the starting point of the development of such systems, extremely simple molecular structures could have spontaneously formed with a certain "stock", an excess of negentropy (orderliness). Protobionts should have a number of characteristics for a successful existence emerging structures. Their design probably included molecules that interact with environmental factors - receptors and molecules that actively influence the environment - effectors.

In addition, the complexes should include molecules integrating receptors with effectors and having the ability to change their own state to "associatively memorize" the states that increase the stability of all such molecular structure. In this case, the response to environmental factors should depend on the direction of the gradient. With the next coincidence of environmental factors, the "saved state" of the molecular system can "predict" future events.

That is, in fact, such a simple molecular construction acquires prognostic properties. This allowed such systems with great probability avoiding unfavorable factors or acquiring additional energy and information. The increase in such complexes during the merger (mutual absorption) led to an increase in energy reserves, and negentropy. The increase in the number of receptors and effectors broadened the spectrum of the perceived environmental factors and possible reactions to their combination.

4 Conclusion

Thus, the paradigm of the mechanisms of the work of BIS should include a principles negentropy based on forecasting. With the interchange of energy and information in spontaneously arising simple molecular system with the medium, there is probably a zone of thermodynamic stability of such molecular structures. Later, biological systems of the following levels are formed from these elementary blocks. The selection of such structures by the environment increases their stability. The use of such a paradigm based on an understanding of the physical essence and evolutionary origin of the emergence of the basic properties of BIS will allow solving the tasks of reengineering, creating new architectures for information devices, and correcting the work of biological systems in pathologies. Creation of models of biological information systems is possible by forming algorithms of the simplest logical, in fact, devices that implement the negentropic principle. The construction of higher-level systems from such devices can be based on knowledge of the evolution of biosystems. In fact, modeling evolution in the direction known from the history of this process. The possibility of creating a theory of the stability of such systems due to information prognostic processes, their models at this stage is conditioned by the existing huge array of analytical materials. This requires theoretical and experimental analysis and convergence from the level of disparate data to the level of knowledge, theories and a complex of technologies. Creation of a model of the molecular information organization of basic systems (molecular-cellular level) will allow forming a new theory of the operation of BIS. The use of such a paradigm for analyzing systems at higher levels is possible taking into account the continuity of the principles of biological systems from level to level, the laws of emergence. Such a theory, as in its time, for example, the formation of ideas about the molecular basis for the transfer of genetic information, although it does not solve all problems, but will allow to carry out purposefully the basic tasks of bioneuroinformatics. We can hope that the problem of constructing the concept of a general theory of information systems through their evolutionary self-organization can be solved with real effort, resources and time.

The development of a new paradigm naturally requires further research and convergence of efforts of specialists from a wide range of areas of science, the integration of humanitarian and natural-science approaches in the study of information processes in biological systems.

References

1. Stern, P.: Neuroscience: In search of new concepts. Science **358**(6362), 464–465 (2017). http://science.sciencemag.org/content/358/6362/464
2. Schrödinger, E.: What Is Life? The Physical Aspect of the Living Cell. University Press, Cambridge (1944)
3. Brillouin, L.: Science and Information Theory. Academic Press, Cambridge (1956)
4. Krushinsky, A.: The cost of problem solving: biophysical background and probable evolutionary consequences. Russ. J. Cogni. Sci. **2**(1), 52–61 (2015)
5. Hebb, D.O.: The Organization of Behavior. Wiley, New York (1949)
6. Caporale, N., Dan, Y.: Spike timing-dependent plasticity: a Hebbian learning rule. Annu. Rev. Neurosci. **31**, 25–46 (2008). https://doi.org/10.1146/annurev.neuro.31.060407.125639
7. Lehn J.-M.: Supramolecular Chemistry: Concepts and Perspectives. Wiley-VCH, Morrisville (1995)
8. Hirst, A.R., Escuder, B., Miravet, J.F., Smith, D.K.: High-tech applications of self assembling supramolecular nanostructures gel phase materials: from regenerative medicine to electronic devices. Angewandte Chemie Int. Edition **47** (2008). https://doi.org/10.1002/anie.200800022

Author Index

B. Kryzhanovsky et al. (Eds.): NEUROINFORMATICS 2018, SCI 799, pp. 343–344, 2019.
https://doi.org/10.1007/978-3-030-01328-8

Printed in the United States
By Bookmasters